ᐱ

Springer Geography

The Springer Geography series seeks to publish a broad portfolio of scientific books, aiming at researchers, students, and everyone interested in geographical research. The series includes peer-reviewed monographs, edited volumes, textbooks, and conference proceedings. It covers the entire research area of geography including, but not limited to, Economic Geography, Physical Geography, Quantitative Geography, and Regional/Urban Planning.

More information about this series at http://www.springer.com/series/10180

Jinyan Zhan

Editor

Impacts of Land-use Change on Ecosystem Services

 Springer

Editor
Jinyan Zhan
School of Environment
Beijing Normal University
Beijing
China

ISSN 2194-315X   ISSN 2194-3168   (electronic)
Springer Geography
ISBN 978-3-662-48007-6   ISBN 978-3-662-48008-3   (eBook)
DOI 10.1007/978-3-662-48008-3

Library of Congress Control Number: 2015946089

Springer Heidelberg New York Dordrecht London

Printed on acid-free paper

Springer-Verlag GmbH Berlin Heidelberg is part of Springer Science+Business Media
(www.springer.com)

Preface

Ecosystem is the life support system and the foundation of human survival and socioeconomic development. Land use changes directly and indirectly affect the sustainable development of ecosystem services. Optimization of land use management based on impact assessment is a hot spot in the academic circles. How land use changes affect the ecosystem services, and how to conduct the evaluation of the interactive effects between them are some of basic scientific questions in this field under investigation. This book aims to uncover the linkage among the land use, ecosystem services, and also human well-being, to describe the effects of land use changes on ecosystem services and human well-being, and to develop the models to quantitatively analyze the impacts in typical areas. These regions play significant roles in the sustainable development of China and even the world. For instance, the Three River Source Area is regarded as the "China water tower" and the North China Plain is one of the major grain production bases of China. Based on the influence mechanism of land use changes on ecosystem services and human well-being, this book serves as valuable decision-making references for land use management in terms of conserving ecosystem services as well as improving human well-being.

Quantitative analysis for the impacts of land use changes on ecosystem services is crucial for shaping the ideas of sustainable development. This book starts with revisiting the relevant studies on impact assessments of land use changes on key ecosystem services in Chap. 1. The impact assessment of land use conversions on land degradation and agricultural productivity was introduced in Chap. 2. Chapter 3 mainly evaluates the ecosystem services and conducts the impact assessment of land use conversion and land use management measures on the economic valued ecosystem services. Subsequently, Chap. 4 narrows down the impact assessments of land use changes and climate changes on ecosystem services in terms of the grassland, including the changes in grassland productivity due to land use and climate changes, as well as the changes in economic returns of livestock production in grassland. Chapter 5 focuses on the impact assessment on water and heat fluxes of terrestrial ecosystem due to land use changes. Finally, Chap. 6 investigates land use conversion and climate change-oriented measures, and then it introduces strategies on conserving sustainable development and building resilience of ecosystem services.

The authors claim full responsibility for any errors appearing in this work.

May 2015   Jinyan Zhan

Acknowledgments

This book was supported by National Natural Science Foundation of China for Distinguished Young Scholar (Grant No. 71225005). First and foremost, I appreciate the contributions of the whole research team. Each member played an important role in terms of the data collection and database buildings related to the spatial and temporal land use and land cover, the climate, and the ecosystem services. I am impressed by the responsibility and contributions of them to this program in the whole process. Moreover, I appreciate the team leaders for their creative thoughts and advice.

Further, the major research plan of the National Natural Science Foundation of China (Grant No. 91325302), the Key Project funded by the Chinese Academy of Sciences (Grant No. KZZD-EW-08), the Key Project in the National Science and Technology Pillar Program (2013BAC03B03), and the National Key Program for Developing Basic Science in China (Grant No. 2010CB950900) also support this book in terms of data and models. In addition, during the preparation of this book, a number of scholars and experts in these projects gave us helpful advice and expertise.

Sincerely, I would like to express my gratitude to all the chapter authors, as well as to all anonymous reviewers in *Energies* and *Advances in Meteorology* for their constructive comments, which helped us to improve the book.

Finally, the colleagues and students from Beijing Normal University and the Institute of Geographic Sciences and Natural Resources Research, Chinese Academy of Sciences did a lot of work on map/table polishing and parameters revision, as well as text proofreading. I appreciate all their assistants. I thank all the people for their help directly and indirectly to complete this book.

Contents

Chapter 1
Reviews on Impact Assessments of Land-Use Change on Key Ecosystem Services

Xiangzheng Deng, Zhihui Li, Jikun Huang, Qingling Shi, Yanfei Li, Rongrong Zhang and Juan Huang

Abstract It is commonly acknowledged that land-use change (LUC) and climate change have exerted significant effects on ecosystem services which are essential and vital to human well-being. This chapter conducted a revisit to relevant researches on the impacts of LUC on human well-being via specifically altering the ecosystem provisioning services and climate regulation services. As to the issues related to ecosystem provisioning services, first, the explorations on the influences of LUC on ecosystem provisioning services were reviewed, including the researches on the influences of LUC on agro-ecosystem services and forest and/or grassland ecosystem services. Then, the quantitative identification of the impacts of LUC on ecosystem provisioning services was commented on. In light of enhanced observation and valuation methods, several approaches to ecosystem

X. Deng (✉) · Z. Li · J. Huang
Institute of Geographic Sciences and Natural Resources Research,
Chinese Academy of Sciences, Beijing 100101, China
e-mail: dengxz.ccap@igsnrr.ac.cn

X. Deng · Z. Li · J. Huang
Center for Chinese Agricultural Policy, Chinese Academy of Sciences,
Beijing 100101, China

Z. Li
University of Chinese Academy of Sciences, Beijing 100049, China

Q. Shi
China Center for Economic Studies, School of Economics, Fudan University,
Shanghai 200433, China

Y. Li · R. Zhang
Faculty of Resources and Environmental Science, Hubei University,
Wuhan, Hubei 430062, China

J. Huang
State Key Laboratory of Water Environment Simulation, School of Environment,
Beijing Normal University, Beijing 100875, China

© Springer-Verlag Berlin Heidelberg 2015
J. Zhan (ed.), *Impacts of Land-use Change on Ecosystem Services*,
Springer Geography, DOI 10.1007/978-3-662-48008-3_1

services and improved models for assessing those ecosystem services were analyzed. The major indicators used to uncover the influences of LUC on human well-being were summarized including the increase of inputs and the reduction of outputs in production and the augmented health risk induced by the irrational land uses. Finally, the research gaps were uncovered and several research directions to address these gaps were proposed. As to climate regulation services, the research efforts in the drivers of and their corresponding effects on climate regulation services are briefly identified. Then, we explicitly reviewed the researches on the effects of LUC and climate change on climate regulation services, especially focused on the certain methods and models used to quantify the effects on the major drivers of climate regulation services. After that, the effects of LUC and climate change on human well-being via climate regulation services were revisited and commented accordingly. Finally, we discussed the current research gaps and proposed some research prospects in future studies.

Keywords Ecosystem provisioning services · Climate regulation services · Land-use change · Climate change · Human well-being

Impacts of Land-Use Change on Ecosystem Provisioning Services

Introduction

The relationship between human activities and ecosystems has been discussed for many years by both natural and social scientists. LUC and climate variations and their effects on ecosystems have been core issues of the International Geosphere-Biosphere Program (IGBP) and International Human Dimensions Program on Global Environmental Change (IHDP). As two interacting processes, LUC and climate variations influence each other. On one hand, climate variations affect human activities, which indirectly exert influences on LUC, and on the other hand, LUC caused by humans accelerates influences on climate variations. Meanwhile, all these changes exert impacts on ecosystems together. In recent years, ecosystem services have been considered as an entry point of science to uncover the human and nature coevolution processes. Ecosystem services represent the benefits that living organisms derived from ecosystems to maintain the earth's life support system and emphasize the role of humans in socioecological systems, which include supporting services, regulating services, provisioning services, and cultural services (Reid et al. 2005). As the global population grows and its consumption patterns change, additional land will be required for living space and agricultural production. Then, the knotty question facing global society is how to meet humans' growing demands for living space, food, fuel, and other materials while sustain ecosystem services and biodiversity under LUC and climate variations. Numerous studies have shown that LUC and climate variations affected the structure and function of ecosystems and then affect the supply of ecosystem services (Nelson et al. 2010). LUC might increase the provision and value of some services but decrease others (Antrop 2000). And land-use decisions that intend to maximize a single outcome such as agricultural production or timber production are likely to generate an accompanying decline in the provision of other services.

It is well known that ecosystems provide necessary services for the livelihoods and human well-being through various ecosystem services, which directly or indirectly sustain the quality of human life. While LUC and climate variations have effects on those various ecosystem services in space and time, which are mainly presented through agro-ecosystem and forest and/or grassland ecosystems. As to agro-ecosystem services, it would definitely be weakened if the cultivated land was degenerated and the climate variations were intensified remarkably, which would result in the loss of food production. And it is the same to the forest ecosystem services and the grassland ecosystem services. And nowadays, there is considerable uncertainty about the values of ecosystem services, which are of great importance to be identified (Daily et al. 2009).

Ecosystems play an important role in providing goods through provisioning services, which are the most apparent connection between ecosystem services and human well-being. Provisioning services are manifested in the goods people obtain from ecosystems such as food and fiber; fuel in the form of peat; wood or nonwoody biomass; and water from rivers, lakes, and aquifers. These goods may be provided by heavily managed ecosystems, such as agro-ecosystem and plantation forests. Thus, provisioning services are the focus of human activities. When humans derive overmuch provisioning services from nature, it will put both ecosystems and humans at risk. The effects of LUC and climate variations may increase changes in ecosystem service delivery (Alcamo et al. 2005). As to the impacts of climate variations on human well-being, it can aggravate the situation for food security by increasing risks of crop failure because of the higher frequency of extreme events and progressive changes of climate (Sivakumar et al. 2005). In addition, as some ecosystem service decline, some new human actions, such as the excessive use of fertilizers and pesticides, have had adverse impacts on ecosystems and further on human well-being (Power 2010). Studies on the impacts of LUC and climate variations on ecosystem provisioning services and the impacts of provisioning services changes on human well-being will provide scientific and theoretical basis for global policy making.

Comprehensive understanding and acknowledgements of main research progress about the relationship between human activities and ecosystems and the influence on human well-being via altering the ecosystem services are of great significance to guide future studies and further policy making. Thus, this study integrated previous studies to conduct a revisit to the impacts of LUC and climate variations on the human well-being via altering the ecosystem provisioning services (Fig. 1.1). In this chapter "Exploration on the Functions of LUC on Ecosystem Provisioning Services" reviewed the exploration on the functions of LUC and climate variations on ecosystem provisioning services. Climate variations related to ecosystem services can be expressed by changes in temperature, precipitation, heat flux, some abnormal climate, etc. LUC related to ecosystem

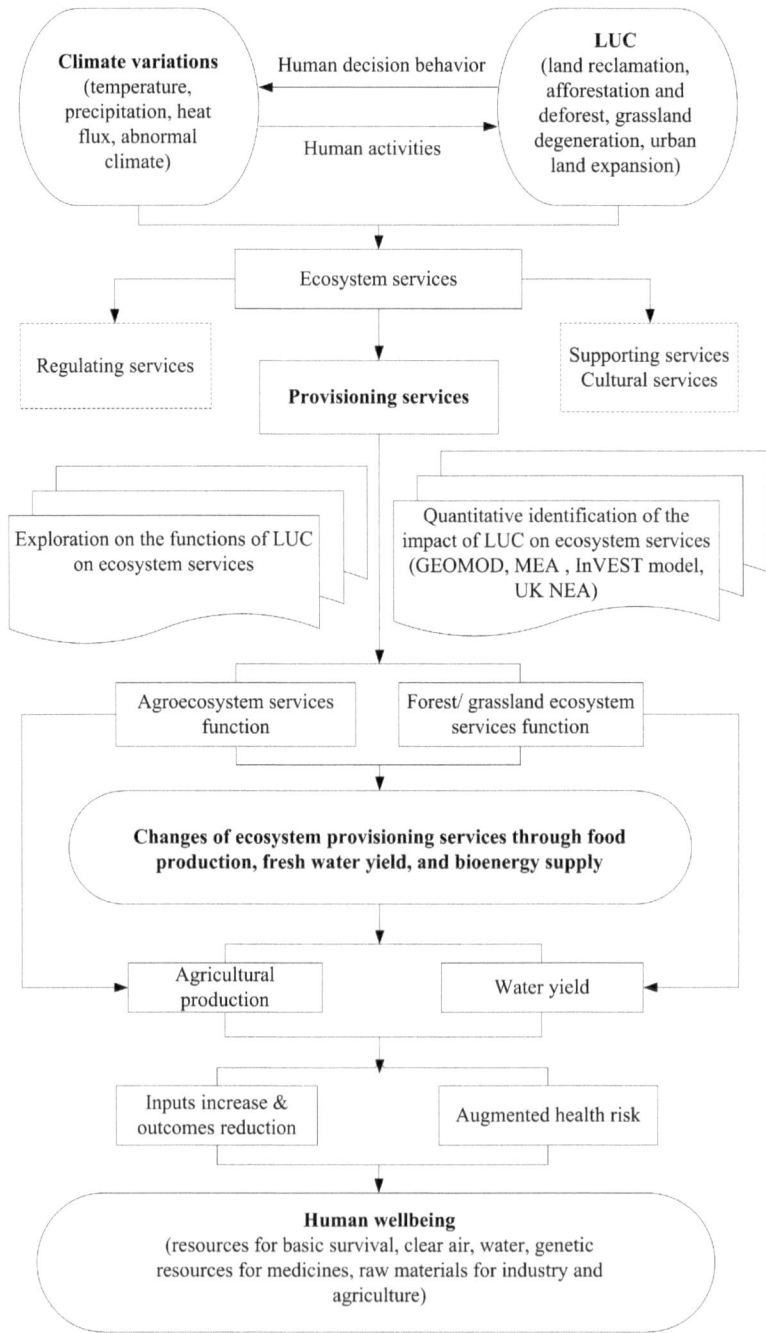

Fig. 1.1 Diagram to trace the impacts of LUC and climate variations on the human well-being via altering the ecosystem provisioning services

services includes cultivated land reclamation, afforestation and deforestation, grassland degeneration, and built-up land expansion. The changes of them would have impacts on ecosystem services, including regulating services, provisioning services, supporting services, and cultural services. In addition, the altering of the provisioning services of agro-ecosystem and forest and/or grassland ecosystems was explored and commented on. In Section "Quantitative Identification of the Impacts of LUC on Ecosystem Services", the quantitative identifications of LUC and climate variation impacts on ecosystem services were reviewed. The enhanced observation and valuation approaches of ecosystem services and improved models for assessing ecosystem services were illustrated. In Section "Major Indicators Identifying Influences of Ecosystem Provisioning Services on Human Well-Being", the major indicators used to uncover the influences of ecosystem service changes on human well-being were revisited via the increased inputs with reduced outputs in production and the augmented health risk induced by the irrational land uses. As to Section "Concluding Remarks", some research gaps were identified, and the research needs and research prospects in further studies were also refined.

Exploration on the Functions of LUC on Ecosystem Provisioning Services

The ecosystems can provide a variety of direct and indirect services to humans and other living organisms, and those services can be affected by climate variations and human activities, especially human-induced LUC (Bangash et al. 2013). Some studies have shown that LUC and climate variations are among the greatest global environmental pressures resulting from anthropogenic activities, which significantly influence the provision of crucial ecosystem services, such as carbon sequestration, water flow regulation, and food and fiber production, at a variety of scales (Reyers et al. 2009). In the researches of the relationship between LUC and ecosystem services, the analyses of the impacts of LUC on ecosystem services were usually conducted from the aspect of land-use quantity and structure changes. Some studies have indicated that the diversification of land use would help to improve ecosystem services (Swift et al. 2004). In addition, combined effects of LUC and climate variations may change ecosystem services, especially the food provision and water yield of agro-ecosystem, and forest and/or grassland ecosystems (Sun et al. 2006).

Influences of LUC on the Aagro-Ecosystem Services

As a kind of specific complex manual–natural ecosystem, agro-ecosystem has not only efficient and direct production function, but also the function of environmental services, tourism services, and aesthetic services (Swinton et al. 2007). Agriculture is a dominant form of land management globally, and agro-ecosystem covers nearly 40 % of the terrestrial surface of the earth (Power 2010).

Agro-ecosystem is faced with severe challenges under the context of global warming and the intensive human-induced LUC.

More and more current studies have shown that LUC especially rapid urbanization has already directly or indirectly affected the food provisioning services of agro-ecosystem (Jaradat and Boody 2012). The quality and quantity changes of agricultural land have potential effects on provision services of agro-ecosystem. However, some studies have shown that agricultural land use has degraded the soil, water, net primary productivity (NPP), and the biological assets in agro-ecosystem to such an extent that the restoration of natural capital and rehabilitation of ecosystem services are needed through changes in land use and management (Foley et al. 2011). A primary reason for this degradation is the failure of agricultural commodity markets to internalize environmental costs associated with land use and management decisions.

The LUC especially the excessive reclamation of cropland and the intensive agriculture land use exerts potential effects on biodiversity conservation of agro-ecosystem, which will influence the stability of provisioning services in agro-ecosystem. Some studies have indicated that the rapid expansion and intensification of row crop production have resulted in the loss of habitat and spatial heterogeneity in agro-ecosystem, which affects the ecosystem provisioning services (Gavier-Pizarro et al. 2012). Land-use conversion from natural lands to croplands, grazing lands, and urban areas has been increased over time, resulted in reduced or modified biodiversity, altered functional processes, and diminished provision of ecosystem goods and services to society globally (DIAz et al. 2007). Some studies also indicated that the intensive agricultural development could change land use, which can further affect regional ecosystem services (Dale and Polasky 2007).

Influences of LUC on the Ecosystem Services from Forest and/or Grassland

Forest and grassland ecosystems are indispensable constituent parts of terrestrial ecosystems which play important roles in global climate variations. Climate variations affect the water yield of forest and/or grassland ecosystems via its direct influence on precipitation and evaporation process of atmosphere hydrologic cycle. LUC, including converting grassland or shrublands to plantations, afforestation, and reforestation, is gaining attention globally and will alter many ecosystem processes, including water yield of forest and grassland ecosystems (Farley et al. 2005; Nosetto et al. 2005). Changes in the extent and composition of forest, grassland, wetland, and other ecosystems have large impacts on the biophysical conditions, which further affect the provision of ecosystem services and biodiversity conservation. The LUC influences the water yield of ecosystems through changing the transpiration, interception, and evaporation, all of which tend to increase when grassland or scrubland is replaced with forests. Transpiration rates are influenced by changes in rooting characteristics, leaf area, stomata response, plant surface albedo, and turbulence (Vertessy et al. 2001).

Much progress has been made to understand the effect of LUC on water yield of forest ecosystems during the past century all over the world, and the results of which generally indicated that LUC has both positive and negative effects on water yield. For example, clear-cutting forests in the USA may result in the increase in annual water yield (Ice and Stednick 2004). The vegetation restoration will have positive effects on watershed health by reducing soil erosion and non-point source pollution, enhancing terrestrial and aquatic habitat, and increasing ecosystem carbon sequestration (Sun et al. 2006). Some studies also indicated that vegetation changes, particularly those involving transitions between forests and grasslands dominated covers, often modify evaporative water losses as a result of plant-mediated shifts in moisture access and demand. And massive afforestation of native grassland has strong yet poorly quantified effects on the hydrological cycle (Nosetto et al. 2005). Since forests with well-developed root systems cost plentiful groundwater and soil water, it can save plenty of water to convert forest into short seasonal crops. To plant abundant pasture instead of forest in catchment areas is becoming a widely used method to increase water yield.

Quantitative Identification of the Impacts of LUC on Ecosystem Services

Many ecologists and natural scientists study ecosystem processes to understand ecosystem services across different landscapes via quantifying ecosystem services (Raudsepp-Hearne et al. 2010). Quantification and valuation of services, if linked with payments or incentives, can enhance policies and regulations that properly reward decisions that yield public benefits. It is well known that ecosystems are essential to the existence of humans, while ecosystem services are typically not priced correctly at their value because of absence of markets for ecosystem goods and services and inadequate or nonexistent information about the value of goods and services. There are many studies evaluating the impacts of LUC on ecosystem services in both developed and developing countries (Su et al. 2012). To sum up the quantitative researches so far, we found that there are two kinds of methods to value the ecosystem services, namely observations based on remote sensing and GIS technology and modeling approaches.

Enhanced Observation and Valuation Approaches of Ecosystem Services

The historical ecosystem service value could be reflected by remote sensing along with a GIS-based model (GEOMOD) since LUC is of the upper most driving forces of regional ecosystems and has huge impacts on ecosystem service value. Remote sensing provides reliable area-wide data for quantifying and mapping ecosystem services at comparatively low costs, and with the fast, frequent, and continuous observations for monitoring. GEOMOD model is a kind of method and technique to allocate LUC spatially and to evaluate its impact on ecosystems simply and transparently.

The selection of indicators to be used in the analyses of potential impacts of LUC is the main challenge after obtaining the remote sensing data. The valuation of different ecosystem services and the spatial–temporal monitoring of their respective changes can provide useful indicators of the potential impacts of LUC. Generally speaking, as the ecosystem service values were different for each land-use category (Estoque and Murayama 2012), Costanza et al. (1997) first attempted to estimate the ecosystem service value coefficients (Costanza et al. 1997), and then, many researches use the same approach in order to quantify and map the ecosystem service values at global or regional scales (Turner et al. 2007). Although there are many potential conceptual and empirical problems and limitations to estimate the ecosystem service values (Nelson et al. 2009), the magnitude of the estimated ecosystem service value changes in the LUC is substantial. Thus, it may still be possible to draw general inferences about the effect of the perceived LUC on the estimated ecosystem service values.

Improved Models for Assessing Ecosystem Services

More and more ecosystem service values were assessed by models around the world in recent decades, and it seems to have become a trend to assess various ecosystem services with models. To avoid the weakness of common assessment models used before, some improved approaches or models were formed to assess the ecosystem service values, including the Millennium Ecosystem Assessment (MEA) approach (Chopra et al. 2005), Integrated Valuation of Ecosystem Services and Tradeoffs (InVEST) (Tallis and Polasky 2009) model, and the UK National Ecosystem Assessment (UK NEA) (Assessment 2011).

The MEA was set up in 2000, and it is the first time to provide a comprehensive picture of past, present, and possible future trends in ecosystem services and their values and propose corresponding measures. Many researchers have assessed the ecosystem services at national scale with framework of MEA. They have valued the agricultural ecosystem services, forest ecosystem services, and grassland ecosystem services in China (Xie et al. 2006). For instance, 17 ecosystem services in 18 categories of grassland ecosystem in China have been assessed (Xiao et al. 2003). Exploring the researches between LUC and ecosystem services, we found that most studies analyzed the impacts by analyzing the quantity and structure change of land use. It is generally acknowledged that the provision of ecosystem services depends on biophysical conditions and changes over space and time due to human-induced LUC. Spatial patterns of LUC can be linked to large regions and provide direct measures of human activities (Riitters et al. 2002).

The InVEST model has been widely used in valuating ecosystem service values (Polasky et al. 2011). The model uses maps and tabular data of land use and land management in conjunction with environmental information, such as soil, topography, and climate, to generate spatially explicit predictions of the ecosystem services. InVEST model estimates the provision and value of ecosystem services under alternative land-use scenarios. Economic information about demand for ecosystem services can be combined with biophysical supply to generate predictive maps of services use and values (Xie et al. 2006; Daily et al. 2009). InVEST model also analyzes the impacts of land use and land management on species habitat provision and quality. Thus, the model provides a powerful tool for quantifying and valuing multiple ecosystem services and assessing the impacts of LUC. By varying land use or land management and evaluating the corresponding output with InVEST, we can provide useful information to managers and policy-makers to weigh the trade-offs in ecosystem services, biodiversity conservation, and other land-use objectives.

The UK NEA is the first analysis of the UK's natural environment in terms of the benefits it provides to society and the nation's continuing prosperity. It has been a wide-ranging, multi-stakeholder, cross-disciplinary process, designed to provide a comprehensive picture of past, present, and possible future trends in ecosystem services and their values; it is underpinned by the best available evidence and the most up-to-date conceptual thinking and analytical tools, which can be applied to assess the ecosystem service values aimed to describe the changes of key drivers that affect the UK's ecosystems, including changes in land use and climate. The UK NEA distinguished between the ecosystem processes and intermediate ecosystem services and the final ecosystem services that directly deliver welfare gains and/or losses to people. This distinction is important to avoid double counting in the valuation of ecosystem services (Fisher et al. 2008). As the researches go on, the UK NEA would have a broad application prospects to assess the impacts of human activities on ecosystem services.

At any rate, ecosystem services play an important role in maintaining the balance of global ecosystems and improving human living environment. Quantifying and mapping those ecosystem services is necessary to periodically determine the response of ecosystem services to global change, such as LUC and climate variations. Those approaches and models, though being widely used to quantify and map the ecosystem service values around the world, still have potential to be improved in order to get more accurate assessment results. In addition, so far there is still no assessment system and method that has been commonly approved by researchers. Therefore, it is still a hot issue to study the assessment of ecosystem service values to clarify the relationship between LUC and ecosystem services.

Major Indicators Identifying Influences of Ecosystem Provisioning Services on Human Well-Being

Ecosystem services are essential for the maintenance of human well-being, and the links between ecosystem services and human well-being are complex, diverse, and complicated to assess properly with the consideration of different spatial and temporal scales (Pereira et al. 2005). Healthy ecosystems provide services that are the foundation for human well-being including the provision of resources for basic survival, such as clean air, water, and genetic resources for medicines, along with the provision of raw materials for industry and agriculture (Daily et al. 1997). Thus, the degradation and loss of ecosystem services have negative effect on human well-being. On the one hand, the degradation and loss of ecosystem provisioning services will increase the inputs in production to recover reduced outputs in the ecosystems. On the other hand, the degradation and loss of provisioning ecosystem services will increase the human health risk.

Increased Inputs with Reduced Outputs in Production

Ecosystems are changed by the LUC and climate variations, all of which may reduce or increase the supply of ecosystem services temporarily or permanently. Some evidences have shown that the climate variations and human activities especially LUC have changed agricultural and natural ecosystems more rapidly and extensively over the last 50 years (Reid et al. 2005).The MA reported that 15 of the world's 24 ecosystem services are in decline, which have affected human well-being and threaten the survival of other species. The declining ability of the earth's systems to meet the needs of a growing population and sustain the life support systems of the planet is a very urgent and serious issue. Due to the impacts of climate variation and LUC on agro-ecosystem services, the outputs humans obtained in agricultural production will decrease, including the food production and water yield. Therefore, humans would input more production factors such as fertilizers and pesticides in the process of food production and invest more in search of more water resources to sustain the continual ecosystem provision of human well-being.

The development of novel chemicals and new technologies during the last century has supported the modern agricultural revolution, resulting in an increase in food production and harvest rates (Galis et al. 2013). Some studies have indicated that the application of fertilizers (such as the nitrogen application) and pesticides in some regions increased rapidly to meet the demands for greater food production needs (Nelson et al. 2006). The increasing use of these chemicals and technologies removed the constraint of nutrient limitation for crop growth, as well as competitive pests and weeds, resulting in increased outputs. However, on the other side, these increased outputs are at the cost of the reducing ecosystem services, especially the provisioning services. Thus, human well-being would be weakened. With the ecosystem services changing, governments have supplied subsidies and grants to adjust personal and business behavior.

Augmented Health Risk Induced by the Irrational Land Uses

Ecosystem services can support the fundamental need of human well-being in a variety of ways. The changes of ecosystem services would both directly and indirectly put the humans' health at risk through the insufficient provision of food and freshwater, inorganic chemicals and persistent organic chemical pollutants in food and water, and infectious disease caused by ecosystem service loss. Besides, the indirect health risk was caused by the irrational land uses.

The insufficient accesses to the ecosystem provisioning services of food production and water yield are particularly important factors leading to the health risks in human well-being. Some studies have indicated that a lack of access to the ecosystem provisioning services of food causes far more than physical harm, and it may put thousands of millions of people in mental and physical potential risks by reducing intelligent and physical growth, in some cases from the moment of human conception (Reid et al. 2005). Undernutrition was recently assessed as an underlying cause of death each year worldwide, which is particularly common in sub-Saharan Africa and south Asia, especially in India (Caulfield et al. 2004). Vegetation especially the grass and forest are important for the interception of water. Some studies have shown that about 1 billion people were affected by land degradation caused by soil erosion, waterlogging, or salinity of irrigated land.

Humans are also at risk due to inorganic chemicals and persistent organic chemical pollutants in food and water, and the infectious disease caused by ecosystem service loss. Some studies indicated human actions, for example, releasing toxic chemicals into the environment, or using pesticide and chemical fertilizer, will pollute the water and food, which can have adverse effects on various organ systems (Aktar et al. 2009). Some evidence indicated that some chemicals from pesticide and chemical fertilizer have increased the microbial contamination of drinking water, which has led to the infectious diseases accounting for approximately 6 % of all deaths globally. The pattern and extent of change in incidence of particular infectious disease depends on the particular ecosystems affected, such as type of LUC. Some studies showed that climate change and some LUC, such as deforestation, might alter infectious disease patterns (Patz et al. 2003; Sehgal 2010). There have been lots of studies to investigate the influence of the increased income and the health risk of human well-being, which will be helpful to profoundly understand the influence mechanism and extent of the degradation of ecosystem services on human well-being.

Concluding Remarks

Based on current researches about the effects of LUC and climate variations on ecosystem services, we mainly focus on ecosystem provisioning services and the influence of the changes in provisioning services on human well-being. Firstly, we explored the researches on identification and quantification of the impacts of LUC on ecosystem service values and latter examined how the impact on ecosystem provisioning services affects human well-being through analyses of the increased inputs and the reduced outputs of agricultural production and the augmented health risk of humans.

So far, there are still some researches to be done to uncover the impacts of LUC and climate variations on human well-being via ecosystem provisioning services. First of all, the current researches are focused on the ecosystem service values, but the mechanisms through which LUC and climate variations influence ecosystem services are still not well understood. However, this is of great significance to the sustainable development of human well-being. Secondly, there would be uncertainty involved when the remote sensing data were used for quantifying and mapping ecosystem services. Therefore, validating the reliability of the results obtained by using remote sensing in quantifying and mapping ecosystem services needs to be done in further researches. As to the impacts of ecosystem service changes on human well-being, more researches on quantification of these impacts need to be done to make the research more comprehensive.

It has been shown by research practice that the assessment of ecosystem services was useful to the setting of strategies and policies, with potentially far-reaching influence on human activities. Thus, further researches are needed to focus on the following three issues. First, there is a need to formulate a set of thorough and normative method to assess ecosystem service values and improve the accuracy of assessment results. Second, an in-depth process-based analysis of the relationship between human activities and ecosystem service function is needed. Third, there is an urgent need to promote the application of ecosystem service values in various aspects of production, livelihood, and government decision making and eventually serve for human well-being.

Impacts of Land-Use Change on Climate Regulation Services

Introduction

Land-use change (LUC) and climate change are two major factors that result in the changes of ecosystem services (Schröter et al. 2005). Along with the socio-economic development and emerging ecological environmental problems, global changes and ecosystem services are becoming the research hot topics. The relationships among LUC, climate change, and ecosystem services are interlaced and complex, in which temporal and spatial variations in human-induced LUC and climate variability can result in the difference of ecosystem services (Chen et al. 2013).

Natural ecosystem delivers a lot of benefits to human beings, and these benefits are known as ecosystem services. According to the Millennium Ecosystem Assessment (2005), these ecosystem services include provisioning services such as provision of food, water, timber, fiber, and genetic resources; regulating services such as the regulation of climate, floods, disease, and water quality as well as waste treatment; cultural services such as recreation, aesthetic enjoyment, and spiritual fulfillment; and supporting services such as soil formation, pollination, and nutrient cycling (Reid et al. 2005). Among all these services, supporting services and regulating services underpin the delivery of other service categories (Kumar 2010). What is more, there often exist trade-offs between different services when humans make management choices, which can change the type, magnitude, and relative mix of services provided by ecosystems (Rodriguez et al. 2006). However, people generally prefer provisioning and cultural services over regulating services (Carpenter et al. 2006) and thus tend to undervalue regulating services. Consequently, decision-makers often ignore these regulating services in ways that will seriously undermine the long-term existence of provisioning services (Kumar 2010). The regulating services provided by ecosystems are diverse, among which climate regulation is a final ecosystem service. The ecosystems regulate climate through biogeochemical and biogeophysical processes, as sources or sinks of greenhouse gases (GHGs) and as sources of aerosols all of which affect temperature and cloud formation (Bonan 2008; Fowler et al. 2009).

The processes involved in climate regulation include the following: (1) CO_2 in the atmosphere was absorbed through photosynthesis; (2) evapotranspiration from soils and plants controls the amount of water vapor entering the atmosphere, thus regulating cloud formation and the radiative properties of the atmosphere; (3) the change of the albedo of different land surfaces can affect the climate; for example, the change in vegetation can have a cooling or heating effect on the surface climate and may affect precipitation; (4) the regulation of aerosols comes from soil erosion or vegetation through vegetation scavenging, which affects radiative heating of the atmosphere, surface albedo, and so forth (Bonan 2008).

There are many direct and indirect drivers that can affect the process of climate regulation services. For the systematic understanding of the effect mechanism and quantitative evaluation of the effects on climate regulation, it is important to clarify the major drivers and quantitatively analyze the effects induced by LUC and climate change on those drivers through both biogeochemical and biophysical processes. Comparatively, a direct driver more unequivocally influences ecosystem processes, while an indirect driver operates more diffusely by altering one or more direct drivers; that is, direct drivers have much more explicit effects on ecosystem processes (Nelson et al. 2005) and usually cause physical change that can be identified and monitored (Ash et al. 2008). As to climate regulation services, the indirect drivers mainly include demographic drivers (population growth and distribution, migration, ethnicity, etc.), economic drivers (economic growth and consumer choice, market force, industry size, globalization, etc.), and sociopolitical drivers (legislation, regulation, etc.). The direct drivers are listed in Table 1.1, among which land-use drivers are the most important in the ecosystem context and can be identified as the main drivers of climate regulation (Anderson-Teixeira et al. 2012), while over longer term, climate change will also have feedback to climate regulation services (Pete et al. 2011).

In this study, we focus on choosing close researches to explore how LUC and climate change affect the climate regulation through biogeochemical and biogeophysical processes, respectively. Then, incorporating with the effects on climate regulation services, researches on impact assessment for human well-being

Table 1.1 Direct drivers of changes in climate regulation and their corresponding effects, adapted from (Pete et al. 2011)

Category of drivers	Change of drivers	Effect on climate regulation services
Habitat change: land and sea use	Productive area: expansion, conversion, abandonment (agriculture, forestry)	Affects carbon sinks and existing stores, greenhouse gas (GHG) emissions, albedo and evapotranspiration, shade and shelter
	Mineral and aggregate extraction (peat)	Affects soil carbon stores, GHG emissions
	Urbanization and artificial sealed surfaces	Affects soil carbon stores, albedo, shade, shelter, local temperatures, and humidity
Pollution and nutrient enrichment	Pollution emissions and deposition	Affects aerosol sources (soot)
	Nutrient and chemical inputs	Affects GHG emissions
Harvest levels/resource consumption	Livestock stocking rates	Affects GHG emissions
Climate variability and change	Temperature and precipitation	Affects existing carbon stores, evapotranspiration
	CO_2 and ocean acidification	Affects existing carbon stores, GHG emissions, aerosol sources
	Sea-level change	Affects existing carbon stores

were further revisited. This study firstly describes the climate regulation services and the needs and significance to study climate regulation services, then examines how LUC and climate change affect climate regulation services through review of the researches on major drivers, and after that outlines how changes of the role of ecosystem services in regulating climate affect human well-being through investigation of four major aspects (economic value, extreme weather, food security, and human health) that are closely related to human well-being. This review study intends to provide a reference for the future research on LUC, climate change, climate regulation services, and human well-being. The framework for the review about the effects induced by LUC and climate change on climate regulation and further impact assessment for human well-being is shown in Fig. 1.2.

LUC-Induced Effects on Climate Regulation Services

As mentioned above, LUC plays an important role in climate regulation. Anthropogenic land use has been and will continue to be a major driver of the changes in climate system (Anderson-Teixeira et al. 2012). And there have been many observations and simulations revealing that LUC exerts effects on climate regulation services through biogeochemical and biogeophysical processes.

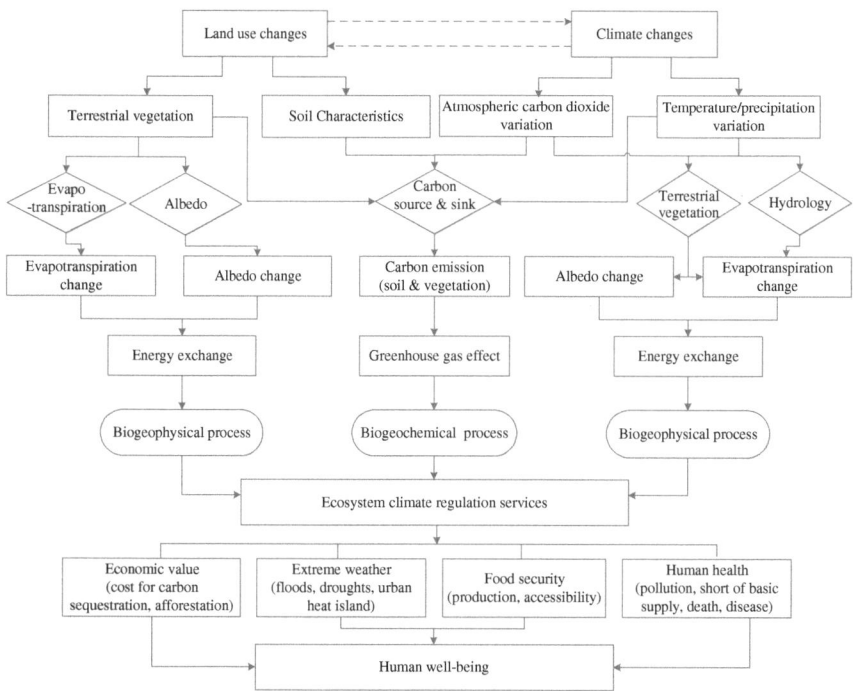

Fig. 1.2 Framework for integrating land-use-induced effects on climate regulation services into impact assessment for human well-being

Cumulative Effects Through Biogeochemical Processes

In terms of biogeochemical processes, LUC mainly affects the climate regulation services through the emission and sequestration of GHGs, especially through altering CO_2 flux. The total amount of carbon stored in terrestrial biosphere is an important factor in climate regulation (McGuire et al. 2001). Terrestrial ecosystems contribute to climate regulation primarily through carbon dynamics. Plants absorb CO_2 through photosynthesis, storing carbon in vegetation and soils, and the carbon accumulated in soil and biomass represents a pool of carbon which is greater than the atmospheric carbon pool (Lal 2004a). Deforestation, forest degradation, and other land-use practices accounted for approximately 20 % of global anthropogenic CO_2 emissions during the 1990s (IPCC 2007). When carbon is released from the earth during cultivation, deforestation, fire, and other land-use practices, it binds with other chemicals to form GHGs in the atmosphere and accelerates global climate change (Lal 2004b). The conservation of carbon sinks or pools is therefore important to mitigate GHGs levels. Thus, it is of great significance to investigate the effect of LUC on climate regulation through its biogeochemical process, that is, through the effects on the cycle of carbon. And LUC can change the release of carbon to the atmosphere mainly through the disturbance on terrestrial vegetation and soils.

Biogeochemical Process Related to Terrestrial Vegetation

Terrestrial vegetation is a large carbon sink which plays an important role in the global carbon cycle and is valued globally for the services it provides to society. Vegetation classifications have been related to climate variables and used in the assessment of possible global response to climate change (Smith et al. 1992). Olson et al. (1983) built up a computerized database to document the map of vegetation and corresponding carbon density for natural and modified complexes of ecosystems. The map provides a basis for making improved estimates of vegetation areas and carbon quantities (Olson et al. 1983), illustrating that different types of vegetation have various ability of carbon sequestration. LUC can have great effects on the structure of terrestrial vegetation (Reidsma et al. 2006); for example, once the forest is being converted to cultivated land, the biodiversity will decline. The distribution of sources and sinks of carbon among the world's terrestrial ecosystem is uncertain, through deforestation, urbanization, expansion of cultivated land, and other land-use practices; LUC variously alter the land surface and species compositions and exert various effects on the carbon cycle.

It is relatively difficult to quantify the process that LUC affects the carbon sink or net source of terrestrial vegetation. Most researches developed and applied different models to record the carbon emission or sequestration resulted from LUC. Firstly, many researchers adopted empirical data to simulate carbon emission and sequestration. Houghton et al. (1983, 2000, 2003) calculated the carbon emission resulting from LUC and the potential for sequestering carbon of different land covers mainly based on the parameters set for each vegetation type in each regime (Houghton et al. 1983, 2000, 2003), and the model was called the "bookkeeping" terrestrial carbon model. Further, Houghton et al. (2001) documented a numeric data package that consists of annual estimates of the net flux of carbon between terrestrial ecosystems and the atmosphere resulting from deliberate LUC, especially forest clearing for agriculture and the harvest of wood for wood products or energy from 1850 through 1990 (Houghton et al. 2001). DeFries et al. (2002) further apply the "bookkeeping" model in conjunction with multiple sources of remote sensing data to estimate carbon fluxes from tropical LUC (DeFries et al. 2002). The "bookkeeping" model tracks the amount of carbon released to the atmosphere from clearing and decay of plant material and adds the amount of carbon accumulated as vegetation grows back. However, the parameters in the "bookkeeping" model are mainly based on empirical data and lack mechanism process. Then, some researches started to combine process-based ecosystem models with the "bookkeeping" model, which can associate the land-use information with its related process and thus enhanced the accuracy of the estimation. For example, one type of the process-based ecosystem model, the Terrestrial Ecosystem Model (TEM), in which the effect of LUC was characterized by the specific parameters of the function of different vegetation types, can be combined with the "bookkeeping" (McGuire et al. 2001). And DeFries et al. (1999) predicted the effect of LUC on carbon cycle based on the vegetation distribution information derived from remote sensing data and combined with Carnegie–Ames–Stanford Approach

Model (CASA) (DeFries et al. 1999). What is more, Sitch et al. (2003) also coupled the "bookkeeping" model with Lund-Potsdam-Jena dynamic global vegetation model (LPJ) (Sitch et al. 2003). Particularly, McGuire et al. (2001) combined four models, that is, TEM, High-Resolution Biosphere Model (HRBM), LPJ, and Integrated Biosphere Simulator (IBIS) model (McGuire et al. 2001). And Levy et al. (2004) combine the HYBRID vegetation dynamic model with the "bookkeeping" model to estimate the carbon dynamic resulted from LUC (Levy et al. 2004). Differently, Schröter et al. (2005) used a range of ecosystem models and scenarios of climate and LUC to assess the vulnerability of ecosystem services, and the simulation results showed that LUC affected terrestrial carbon sink positively through decreases in agricultural land and increased afforestation during the twenty-first century. Different models have different results about the amount of the carbon emission induced by LUC, but all the results show that global deforestation can lead to large carbon emission.

Among all the vegetation types, forest as one of the productive area is a major part of terrestrial ecosystem. Global deforestation has great effect on the carbon emission (Woodwell et al. 1983). Using forest inventory data and long-term ecosystem carbon studies, Pan et al. (2011) estimated a total forest sink of petagrams of carbon per year (Pg C $year^{-1}$) globally for 1990–2007, and tropical deforestation around the world contributes to a gross emission of Pg C $year^{-1}$, which was equal to 40 % of global fossil fuel emissions (Pan et al. 2011). Combined with spatial explicit information on changes in forest area which derived from satellite observations and terrestrial carbon models, many researches focused on the tropics to detect whether tropic deforestation acts as carbon sink or net source and identified that tropics acted as carbon sink (DeFries et al. 2002). Explicitly, Arora et al. (2009) assessed the biogeochemical behavior of the CCCma earth system model (CanESM1) against observations for 1850–2000 and then compared simulated atmospheric CO_2 concentration with available observations and observation-based estimates; the simulation results showed that forests' different photosynthesis ability and CO_2 emissions from LUC resulted in different density of CO_2 in atmosphere, which indicated that the tropics were large carbon sinks (Arora et al. 2009).

Currently, based on model simulation, it was acknowledged that LUC can affect the carbon sink or source. Firstly, many researches were based on empirical data and statistical models which can use only observation data about the land cover without knowing conversions among different land cover types. While with the application of remote sensing and developed ecosystem models, the carbon emission induced by LUC can be more accurately estimated. However, most models also cannot take some important factors into account, such as the feedback of regional climate change, the water cycle before and after LUC, and the physical structure of soil. More efforts should be devoted to develop models that can comprehensively integrate all the factors to do simulations.

Biogeochemical Process Related to Soil Carbon

The changes of LUC can affect the sink of carbon not only through the sequestration of vegetation covering the land, but also through the changes of soil characteristic of different land cover types. It has been identified that different land cover types' potentials of soil carbon are various. For example, tropical forests were estimated to store approximately 206 Pg C in the soil globally, which was relatively less than half of that of boreal forests' soil (Dixon et al. 1994). And explicitly, wet and moist tropical forests tend to have greater soil carbon pools per unit area than tropical dry forests due to higher rates of NPP (Brown and Lugo 1982). The potential of cropland soils (Kimble et al. 1998) and grazing land soils to sequester carbon and mitigate the greenhouse effect is different. Dawson and Smith (2007) reviewed effects of LUC and climate change on soil carbon losses and found that the soil carbon emissions of agriculture land (cropland and grassland), forestry land, and peatlands/wetlands explicitly have different ability of carbon sequestration, and human's land-use management is critical to regulate carbon emission.

LUC plays an important role in determining whether soil is a sink or source of atmospheric CO_2, since it can change the amount of soil carbon sequestration. For example, deforestation can result in an initial loss of carbon from soils due to increased decomposition rates, erosion, and reduced inputs of vegetation residues, namely soil organic matters (SOM) (Lugo and Brown 1993). Ayanaba (1976) reported that deforestation for cultivated land leads to the reduction of soil total carbon content (Ayanaba 1976). As most researches about the effect of deforestation on soil carbon only focused on one kind of land conversion while did not simultaneously take all the land cover types into consideration, Powers (2004) presented better understanding of the effects of LUC on regional and global ecosystem process by considering all of the land-use conversions that occur in the landscape and detect the difference among the effects on soil carbon of each land-use conversion based on statistical analysis (Powers 2004). Scott et al. (2002) designed a soil carbon monitoring system for New Zealand using country-specific land use and soil carbon information, which can be used to estimate both soil carbon stocks at a single point in time and also future soil carbon stocks in response to LUC (Scott et al. 2002).

LUC affects the soil carbon content mainly through regulating soil organic carbon (SOC) pools. The dynamic changes of SOC have a strong effect on atmospheric composition and the rate of climate change, thus playing an important role in climate regulation. There have been many studies quantifying the effects of LUC on SOC, and since the carbon dynamics in soil are a long-term process which is relatively difficult for doing field researches and observations, most of the studies were based on model simulations. The major SOM models include SOM model (SOMM), Institute of Terrestrial Ecology, Edinburgh (ITE), Verberne, Rothamsted C model (ROTHC), Carbon-Nitrogen-Dynamics (CANDY), DeNitrification and DeComposition (DNDC), CENTURY, DAISY, and NCSOIL. Smith et al. (1997) compared these nine models with data from seven long-term experiments representing multiple vegetation and management conditions (Smith et al. 1997); the results showed that a comparison of the overall performance of models across all datasets reveals that the model errors of one group of models (ROTHC, CANDY, DNDC, CENTURY, DAISY, and NCSOIL) did not differ significantly from each other. Another group (SOMM, ITE, and Verberne) did not differ significantly from each other but showed significantly larger model errors than did models in the first group. Among all the nine models, ROTHC and CENTURY are two of the most widely used and tested SOM models. Shrestha et al. (2009) used the CENTURY model to simulate the changes in SOC pool over 100 years (1950–2050) under managed dense Shorea forest, rainfed upland, and irrigated low land in a midhill watershed of Nepal (Shrestha et al. 2009). And Dieye et al. (2012) analyzed the sensitivity of the GEMS SOC model in response to LUC (Dieye et al. 2012); the GEMS model was also developed from the CENTURY model (Metherell 1993). Barancikova et al. (2010) and Francaviglia et al. (2012) used the ROTHC model to simulate the changes in agricultural land's SOC pools (Barančíková et al. 2010; Francaviglia et al. 2012). SOC cannot only be affected by LUC, but also by climate change, which will be revisited in Section "Biogeochemical Process of the Effect on Climate Regulation Services".

Combined Biogeochemical Process of Terrestrial Vegetation and Soil

As mentioned above, most researches focused on the calculation of carbon emission resulted from the terrestrial vegetation changes with different models and methodologies, based on the historical observations or scenario analysis. And also, there are researches trying to figure out the change of carbon emission as the change of soil characteristic of different land cover types resulted from land-use management practices. However, it is also important to combine the two biogeochemical processes together to detect the effect of LUC on carbon cycle and further analyze the effect on climate regulation services.

As to this point of view, Woomer et al. (2004) defined the total carbon in the ecosystem as the sum of the woody biomass, herbaceous biomass, root, litter, and soil carbon pools, and using an inventory procedure involving satellite images which reveal historical LUC and recent field measurements of standing carbon stocks occurring in soil and plant; they estimated Senegal's terrestrial carbon stocks in 1965, 1985, and 2000 (Woomer et al. 2004). In addition, the changes of carbon emission from vegetation and soil resulted from LUC and climate change can also counteract with each other. Synthetically considering the carbon emission from vegetation and soil induced by LUC and climate change, based on a series of ecosystem models and scenarios of LUC and climate change to evaluate the ecosystem service supply and vulnerability, Schröter et al. (2005) confirmed that Europe's terrestrial biosphere acted as a net carbon sink, with decreased agricultural land and increased afforestation resulting in the increasing sequestrated amount of carbon, while soil carbon losses due to warming would balance the carbon sequestrated by terrestrial biosphere by 2050 and lead to carbon releases by the end of this century (Schröter et al. 2005). In the (InVEST) model, a simplified carbon cycle that maps and quantifies the amount of stored and sequestered carbon based on five carbon pools (aboveground biomass, belowground biomass, soil, dead organic matter, and harvested wood products) is incorporated. Leh et al. (2013) used average literature values in the InVest model to calculate the carbon as part of the ecosystem services in major land classes according to the proportion of land-use/land cover (LULC) area in Ghana and Cote d'Ivoire, West Africa (Leh et al. 2013); the model is of great significance to evaluate the ecosystem services.

In summary, effect on climate regulation through biogeochemical process mainly focuses on the carbon emission and sequestration both in terrestrial vegetation and in soil; from comprehensive and global perspective, it is important to regulate the carbon cycle to control global warming and there should be more researches to combine the whole biogeochemical processes of LUC's effect on carbon cycle together to provide references for decision-makers to make land-use management practices.

Overall Effects Accounting Through Biogeophysical Processes

LUC is a general term for the human modification of earth's terrestrial surface, and the albedo and evapotranspiration of different surfaces vary with land cover types. The biogeophysical effect mechanisms of LUC on climate regulation are mainly through changing the albedo and evapotranspiration, which are closely related to energy flux between land surface and atmosphere (West et al. 2010).

Biogeophysical Process Related to Albedo

Albedo represents the fraction of incoming radiation reflected by the surface. A reduction in albedo means that a larger fraction of the incoming radiative energy is absorbed by the surface, further results in warming. The effect of albedo changes on regional and local climates is a research hot spot, especially changes in climate in response to changes of vegetated land cover and built-up areas. These changes alter surface heat balance not only by changing surface albedo, but also by altering evaporative heat transfer caused by evapotranspiration from vegetation and by changes in surface roughness.

Land surface with different vegetation types possesses different albedos; for example, forests typically have lower albedos than bare ground, grassland, and cropland and therefore absorb more incoming solar radiation (Bonan 2008; Anderson-Teixeira et al. 2012), indicating that different vegetation types participated in the land surface play a role in regulating surface energy, thus affecting surface climate. Based on different methodologies and models, there have been many researches focusing on the net radiation associated with the change of albedo resulted from LUC. In the 1970s, Charney (1975) firstly pointed out that albedo changes play an important role in global climate based on global climate model (GCM) and put forward the biogeophysical feedback mechanism (Charney 1975).

The mechanism explains that vegetation reduction due to drought will lead to increasing the albedo, decreasing the net radiation of the surface and corresponding sensible and latent heat, and further weakening the convergence upward movement, thus resulting in the reduction of cloud and precipitation. It can be seen that albedo plays an important role in surface net radiation and energy exchange. The radiative forcing due to surface albedo change can be simulated with the radiative transfer scheme of the third Hadley Centre Atmosphere Model (HadAM3). Betts (2000) applied HadAM3 to simulate the radiative forcing associated with changes in surface albedo and revealed that decreasing albedo exerts a positive radiative forcing on climate (Betts 2000). Snyder et al. (2004) investigated the participation of different vegetation types within the physical climate system based on CCM3-IBIS (the Community Climate Model coupled to the IBIS), a coupled atmosphere–biosphere model; analyzed the effects of six different vegetation biomes (tropical, boreal, temperate forests, savanna, grassland and steppe, and shrubland/tundra) on the climate through their role in modulating the biogeophysical exchanges of energy, water, and momentum between the land surface and the atmosphere, and explained the role of the albedo in modifying the surface radiation budget (Snyder et al. 2004).

Biogeophysical Process Related to Evapotranspiration

Evapotranspiration causes local cooling due to latent heat transfer from the surface to the atmosphere. Evapotranspiration can also influence cloud cover, which, in turn, affects the amount of energy reaching the surface. Clearing vegetation reduces evapotranspiration and associated latent heat flux, as without the vegetation, energy normally used to evaporate water instead of heating the land (Anderson-Teixeira et al. 2012). Evapotranspiration is an important ecosystem integrity indicator and is strongly related to land cover types. The process of evapotranspiration consumes energy and therefore has a cooling effect and a positive effect on microclimate regulation. To quantify the climate regulation and assess the effect of LUC on climate regulation, Kroll et al. (2009) used thermal emissions of different land-use classes with different land surface emissivity derived from remote sensing data and land cover information of MOLAND classification based on PLUREL scenario (Nilsson and Nielsen 2011). And Twine et al. (2004) analyzed the effects of LUC on the energy and water balance of the Mississippi River basin using the IBIS model, in which the forest cover was assumed to be completely converted to crop cover which resulted in decreasing annual average net radiation and evapotranspiration (Twine et al. 2004). In order to assess the effect of LUC on evapotranspiration, Liu et al. (2008) used Dynamic Land Ecosystem Model (DLEM) in conjunction with spatial data of LUC to estimate the LUC's effects on the magnitude, spatial and temporal variations of evapotranspiration in China, and results showed that deforestation averagely increased evapotranspiration by 138 mm/year and urban sprawl generally decreased evapotranspiration by 98 mm/year during 1900–2000, and so forth (Liu et al. 2008). Yang et al. (2012) first used a knowledge-based decision tree (K-DT) classification technique to detect LUC which was characterized with deforestation and expansion of farmland, barren land, and residential land; then, a two-source potential evapotranspiration (PET) model was used to estimate the PET response to LUC; and the result showed a decreasing trend of PET at Shalamulun River watershed in China (Yang et al. 2012). And more efforts on quantifying the biogeophysical regulation of climate by ecosystems have largely focused on regional analyses, using GCMs that include land surface models coupled with atmospheric circulation models, and these regional-scale analyses have focused on areas with strong land–climate feedbacks, including the Amazon (Shukla et al. 1990) and boreal regions (Bonan et al. 1992).

As mentioned above, current approaches for quantifying the biogeophysical regulation of climate by ecosystems (especially through effects on atmospheric heat and moisture) were mainly concentrated on highly complex models, especially atmospheric models, such as HadAM3 and GCMs; thus, few nonmeteorologists have access (or the expertise) to run and interpret highly complex systems; then, West et al. (2010) put forward an alternative approach for quantifying climate regulation by ecosystem; they developed a simple climate regulation index that could quickly produce approximations of biogeophysical regulation of climate by terrestrial ecosystems, in which the potential effects of LUC on biogeophysical climate regulation were estimated based on comparing a natural vegetation scenario to a bare ground scenario (West et al. 2010). However, conversion of natural land to other land cover types can have different results; thus, the model has potential to be improved.

In some sense, researches of effect on climate regulation of LUC through biogeophysical process are of great significance and mainly focused on the net radiation, energy balance, and so forth, which is related to albedo change and evapotranspiration rate. And most researchers simulate the effects with complex atmospheric models, biosphere models, and combined models, and the effect mechanism is relatively similar, while more attention should be paid to how to integrate and simplify the whole process to support most policy-makers, nonmeteorologists, and other related people decision making.

Effects on Climate Regulation Services Due to Climate Change

According to the forth report of Intergovernmental Panel on Climate Change (IPCC AR4 2007), there is more than 90 % probability that human activities have affected the climate, mainly through two approaches: fossil fuel burning and land-use/land cover change. LUC can not only directly exert effect on the drivers of climate regulation, but also indirectly through its impact on climate change (Deng et al. 2013). And climate change (temperature and precipitation variation and CO_2 variation) is also likely to put many ecosystem services that humans derive from lands, waters, and so forth at risk. Over the long term, climate change will feed back to climate regulation services also through both biogeochemical and biogeophysical processes.

Biogeochemical Process of the Effect on Climate Regulation Services

The amount of carbon in atmosphere is important to climate regulation services. Climate variations, such as the changes of temperature, precipitation, and atmospheric CO_2 content, affect existing carbon stores through biogeochemical process. Friedlingstein et al. (2006) used eleven coupled climate–carbon cycle models with a common protocol to study the coupling between climate change and the carbon cycle, through the feedback analysis; their results indicated that future climate change will reduce the efficiency of the earth system to absorb the anthropogenic carbon perturbation (Friedlingstein et al. 2006). And as to carbon accumulation, soil carbon is a major component of global carbon inventory which interacts with atmospheric CO_2. Rising atmospheric CO_2 concentrations will increase radiative forcing and is expected to increase soil temperatures and accelerate decomposition rates of SOC, which, in return, will increase CO_2 accumulation rate in the atmosphere (Jobbagy and Jackson 2000).

Regarding the temperature changes, which are major elements affecting the photosynthesis and decomposition of SOM, Jenkinson et al. (1991) used the Rothamsted model for the turnover of organic matter in soil to calculate the amount of CO_2 that would be released from the world stock of SOM if temperatures increase as predicted and the annual return of plant debris to the soil being held constant. The results showed that if world temperatures rise by 0.03 °C per year, the additional release of CO_2 from SOM over the next 60 years will be 61×10^{15} gC (Jenkinson et al. 1991). The temperature's effect on soil carbon losses is also confirmed by recent experimental and modeling studies (Knorr et al. 2005), and lots of experimental studies overwhelmingly indicate that increased SOC decomposition resulting from higher temperature led to increased CO_2 emissions from soils (Sanderman et al. 2003). In addition to carbon emission, soil carbon sequestration can be regulated by temperature change (Davidson and Janssens 2006).

As to moisture or precipitation conditions, in mature tropical forests, soil carbon pools tend to decrease exponentially as the ratio of temperature to precipitation increases, corresponding to a gradient from wet to dry forests. Albani et al. (2006) used ecosystem demography (ED) model, through comparison between the patterns of variability in net ecosystem productivity (NEP) and the patterns of temperature and precipitation variability from 1948 to 2003, and revealed that the periods of carbon losses during the 1960s were caused by anomalously dry conditions, while the carbon losses that occurred in the late 1990s were due to a combination of reduced precipitation and anomalously warm temperatures in the eastern USA (Albani et al. 2006). What is more, to better quantify climate change's effect on soil carbon, Ise and Moorcroft (2006) used a mechanistic decomposition model in which the effects of temperature and moisture are multiplicative, to estimate the global-scale temperature and moisture dependencies of SOC decomposition, and the results indicated that modeling of temperature and moisture dependencies of SOC decomposition in global-scale models should consider effects of scale which were not considered in other models. And as mentioned in Section "Cumulative Effects Through Biogeochemical Processes", SOC can also be affected by climate change. In many SOM models, climate data are important part as the input data into the models, and SOC changes in response to climate change under different climate scenarios which derived from GCMs and other climate models were usually analyzed (Xu et al. 2011). What is more, as the biogeochemical process of carbon sequestration is complex, it is difficult to take into account of all influencing factors; Zhan et al. (2012) used a panel data model and decomposition analyses to figure out the subtle effects of climatic, demographic, geographic, and economic factors, which revealed the importance of climatic factors in carbon sequestration (Zhan et al. 2012).

Biogeophysical Process of the Effects on Climate Regulation

Human-induced climate change is also important factors affecting the climate regulation services. Along with the industrialization and other human activities, the global climate showed a warming trend, which in the long term will return feedback to climate regulation services through biogeophysical process.

Effect on Evapotranspiration

Climate change can exert effects on the hydrological cycle and alter the evapotranspiration that both with implication for ecosystem services and feedback to regional and global climate. Land evapotranspiration is a central process in the climate system and a nexus of the water, energy, and carbon cycles. As evapotranspiration cannot be measured directly at the scale of climate observations and climate predictions, therefore most researches generally applied hydrological models to estimate the effect of climate change on evapotranspiration; among the models, different PET models are often used. To accurately detect the effect of climate change on PET, more and more researchers turned over to analyze the sensitivity of these models to climate change (Lu et al. 2005).

Effect on Albedo Due to Changes of Terrestrial Vegetation

Climate change also exerts effect on climate regulation via the change of vegetation which plays an important role in climate regulation through albedo and evapotranspiration change. The response of terrestrial vegetation to climate change has long been a research hot spot; there are many studies about understanding and quantifying climate-induced vegetation change (Solomon 1986).

Regarding temperature and precipitation changes, Overpeck et al. (1990) used the FORENA model to simulate the effects of climate change on four forest types in eastern North America with three types of climate change experiments; the results indicated that climate change would induce increases in forest and vegetation disturbance, which can significantly alter the total biomass and compositional response of forests to future warming (Overpeck et al. 1990). As anthropogenic increases in the atmospheric concentration of CO_2 and other GHGs are predicted to cause a warming of the global climate by modifying radiative forcing, Peters (1985) had a widespread hypothesis that global warming resulted from greenhouse effect can diminish biological diversity and shift the terrestrial pattern (Peters 1985). Moreover, corresponding climate change with enriched CO_2 may also alter the density of vegetation cover, thus modifying the physical characteristics of the land surface to provide climate feedback. In a vegetation/ecosystem modeling and analysis project (1995), three biogeochemical models (BIOME-BGC (BioGeochemistry Cycles), CENTURY, and TEM) and three biogeophysical models [BIOME2, Dynamic Global Phytogeography Model (DOLY), and Mapped Atmosphere-Plant Soil System (MAPSS)] were applied, respectively, to simulate the geographic distribution of vegetation types under doubled CO_2 and a range of climate scenarios (temperature and precipitation variations), and the results indicated that carbon concentration and temperature and precipitation variations can obviously change the distribution of vegetation types (Melillo et al. 1995). Betts et al. (1997) used a general circulation model iteratively coupled with an equilibrium vegetation model to quantify the effects of both physiological and structural vegetation feedbacks on a doubled CO_2 climate, with the output vegetation structure mainly represented by Leaf Area Index (LAI). The results showed that changes in vegetation structure can significantly feed back to regional-scale climate (Betts et al. 1997). The vegetation structure feeds back to climate mainly through two opposite effects: increased LAI tended to warm the land surface by lowering its albedo and cool the land surface by enhancing evaporation, while decreased LAI tended to cool the surface via increased albedo and warm the surface via reduced evaporation. To understand how vegetation growth responded to the climatic variations, Piao et al. (2006) took the LAI as an indicator to represent the vegetation activity and based on a mechanism terrestrial carbon model, to detect effect of current climate change on vegetation growth in the Northern Hemisphere, and results showed that the vegetation change in different areas can be explained by temperature and precipitation, respectively (Piao et al. 2006).

Most of the researches focused on terrestrial vegetation ecosystem, while Lovelock and Kump (1994) suggested that if living organism participates in climate regulation in an active and responsive way, they do most probably as part of a tightly coupled system, which includes the biota, the atmosphere, the ocean, and the crustal rock. Within such system, the growth of organisms changes environmental conditions and environmental changes feed back to the growth. Thus, they conducted qualitative analysis of the effects of temperature change on the feedbacks induced by changes in surface distribution of both marine algae and land plants, as to detect how the planetary area occupied by these two ecosystems varies with temperature (Lovelock and Kump 1994). There are many kinds of ecosystems on the earth, among which terrestrial ecosystem is much more connected with human beings, while it is also of great significance to do researches that combine other ecosystems with terrestrial ecosystem to comprehensively detect the impact mechanism of LUC and climate change.

Impact Assessment for Human Well-Being Due to Land-Use-Induced Climate Regulation Services Changes

According to the Millennium Ecosystem Assessment (2005), human well-being is assumed to have multiple constituents, including the basic material for a good life (adequate livelihoods, sufficient nutrient food, shelters, access to goods, etc.); health (feeling well, access to clean air and water, etc.); security (secure resource access, personal safety, and security from disasters, etc.); and freedom of choice and action (Reid et al. 2005). The services provided by ecosystems for all those unalterable needs indicate that ecosystem services are essential to human well-being. Climate regulation relates to the maintenance of a favorable climate, both at local and at global scales, which has important implications for health, crop productivity, and other human activities. In this study, we reviewed how changes in the role of ecosystem services in regulating climate affect human well-being, mainly including economy benefits, food security, human health, and a healthy environment.

Economic Value Determined by Climate Regulation

Ecosystem services and the natural capital stocks that produce the services are critical to support the earth's life system, contribute to human well-being both directly and indirectly, and therefore represent part of the total economic value of the planet. Costanza et al. (1997) firstly developed the Ecosystem Value System (ESV) calculation system to estimate the economic value of 17 ecosystem services (including climate regulation services) for 16 biomes, in which land use is one of the major factors (Costanza et al. 1997). Based on the ESV calculation system, some researchers examine the potential effects of the past and future LUC on the ecosystem services, and the results showed that the changes of ecosystem service value (among which climate regulation accounted for 10 % of the change) in Baguio City about 98 % were due to forest cover loss for 1988–2009.

Forest ecosystems play an important role in climate regulation through trapping moisture and cooling the earth's surface, thus regulating precipitation and temperature. Costanza et al. (1997) found that forests yield US$450 per hectare per year in terms of climate regulation benefits (Costanza et al. 1997). Explicitly, in the urban context which is directly related to human well-being, urban and community forests can strongly influence the physical/biological environment and mitigate many effects of urban development by moderating climate; conserving energy, CO_2, and water; improving air quality; controlling rainfall runoff; and so forth. As in the case of Tucson, Arizona, each tree would give benefits in the range of US$20.75 annually by reducing cooling costs for buildings. Dwyer et al. (1992) calculated that 100 million mature trees in US cities could reduce annual energy costs by US$2 billion based on computer simulation estimate (Dwyer et al. 1992).

LUC and climate change can exert effect on the capacity of ecosystem to absorb carbon, which will result in the loss of climate regulation services and further loss of socioeconomic benefits. To estimate the value of carbon sequestration in Uganda's protected area, Howard (1995) used two different approaches: First, based on figures of the damage that would occur if the land was converted and carbon released in the atmosphere, the value of Uganda's protected areas as a carbon sink was estimated at US$245 million, which amounts to a US$17.4 million annually; second, the replacement cost approach was also used to estimate the cost of replacing this carbon sink functions through an afforestation scheme and was valued at US$20.3 million annually (Howard 1995). The results indicated that along with the degradation or loss of climate regulation services in carbon sequestration, much cost will be input to ecological engineering, such as afforestation and construction of environmental protection area.

Effects of Extreme Weather Events

In response to ecosystem service changes, extreme weather events directly affect human well-beings, leading to drastic disasters which can influence most aspects of humans, such as living environment, agriculture production, and health. Regional climatic conditions are influenced by changes in ecosystems and landscapes, especially deforestation and desertification. Human-induced alteration of atmospheric composition (the greenhouse effect) also affects climatic conditions. On a large scale, the long-term ongoing GHGs effect will lead to increasing risks of deaths in extreme weather events (heat waves, floods, droughts, etc.) (Chappell et al. 2008). These events have local and sometimes regional effects, directly through deaths and injuries and indirectly through economic disruption, infrastructure damage, and population displacement. In turn, this may lead to increased incidence of certain communicable diseases as a result of overcrowding, lack of clean water and shelter, poor nutritional status, and adverse effects on mental health.

Climate regulation of hydrology is a research area, in which natural climate system affects hydrology and further exerts effect on the runoff, rainfall cyclicity, water quality, soil development, and so forth. As a result, there are marked variations in the likelihood of flooding and droughts, the intensity of erosion and nutrient cycling, and the demand for irrigation water and water supplies, which further will bring substantial loss of human well-being. And as to urban ecosystem, climate change caused by increased anthropogenic emissions of CO_2 and other GHGs is a long-term climate hazard with the potential to alter the intensity, temporal pattern, and spatial extent of the urban heat island in metropolitan regions (Rosenzweig et al. 2005). The microclimate regulation of all natural ecosystem in the urban areas contributes to reducing the urban heat island effect, as urban green increases evapotranspiration and therefore has a cooling effect; from the perspective of economics, vegetation can also decrease energy use for heating and air conditioning substantially in urban areas by shading house in summer and reducing wind speed in winter (Bolund and Hunhammar 1999).

Response of Food Security to Climate Regulation

The accessibility of food is highly dependent on suitable climatic conditions, and the sustainable production of food and raw materials is vulnerable to changes in temperature, precipitation, and concentrations of CO_2. Though the role of climate regulation in food security has not been fully estimated, the climate variability which can result from LUC has been examined as the principal source of fluctuations in global food production in different areas, such as the arid and semi-arid tropical countries of the developing world and the northern China (Shi et al. 2013). Specifically speaking, soil carbon sequestration can be regulated by climate change, especially the changes result from temperature change (Davidson and Janssens 2006) and LUC. As land is the basic requirement for cultivation of crops, a region's potential for food production in agricultural areas depends directly on the fertility of available arable lands (Lautenbach et al. 2011), and soil carbon sequestration is a key element for ecosystem biogeochemical carbon cycle and contributes to the formation of fertility in soil, which helps to maintain net productivity, improve water quality, and restore degraded soils and ecosystems (Lal et al. 2007). Thus, along with the climate regulation services changes, the variability of soil carbon sequestration results in the change of food production and accessibility.

Consequence on Human Health

In the MEA, it was stated that stresses on freshwater sources, food-producing systems, and climate regulation could cause major adverse health effects, and each ecosystem service is sensitive to climate conditions which will be affected by LUC and climate change. And the change of ecosystem services will affect the human well-being and health of human (Chappell et al. 2008). Climate regulation is one of the services the ecosystems provide to regulate climate conditions, mainly to ensure that people live an environmentally clean and safe life (Gedney and Valdes 2000). The degradation of climate regulation will decrease the ability of ecosystem to avoid the adverse effect of climate change and affect human health through both direct effect (such as increasing mortality, diseases from extreme weather) and indirect effects (such as climate-induced changes in the distribution of productive ecosystems and the availability of food, water, and energy supplies). Although climate change may have some benefits to human health, most are expected to be negative especially in urban areas. As human migrates from rural areas to cities, more than half of the world's population now live in high-density urban areas, many of which are poorly supplied with either ecosystem or human services (Chappell et al. 2008). And in these urban areas, a reduction in climate regulation services could exacerbate climate stress for large numbers of people, reducing well-being and increasing death rates directly through higher summer temperatures and so forth. Furthermore, climate regulation services of soil carbon sequestration can help mitigate climate change by offsetting emissions of fossil fuels and reducing GHGs effects, and so forth, which is of great significance to improve environment for healthy lives (Lal et al. 2007).

Concluding Remarks

LUC and climate change will alter the supply of ecosystem services that are vital for human well-being. The processes through LUC and climate change influence human well-being via exerting effects on climate regulation services are complex and complicated, involving both biogeochemical and biogeophysical processes. From the perspective of the drivers of and their corresponding effects on climate regulation services, this study mainly revisited the closely related researches to explore how LUC and climate change exert effects on the major drivers, such as carbon emission and sequestration, albedo, and evapotranspiration which are key elements to regulate surface energy exchange and further play an important role in climate regulation services. And, along with the changes of climate regulation services resulted from LUC and climate change, different aspects of human well-beings will be impacted which have also been assessed by many researchers. Generally speaking, LUC and climate change have great effects on climate regulation services. Globally, deforestation, degradation of forest and grassland, land conversion from forest to cultivated land, urban expansion, carbon emission, and so forth all will probably lead to changes in climate regulation services.

So far, lots of researches have been conducted and significant progresses have been made related to each aspect. However, there is still room for improvement. First, there have been many methodologies and models developed or enhanced to investigate the effect mechanism of the drivers of climate regulation services via biogeochemical and biogeophysical processes, while there are less researches that combine both the processes in the models; thus, more research attention should be paid to the effect mechanisms that take both biogeochemical and biogeophysical effects into considerations. Second, in future research, improved models which can combine the effects of LUC and climate change on the climate regulation as a whole should be developed. In addition, as human well-being changes along with the variation of climate regulation resulted from the effect of LUC and climate change, further anthropogenic activities and economic development related to human well-being will respond to LUC and climate change. However, there are relatively less researches that have taken the responses into consideration when assessing the effects, so future research should also focus on these potential societal responses.

References

Aktar W, Sengupta D, Chowdhury A (2009) Impact of pesticides use in agriculture: their benefits and hazards. Interdisc Toxicol 2(1):1–12

Albani M, Medvigy D, Hurtt GC, Moorcroft PR (2006) The contributions of LUC, CO_2 fertilization, and climate variability to the Eastern US carbon sink. Glob Change Biol 12(12):2370–2390

Alcamo J, Van Vuuren D, Cramer W, Alder J, Bennett E, Carpenter S, Christensen V, Foley J, Maerker M, Masui T (2005) Changes in ecosystem services and their drivers across the scenarios. Ecosyst hum well-being 2:297–373

Anderson-Teixeira KJ, Snyder PK, Twine TE, Cuadra SV, Costa MH, DeLucia EH (2012) Climate-regulation services of natural and agricultural ecoregions of the Americas. Nat Clim Change 2(3):177–181

Antrop M (2000) Changing patterns in the urbanized countryside of Western Europe. Landscape Ecol 15(3):257–270

Arora VK, Boer GJ, Christian JR, Curry CL, Denman KL, Zahariev K, Flato GM, Scinocca JF, Merryfield WJ, Lee WG (2009) The effect of terrestrial photosynthesis down regulation on the twentieth-century carbon budget simulated with the CCCma earth system model. J Clim 22(22):6066–6088

Ash N, Lucas N, Bubb P, Iceland C, Irwin F, Ranganathan J, Raudsepp-Hearne C (2008) Framing the link between development and ecosystem services. Ecosystem services: a guide for decision makers. World Resouces Institute, Washington

Assessment, U. N. E (2011) The UK national ecosystem assessment: synthesis of the key findings. UNEP-WCMC Cambridge, UK

Ayanaba A (1976) The effects of clearing and cropping on the organic reserves and biomass of tropical forest soils. Soil Biol Biochem 8(6):519–525

Bangash RF, Passuello A, Sanchez-Canales M, Terrado M, Lopez A, Elorza FJ, Ziv G, Acuna V, Schuhmacher M (2013) Ecosystem services in mediterranean river basin: climate change impact on water provisioning and erosion control. Sci Total Environ 458:246–255

Barančíková G, Halás J, Guttekova M, Makovnikova J, Novakova M, Skalský R, Tarasovičová Z (2010) Application of ROTHC model to predict soil organic carbon stock on agricultural soils of Slovakia. Soil Water Res 5(1):1–9

Betts RA (2000) Offset of the potential carbon sink from boreal forestation by decreases in surface albedo. Nature 408(6809):187–190

Betts RA, Cox PM, Lee SE, Woodward FI (1997) Contrasting physiological and structural vegetation feedbacks in climate change simulations. Nature 387(6635):796–799

Bolund P, Hunhammar S (1999) Ecosystem services in urban areas. Ecol Econ 29(2):293–301

Bonan G (2008) Ecological climatology: concepts and applications, 2nd edn. Cambridge University Press, Cambridge

Bonan GB, Pollard D, Thompson SL (1992) Effects of boreal forest vegetation on global climate. Nature 359(6397):716–718

Brown S, Lugo AE (1982) The storage and production of organic matter in tropical forests and their role in the global carbon cycle. Biotropica, pp 161–187

Carpenter SR, Bennett EM, Peterson GD (2006) Scenarios for ecosystem services: an overview. Ecol Soc 11(1):29

Caulfield LE, de Onis M, Blössner M, Black RE (2004) Undernutrition as an underlying cause of child deaths associated with diarrhea, pneumonia, malaria, and measles. Am J Clin Nutr 80(1):193–198

Chappell NA, Chen YD, Suhaimi J, Bonell M (2008) Climate regulation of humid tropical hydrology, pp 172–177

Charney JG (1975) Dynamics of deserts and drought in the Sahel. Q J Roy Meteorol Soc 101(428):193–202

Chen X, Bai J, Li X, Luo G, Li J, Li BL (2013) Changes in land use/land cover and ecosystem services in central Asia during 1990–2009. Curr Opin Environ Sustain 5(1):116–127

Chopra K, Leemans R, Kumar P, Simons H (2005) Ecosystems and human well-being: policy responses. Island Press, Washington

Costanza R, d'Arge R, de Groot R, Farber S, Grasso M, Hannon B, Limburg K, Naeem S, O'neill RV, Paruelo J, Raskin RG, Sutton P, van den Belt M (1997) The value of the world's ecosystem services and natural capital. Nature 387(6630):253–260

Daily GC, Alexander S, Ehrlich PR, Goulder L, Lubchenco J, Matson PA, Mooney HA, Postel S, Schneider SH, Tilman D (1997) Ecosystem services: benefits supplied to human societies by natural ecosystems. Ecological Society of America, Washington

Daily GC, Polasky S, Goldstein J, Kareiva PM, Mooney HA, Pejchar L, Ricketts TH, Salzman J, Shallenberger R (2009) Ecosystem services in decision making: time to deliver. Front Ecol Environ 7(1):21–28

Dale VH, Polasky S (2007) Measures of the effects of agricultural practices on ecosystem services. Ecol Econ 64(2):286–296

Davidson EA, Janssens IA (2006) Temperature sensitivity of soil carbon decomposition and feedbacks to climate change. Nature 440(7081):165–173

Dawson JJ, Smith P (2007) Carbon losses from soil and its consequences for land-use management. Sci Total Environ 382(2):165–190

De Fries RS, Field CB, Fung I, Collatz GJ, Bounoua L (1999) Combining satellite data and biogeochemical models to estimate global effects of human-induced land cover change on carbon emissions and primary productivity. Glob Biogeochem Cycles 13(3):803–815

DeFries RS, Houghton RA, Hansen MC, Field CB, Skole D, Townshend J (2002) Carbon emissions from tropical deforestation and regrowth based on satellite observations for the 1980s and 1990s. Proc Nat Acad Sci 99(22):14256–14261

Deng X, Zhao C, Yan H (2013) Systematic modeling of impacts of land use and land cover changes on regional climate: a review. Adv Meteorol 2013:11

DIAz S, Lavorel S, McIntyre S, Falczuk V, Casanoves F, Milchunas DG, Skarpe C, Rusch G, Sternberg M, Noy-Meir I (2007) Plant trait responses to grazing–a global synthesis. Glob Change Biol 13(2):313–341

Dieye A, Roy D, Hanan N, Liu S, Hansen M, Toure A (2012) Sensitivity analysis of the GEMS soil organic carbon model to land cover land use classification uncertainties under different climate scenarios in senegal. Biogeosciences 9:631–648

Dixon RK, Brown S, Houghton REA, Solomon A, Trexler M, Wisniewski J (1994) Carbon pools and flux of global forest ecosystems. Science 263(5144):185–189 (Washington)

Dwyer JF, McPherson EG, Schroeder HW, Rowntree RA (1992) Assessing the benefits and costs of the urban forest. J Arboric 18:227

Estoque RC, Murayama Y (2012) Examining the potential impact of land use/cover changes on the ecosystem services of Baguio city, the Philippines: a scenario-based analysis. Appl Geogr 35(1):316–326

Farley KA, Jobbágy EG, Jackson RB (2005) Effects of afforestation on water yield: a global synthesis with implications for policy. Glob Change Biol 11(10):1565–1576

Fisher B, Turner K, Zylstra M, Brouwer R, de Groot R, Farber S, Ferraro P, Green R, Hadley D, Harlow J, Jefferiss P, Kirkby C, Morling P, Mowatt S, Naidoo R, Paavola J, Strassburg B, Yu D, Balmford A (2008) Ecosystem services and economic theory: integration for policy-relevant research. Ecol Appl 18(8):2050–2067

Foley JA, Ramankutty N, Brauman KA, Cassidy ES, Gerber JS, Johnston M, Mueller ND, O'Connell C, Ray DK, West PC (2011) Solutions for a cultivated planet. Nature 478(7369):337–342

Fowler D, Pilegaard K, Sutton M, Ambus P, Raivonen M, Duyzer J, Simpson D, Fagerli H, Fuzzi S, Schjørring JK (2009) Atmospheric composition change: ecosystems–atmosphere interactions. Atmos Environ 43(33):5193–5267

Francaviglia R, Coleman K, Whitmore AP, Doro L, Urracci G, Rubino M, Ledda L (2012) Changes in soil organic carbon and climate change–application of the ROTHC model in agro-silvo-pastoral Mediterranean systems. Agric Syst 112:48–54

Friedlingstein P, Cox P, Betts R, Bopp L, Von Bloh W, Brovkin V, Cadule P, Doney S, Eby M, Fung I (2006) Climate-carbon cycle feedback analysis: results from the C4MIP model intercomparison. J Clim 19(14):3337–3353

Galis A, Marcq C, Marlier D, Portetelle D, Van I, Beckers Y, Thewis A (2013) Comprehensive reviews in food science and food safety

Gavier-Pizarro GI, Calamari NC, Thompson JJ, Canavelli SB, Solari LM, Decarre J, Goijman AP, Suarez RP, Bernardos JN, Zaccagnini ME (2012) Expansion and intensification of row crop agriculture in the Pampas and Espinal of Argentina can reduce ecosystem service provision by changing avian density. Agric Ecosyst Environ 154:44–55

Gedney N, Valdes PJ (2000) The effect of Amazonian deforestation on the northern hemisphere circulation and climate. Geophys Res Lett 27(19):3053–3056

Houghton RA (2003) Revised estimates of the annual net flux of carbon to the atmosphere from changes in land use and land management 1850–2000. Tellus Ser B Chem Phys Meteorol 55(2):378–390

Houghton RA, Hobbie JE, Melillo JM, Moore B, Peterson BJ, Shaver GR, Woodwell GM (1983) Changes in the carbon content of terrestrial biota and soils between 1860 and 1980—a net release of CO_2 to the atmosphere. Ecol Monogr 53(3):235–262

Houghton RA, Hackler JL, Lawrence KT (2000) Changes in terrestrial carbon storage in the United States 2: the role of fire and fire management. Glob Ecol Biogeogr 9(2):145–170

Houghton RA, Hackler JL, Cushman RM (2001) Carbon flux to the atmosphere from land-use changes: 1850–1990. Carbon Dioxide Information Center, Environmental Sciences Division, Oak Ridge National Laboratory, Oak Ridge

Howard PC (1995) The economics of protected areas in Uganda: costs, benefits and policy issues. University of Edinburgh, Edinburgh

Ice GG, Stednick JD (2004) A century of forest and wildland watershed lessons

IPCC (2007) Climate change 2007: the physical science basis. Agenda 6(07):333

Ise T, Moorcroft PR (2006) The global-scale temperature and moisture dependencies of soil organic carbon decomposition: an analysis using a mechanistic decomposition model. Biogeochemistry 80(3):217–231

Jaradat AA, Boody G (2012) Modeling agroecosystem services under simulated climate and land-use changes. Int Sch Res Not 2011

Jenkinson D, Adams D, Wild A (1991) Model estimates of CO_2 emissions from soil in response to global warming. Nature 351(6324):304–306

Jobbagy EG, Jackson RB (2000) The vertical distribution of soil organic carbon and its relation to climate and vegetation. Ecol Appl 10(2):423–436

Kimble JM, Follett RF, Cole CV, Lal R (1998) The potential of US cropland to sequester carbon and mitigate the greenhouse effect. Ann Arbor Press, Chelsea, X **157504112**: 128

Knorr W, Prentice I, House J, Holland E (2005) Long-term sensitivity of soil carbon turnover to warming. Nature 433(7023):298–301

Kumar P (2010) Guidance manual for the valuation of regulating services, United Nations Environment Programme, Nairobi

Lal R (2004a) Soil carbon sequestration impacts on global climate change and food security. Science 304(5677):1623–1627

Lal R (2004b) Soil carbon sequestration to mitigate climate change. Geoderma 123(1):1–22

Lal R, Follett RF, Stewart B, Kimble JM (2007) Soil carbon sequestration to mitigate climate change and advance food security. Soil Sci 172(12):943–956

Lautenbach S, Kugel C, Lausch A, Seppelt R (2011) Analysis of historic changes in regional ecosystem service provisioning using land use data. Ecol Ind 11(2):676–687

Leh MDK, Matlock MD, Cummings EC, Nalley LL (2013) Quantifying and mapping multiple ecosystem services change in West Africa. Agric Ecosyst Environ 165:6–18

Levy PE, Friend AD, White A, Cannell MGR (2004) The influence of land use change on global-scale fluxes of carbon from terrestrial ecosystems. Clim Change 67(2–3):185–209

Liu M, Tian H, Chen G, Ren W, Zhang C, Liu J (2008) Effects of land-use and land-cover change on evapotranspiration and water yield in China during 1900–20001. JAWRA J Am Water Resour Assoc 44(5):1193–1207

Lovelock JE, Kump LR (1994) Failure of climate regulation in a geophysiological model. Nature 369(6483):732–734

Lu J, Sun G, McNulty SG, Amatya DM (2005) A comparison of six potential evapotranspiration methods for regional use in the southeastern United States1, Wiley Online Library, Hoboken

Lugo AE, Brown S (1993) Management of tropical soils as sinks or sources of atmospheric carbon. Plant Soil 149(1):27–41

McGuire AD, Sitch S, Clein JS, Dargaville R, Esser G, Foley J, Heimann M, Joos F, Kaplan J, Kicklighter DW, Meier RA, Melillo JM, Moore B, Prentice IC, Ramankutty N, Reichenau T, Schloss A, Tian H, Williams LJ, Wittenberg U (2001) Carbon balance of the terrestrial biosphere in the twentieth century: analyses of CO_2, climate and land use effects with four process-based ecosystem models. Glob Biogeochem Cycles 15(1):183–206

Melillo J, Borchers J, Chaney J (1995) Vegetation/ecosystem modeling and analysis project: comparing biogeography and geochemistry models in a continental-scale study of terrestrial ecosystem responses to climate change and CO_2 doubling. Glob Biogeochem Cycles 9(4)

Metherell AK (1993) Century: soil organic matter model environment: technical documentation: agroecosystem version 4.0. Colorado State University, Colorado

Nelson GC, Bennett E, Berhe AA, Cassman KG, DeFries R, Dietz T, Dobson A, Dobermann A, Janetos A, Levy M, Marco D, Nakićenović N, O'Neill B, Norgaard T, Petschel-Held G, Ojima D, Pingali P, Watson R, Zurek M (2005) Drivers of change in ecosystem condition and services in ecosystems and human well-being: scenarios: findings of the scenarios working group. Island Press, Washington

Nelson GC, Bennett E, Berhe AA, Cassman K, DeFries RS, Dietz T, Dobermann A, Dobson A, Janetos A, Levy MA (2006) Anthropogenic drivers of ecosystem change: an overview. Ecol Soc 11(2)

Nelson E, Mendoza G, Regetz J, Polasky S, Tallis H, Cameron D, Chan KM, Daily GC, Goldstein J, Kareiva PM (2009) Modeling multiple ecosystem services, biodiversity conservation, commodity production, and tradeoffs at landscape scales. Front Ecol Environ 7(1):4–11

Nelson E, Sander H, Hawthorne P, Conte M, Ennaanay D, Wolny S, Manson S, Polasky S (2010) Projecting global land-use change and its effect on ecosystem service provision and biodiversity with simple models. PLoS ONE 5(12):e14327

Nilsson K, Nielsen TS (2011) Peri-urban land use relationships–strategies and sustainability assessment tools for urban-rural linkages. Chin Landscape Archit 27(186):12–17

Nosetto MD, Jobbagy EG, Paruelo JM (2005) Land-use change and water losses: the case of grassland afforestation across a soil textural gradient in central Argentina. Glob Change Biol 11(7):1101–1117

Olson JS, Watts JA, Allison LJ (1983) Carbon in live vegetation of major world ecosystems. Oak Ridge National Lab, USA

Overpeck JT, Rind D, Goldberg R (1990) Climate-induced changes in forest disturbance and vegetation. Nature 343(6253):51–53

Pan YD, Birdsey RA, Fang JY, Houghton R, Kauppi PE, Kurz WA, Phillips OL, Shvidenko A, Lewis SL, Canadell JG, Ciais P, Jackson RB, Pacala SW, McGuire AD, Piao SL, Rautiainen A, Sitch S, Hayes D (2011) A large and persistent carbon sink in the world's forests. Science 333(6045):988–993

Patz J, Githeko A, McCarty J, Hussein S, Confalonieri U, De Wet N (2003) Climate change and infectious diseases. Clim change hum health: risks responses, pp 103–137

Pereira E, Queiroz C, Pereira HM, Vicente L (2005) Ecosystem services and human well-being: a participatory study in a mountain community in Portugal. Ecol Soc 10(2):14

Pete S, Mike A, Helaina B, Paul B, Chris E, Rosemary H, Simon GP, Timothy Q, Amanda T (2011) Regulating services. UNEP-WCMC, Cambridge

Peters RL (1985) The greenhouse effect and nature reserves. Bioscience 35(11):707–717

Piao SL, Friedlingstein P, Ciais P, Zhou LM, Chen AP (2006) Effect of climate and CO_2 changes on the greening of the Northern Hemisphere over the past two decades. Geophys Res Lett 33(23)

Polasky S, Nelson E, Pennington D, Johnson KA (2011) The impact of land-use change on ecosystem services, biodiversity and returns to landowners: a case study in the state of Minnesota. Environ Resour Econ 48(2):219–242

Power AG (2010) Ecosystem services and agriculture: tradeoffs and synergies. Philos Trans Roy Soc B-Biol Sci 365(1554):2959–2971

Powers JS (2004) Changes in soil carbon and nitrogen after contrasting land-use transitions in northeastern Costa Rica. Ecosystems 7(2):134–146

Raudsepp-Hearne C, Peterson GD, Bennett E (2010) Ecosystem service bundles for analyzing tradeoffs in diverse landscapes. Proc Natl Acad Sci 107(11):5242–5247

Reid W, Mooney H, Cropper A, Capistrano D, Carpenter S, Chopra K, Dasgupta P, Dietz T, Duraiappah A, Hassan R (2005) Ecosystems and human well-being: synthesis, MA (Millennium Ecosystem Assessment). Island Press, Washington

Reidsma P, Tekelenburg T, van den Berg M, Alkemade R (2006) Impacts of land-use change on biodiversity: an assessment of agricultural biodiversity in the European Union. Agric Ecosyst Environ 114(1):86–102

Reyers B, O'Farrell PJ, Cowling RM, Egoh BN, Le Maitre DC, Vlok JH (2009) Ecosystem services, land-cover change, and stakeholders: finding a sustainable foothold for a semiarid biodiversity hotspot

Riitters KH, Wickham JD, O'Neill RV, Jones KB, Smith ER, Coulston JW, Wade TG, Smith JH (2002) Fragmentation of continental United States forests. Ecosystems 5(8):0815–0822

Rodriguez JP, Beard TD, Bennett EM, Cumming GS, Cork SJ, Agard J, Dobson AP, Peterson GD (2006) Trade-offs across space, time, and ecosystem services. Ecol Soc 11(1):28

Rosenzweig C, Solecki WD, Parshall L, Chopping M, Pope G, Goldberg R (2005) Characterizing the urban heat island in current and future climates in New Jersey. Glob Environ Change Part B Environ Hazards 6(1):51–62

Sanderman J, Amundson RG, Baldocchi DD (2003) Application of eddy covariance measurements to the temperature dependence of soil organic matter mean residence time. Glob Biogeochem Cycles 17(2):1061

Schröter D, Cramer W, Leemans R, Prentice IC, Araújo MB, Arnell NW, Bondeau A, Bugmann H, Carter TR, Gracia CA (2005) Ecosystem service supply and vulnerability to global change in Europe. Science 310(5752):1333–1337

Scott NA, Tate KR, Giltrap DJ, Smith CT, Wilde RH, Newsome PFJ, Davis MR (2002) Monitoring land-use change effects on soil carbon in New Zealand: quantifying baseline soil carbon stocks. Environ Pollut 116:S167–S186

Sehgal R (2010) Deforestation and avian infectious diseases. J Exp Biol 213(6):955–960

Shi Q, Lin Y, Zhang E, Yan H, Zhan J (2013) Impacts of cultivated land reclamation on the climate and grain production in Northeast China in the future 30 years. Adv Meteorol 2013:8

Shrestha B, Williams S, Easter M, Paustian K, Singh B (2009) Modeling soil organic carbon stocks and changes in a Nepalese watershed. Agric Ecosyst Environ 132(1):91–97

Shukla J, Nobre C, Sellers P (1990) Amazon deforestation and climate change. Science 247(4948):1322–1325

Sitch S, Smith B, Prentice IC, Arneth A, Bondeau A, Cramer W, Kaplan JO, Levis S, Lucht W, Sykes MT, Thonicke K, Venevsky S (2003) Evaluation of ecosystem dynamics, plant geography and terrestrial carbon cycling in the LPJ dynamic global vegetation model. Glob Change Biol 9(2):161–185

Sivakumar MVK, Das HP, Brunini O (2005) Impacts of present and future climate variability and change on agriculture and forestry in the arid and semi-arid tropics. Clim Change 70(1–2):31–72

Smith TM, Leemans R, Shugart HH (1992) Sensitivity of terrestrial carbon storage to CO_2-induced climate change: comparison of four scenarios based on general circulation models. Clim Change 21(4):367–384

Smith P, Smith J, Powlson D, McGill W, Arah J, Chertov O, Coleman K, Franko U, Frolking S, Jenkinson D (1997) A comparison of the performance of nine soil organic matter models using datasets from seven long-term experiments. Geoderma 81(1):153–225

Snyder P, Delire C, Foley J (2004) Evaluating the influence of different vegetation biomes on the global climate. Clim Dyn 23(3–4):279–302

Solomon AM (1986) Transient response of forests to CO_2-induced climate change: simulation modeling experiments in eastern North America. Oecologia 68(4):567–579

Su SL, Xiao R, Jiang ZL, Zhang Y (2012) Characterizing landscape pattern and ecosystem service value changes for urbanization impacts at an eco-regional scale. Appl Geogr 34:295–305

Sun G, Zhou G, Zhang Z, Wei X, McNulty SG, Vose JM (2006) Potential water yield reduction due to forestation across China. J Hydrol 328(3):548–558

Swift MJ, Izac AMN, van Noordwijk M (2004) Biodiversity and ecosystem services in agricultural landscapes—are we asking the right questions? Agric Ecosyst Environ 104(1):113–134

Swinton SM, Lupi F, Robertson GP, Hamilton SK (2007) Ecosystem services and agriculture: cultivating agricultural ecosystems for diverse benefits. Ecol Econ 64(2):245–252

Tallis H, Polasky S (2009) Mapping and valuing ecosystem services as an approach for conservation and natural-resource management. Ann N Y Acad Sci 1162(1):265–283

Turner WR, Brandon K, Brooks TM, Costanza R, Da Fonseca GA, Portela R (2007) Global conservation of biodiversity and ecosystem services. Bioscience 57(10):868–873

Twine TE, Kucharik CJ, Foley JA (2004) Effects of land cover change on the energy and water balance of the mississippi river basin. J Hydrometeorology 5(4):640–655

Vertessy RA, Watson FG, Sharon K (2001) Factors determining relations between stand age and catchment water balance in mountain ash forests. For Ecol Manage 143(1):13–26

West PC, Narisma GT, Barford CC, Kucharik CJ, Foley JA (2010) An alternative approach for quantifying climate regulation by ecosystems. Front Ecol Environ 9(2):126–133

Woodwell GM, Hobbie J, Houghton R, Melillo J, Moore B, Peterson B, Shaver G (1983) Global deforestation: contribution to atmospheric carbon dioxide. Science 222(4628):1081–1086

Woomer PL, Tieszen LL, Tappan G, Toure A, Sall M (2004) Land use change and terrestrial carbon stocks in Senegal. J Arid Environ 59(3):625–642

Xiao Y, Xie G, An K (2003) Economic value of ecosystem services in Mangcuo Lake drainage basin. Ying yong sheng tai xue bao = The journal of applied ecology/Zhongguo sheng tai xue xue hui, Zhongguo ke xue yuan Shenyang ying yong sheng tai yan jiu suo zhu ban 14(5):676–680

Xie G, Xiao Y, Lu C (2006) Study on ecosystem services: progress, limitation and basic paradigm. J Plant Ecol 2:002

Xu X, Liu W, Kiely G (2011) Modeling the change in soil organic carbon of grassland in response to climate change: effects of measured versus modelled carbon pools for initializing the rothamsted carbon model. Agric Ecosyst Environ 140(3):372–381

Yang XL, Ren LL, Singh VP, Liu XF, Yuan F, Jiang SH, Yong B (2012) Impacts of land use and land cover changes on evapotranspiration and runoff at Shalamulun River watershed, China. Hydrol Res 43(1–2):23–37

Zhan J, Yan H, Chen B, Luo J, Shi N (2012) Decomposition analysis of the mechanism behind the spatial and temporal patterns of changes in carbon bio-sequestration in China. Energies 5(2):386–398

Chapter 2
Impact Assessments on Agricultural Productivity of Land-Use Change

Jinyan Zhan, Feng Wu, Zhihui Li, Yingzhi Lin and Chenchen Shi

Abstract Currently, food safety and its related influencing factors in China are the hot research topics, and cultivated land conversion is one of the significant factors influencing food safety in China. In this chapter, we first investigated land degradation induced by land-use change. Land degradation is a complex process which involves both the natural ecosystem and the socioeconomic system, among which climate and land-use change are the two predominant driving factors. To comprehensively and quantitatively analyze the land degradation process, we employed the normalized difference vegetation index (NDVI) as a proxy to assess land degradation and further applied the binary panel logistic regression model to analyze the impacts of the driving factors on land degradation in the North China Plain. The results revealed that increasing in rainfall and temperature would significantly and positively contribute to the land improvement, and conversion from cultivated land to grassland and forest land showed positive relationship with land improvement, while conversion to built-up area will lead to land degradation. Besides, human agricultural intensification represented by fertilizer utilization will help to

J. Zhan (✉) · F. Wu · C. Shi
State Key Laboratory of Water Environment Simulation, School of Environment,
Beijing Normal University, Beijing 100875, China
e-mail: zhanjy@bnu.edu.cn

Z. Li
Institute of Geographic Sciences and Natural Resources Research,
Chinese Academy of Sciences, Beijing 100101, China

Z. Li
Center for Chinese Agricultural Policy, Chinese Academy of Sciences,
Beijing 100101, China

Z. Li
University of Chinese Academy of Sciences, Beijing 100049, China

Y. Lin
School of Mathematics and Physics, China University of Geosciences (Wuhan),
Wuhan 430074, China

© Springer-Verlag Berlin Heidelberg 2015
J. Zhan (ed.), *Impacts of Land-use Change on Ecosystem Services*,
Springer Geography, DOI 10.1007/978-3-662-48008-3_2

improve the land quality. The economic development may exert positive impacts on land quality to alleviate land degradation, although the rural economic development and agricultural production will exert negative impacts on the land and lead to land degradation. Infrastructure construction would modify the land surface and further resulted in land degradation. The findings of the research will provide scientific information for sustainable land management. Second, we predicted land-use conversions in the North China Plain based on the scenario analysis. Scenario analysis and dynamic prediction of land-use structure which involve many driving factors are helpful to investigate the mechanism of land-use change and even to optimize land-use allocation for sustainable development. In this study, land-use structure changes during 1988–2010 in North China Plain were discerned and the effects of various natural and socioeconomic driving factors on land-use structure changes were quantitatively analyzed based on an econometric model. The key drivers of land-use structure changes in the model are county-level net returns of land resource. In this research, we modified the net returns of each land-use type for three scenarios, including business as usual (BAU) scenario, rapid economic growth (REG) scenario, and coordinated environmental sustainability (CES) scenario. The simulation results showed that, under different scenarios, future land-use structures were different due to the competition among various land-use types. The land-use structure changes in North China Plain in the 40-year future will experience a transfer from cultivated land to built-up area, an increase of forestry land, and decrease of grassland. The results will provide some significant references for land-use management and planning in the study area. Third, we simulated shifting patterns of agroecological zones across China. An agroecological zone (AEZ) is a land resource mapping unit, defined in terms of climate, landform, and soils, and has a specific range of potentials and constraints for cropping (FAO 1996). The shifting patterns of AEZs in China driven by future climatic changes were assessed by applying the agroecological zoning methodology proposed by International Institute for Applied Systems Analysis (IIASA) and Food and Agriculture Organization of the United Nations (FAO) in this study. A data processing scheme was proposed to reduce systematic errors in projected climate data using observed data from meteorological stations. AEZs in China of each of the four periods: 2011–2020, 2021–2030, 2031–2040, and 2041–2050 were drawn. It is found that the future climate change will lead to significant local changes of AEZs in China and the overall pattern of AEZs in China is stable. The shifting patterns of AEZs will be characterized by northward expansion of humid AEZs to subhumid AEZs in south China, eastward expansion of arid AEZs to dry and moist semiarid AEZs in north China, and southward expansion of dry semiarid AEZs to arid AEZs in southwest China.

Keywords Land degradation · NDVI · Land-use conversion · Scenario · Land-use simulation · Agroecological zones · North China Plain

Land Degradation Induced by Climate and Land-Use Change in the North China Plain

Introduction

Land degradation has become a critical issue worldwide, especially in the developing countries, which lead to great concerns about food security. To improve livelihoods of human beings and to keep sustainable development of human society, healthy land ecosystems are basically essential elements. However, the key services of good-quality land and their true values have been usually taken for granted and underestimated, leading to serious land degradation, which not only deteriorates the ecosystem services but also hinders regional sustainable development. Land degradation means a significant reduction of the productive capacity of land. It is an interactive process involving various casual factors, among which climate change, land-use/land-cover changes, and human dominated land management play a significant role (Bajocco et al. 2012). Land degradation involves two interlocking complex systems, the natural ecosystem and the human social system, and both changes in biophysical natural ecosystem and socioeconomic conditions will affect the land degradation process (MEA 2005). Natural forces affect land degradation through periodic stresses of extreme and persistent climatic events. Human activities also contribute to land degradation through deforestation, removal of natural vegetation and urban sprawl that lead to land-use/land-cover changes; unsustainable agricultural land-use management practices, such as use and abuse of fertilizer, pesticide and heavy machinery; and overgrazing, improper crop rotation, and poor irrigation practices (WMO 2005).

In comparison with other regions, land degradation afflicts China more seriously in terms of the extent, intensity, economic impacts, and affected population (Bai and Dent 2009). China accounts for 22 % of the world's population, but only 6.4 % of the global land area and 7.2 % of the global cultivated land area. Agricultural production is an important issue that always plagues the national economy and livelihood of citizens (Deng et al. 2008a). However, the rapid population growth and urbanization, unreasonable human utilization, and influence of natural factors have caused degradation of 5.392 million km^2 of land, which accounts for about 56.2 % of the total national area. Only 1.3 million km^2 of land is suitable for cultivation in China, which accounts for about 14 % of the total national land area. The land degradation has involved more than half of the total cultivated land area and further exerts more pressure on the economic benefits of agricultural production and food security. Besides, as the cultivated land degradation will directly affect the potential land productivity, more inputs such as fertilizer and irrigation water will be needed in order to get the same production as well as yield level, which will increase the production cost (Li et al. 2011). In addition, the structural change and pattern succession of the land system resulting from land conversions will undoubtedly lead to changes in the suitability and quality of land and directly influence land productivity. Hence, sustainable productive land management is crucial for the country's long-term agricultural economy.

China is a large country with significant spatial variation of natural/climatic conditions and diverse socioeconomic characteristics. For example, the eastern, central, and western parts of China have different population density, industrial structure, and per capita income, etc. The difference among regions will affect the land use, leading to difference in the way and extent of economic use of land resources. The North China Plain has been one of the most productive agricultural bases in China for thousands of years. It is of great significance to quantitatively measure the intensity and spatial pattern of land degradation and analyze the driving mechanisms, which can provide information for sustainable land management to ensure the food safety in the North China Plain.

In order to provide information for sustainable land management, it is necessary to quantitatively clarify the impacts of major natural and socioeconomic driving forces on land degradation. The recent development of remote sensing techniques (RS) and geographic information system (GIS) techniques has enhanced the capabilities to obtain and handle spatial information on the heterogeneities of land surface characteristics. The integration of RS and GIS technologies has been proven to be an efficient approach and has been successfully used in various investigations for land degradation assessment. Land degradation is a long-term process indicating the loss of ecosystem function and productivity. For quantitative analysis of land degradation, satellite measurements of the normalized difference vegetation index (NDVI), which is the most widely used proxy for vegetation cover and production, has been widely applied in studies of land degradation from the field scale to national and global scales (Wessels et al. 2004; Bai and Dent 2009). Therefore, the objective of this study is to explore the spatial pattern and temporal variation of land degradation based on NDVI in the North China Plain, then to analyze the impact mechanisms of a set of natural and socioeconomic driving forces on land degradation, with the application of the binary panel logistic regression model.

Data Processing and Analyses

Land Uses in the North China Plain

As estimated with the remote sensing data derived from the Landsat TM and ETM+ images in the year of 2008, the cultivated land accounts for about 68 % of the total land area of the North China Plain (Fig. 2.1). There is a semiarid to semihumid warm temperate climate in the North China Plain, with the average annual temperature of 10–15 °C, and annual rainfall of 500–800 mm. It is evident that the North China Plain is suitable for developing rain-fed agriculture and planting a variety of crops and fruit trees. Meanwhile, it is conducive to mechanization and irrigation with its natural conditions of a flat terrain, deep soil, and contiguous

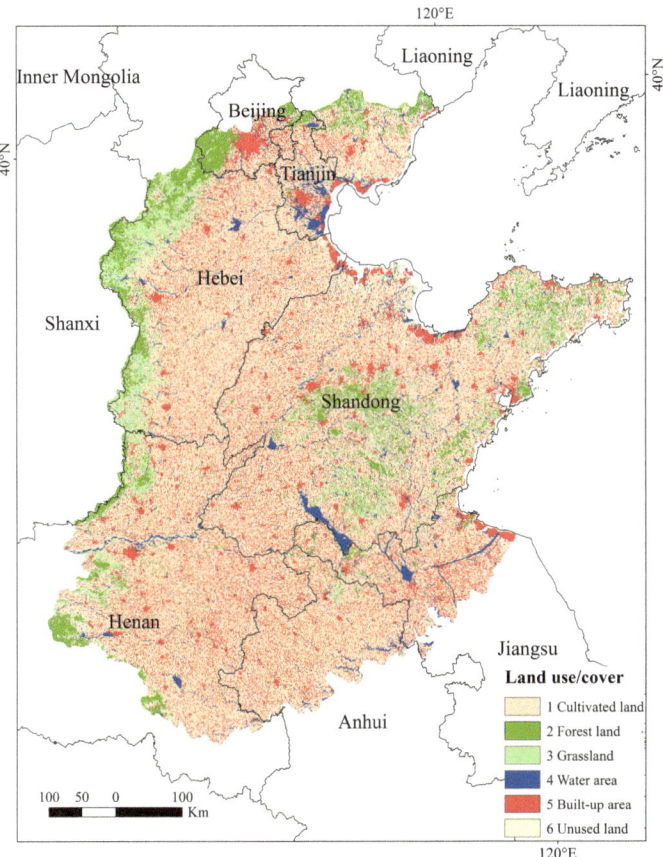

Fig. 2.1 Land-use pattern of the North China Plain in the year of 2008

land concentration. All of these unique conditions have made the North China Plain a major agricultural region and a significant commodity base for grain, cotton, meat, and oil in China.

However, the rapid population growth and urban land expansion have led to serious cultivated land loss and environmental degradation in the North China Plain. Many researches showed that the conversion from cultivated land to other types of land led to the decrease of cultivated land in the North China Plain (Jiang et al. 2011) pointed out that the area of cropland abandonment was much greater than the area of cropland reclamation in the North China Plain (Jiang et al. 2011; Shi et al. 2013). Besides, the quality of cultivated land converted to other land-use types is generally higher than that of the cultivated land converted from other land-use types. As a result, the shrinkage of cultivated land in the North China Plain will exert great impact on the land quality. How to improve land productivity is becoming vital for the improvement of agricultural production. In addition, though the statistic data showed that the grain yields were increasing, it is due to the additional investment into the agricultural production. With this regard, it is an urgent issue to study how to make scientific policy and sustainable land management measures to rationally regulate the land conversion and control land degradation to improve the productivity of the remaining resources in the agricultural sector.

Data Preparation

According to previous studies, driving factors of land degradation are complex. Among which, biophysical causes (include topography that determines soil erosion hazard, and climatic conditions, such as rainfall and temperature) and land-use change (such as deforestation, desertification resulted from unsustainable land management practices) are major direct contributors to land degradation, while the underlying causes of land degradation include population density, economic development, land tenure, and access to agricultural extension, infrastructure, etc. Thus, in order to analyze the situation of agricultural production and impact mechanisms of climate and land-use change integrated with socioeconomic factors on cultivated land degradation in the North China Plain, we collected the NDVI data, land-use datasets, climatic data, socioeconomic statistical data, and other geophysical data of the year of 2000–2008.

NDVI. NDVI is one of the most commonly used indicators for vegetation monitoring (Tucker 1979). Many previous studies have used the NDVI and other measures based on the NDVI as indicators of changes in ecosystem productivity and land degradation, which showed that NDVI is strongly correlated with the green vegetation coverage (Elhag and Walker 2009). Though there are some limitations and criticisms related to the use of this measure, NDVI remains the only dataset available at the global level and the only dataset that reliably provides information about the condition of the aboveground biomass (Nkonya et al. 2011). In this study, we used the remote sensing data of NDVI to measure the land degradation. Before 2006, the NDVI data were obtained from the GIMMS NDVI dataset produced by the GLCF (Global Land Cover Facility) research group at the University of Maryland from July 1981 to December 2006, with a spatial resolution of 8 km by 8 km and a temporal resolution of 15 days. The NDVI data after the year 2006 were derived from SPOT VEGETATION data (ten-day ensembles) which were derived from a dataset developed over the period from April 1998 to December 2010, with a spatial resolution of 1 km by 1 km. Based on the period when the two datasets overlapped (1998–2006), we conducted data processing according to the previous experience of Yin et al. (2014). Firstly, correlation analysis between the maximum monthly NDVI and other factors was performed, following which a linear regression equation was established to extend the Global Inventory Modeling and Mapping Studies (GIMMS) dataset from 2007 to 2008. This helps to eliminate any sensor error in the other two datasets. To ensure the data quality, reliable and internationally recognized data pretreatment processes were used. In order to eliminate cloud contamination effects and the noise caused by other atmospheric phenomena, we included a smoothing method proposed by Chen et al. (2004) based on the Savitzky-Golay filter. Finally, we clipped the NDVI data of the North China Plain during the period from the year of 2000 to 2008.

Land use. The land-use data came from the land-use database of the Resources and Environment Scientific Data Center, Chinese Academy of Sciences (Liu et al. 2010). The database was constructed from remotely sensed digital images by the US Landsat TM/ETM satellite with a spatial resolution of 30 by 30 m. Further the land-use data are upscaled to 1 by 1 km. The data that were analyzed in this study covered the year of 2000 and 2008.

Climatic data. The climatic data, mainly including temperature and rainfall from the year of 2000 to 2008, were derived from the daily records of 117 observation stations covering the entire area of the North China Plain, which were maintained by the China Meteorological Administration. A spline interpolation using the coupling-fitted thin-plate interpolation method was chosen to interpolate the meteorological data. Based on the climatic dependence on topography, climatic variables were interpolated with spatial resolution of 1 by 1 km. At a broader scale, temperature declined roughly with the increasing elevation at a typical standard lapse rate of 6.5 °C per 1 km, so the interpolated air temperature data were adjusted to the sea level according to the altitudinal correction factors based on the digital elevation model (DEM) dataset.

Soil organic matter. The soil fertility map of the study area was clipped from the soil map of China provided by the Institute of Soil Science, Chinese Academy of Sciences. It was based on the maps on the scale of 1:4,000,000 from the Second National Soil Survey of China for Soil Organic Matter (SOM).

Socioeconomic statistical data. In this study, we collected and processed the socioeconomic data of 385 major counties in the North China Plain at a county level for the period of the year of 2000–2008, including the gross domestic production (GDP), population, rural farmers' per capita income, fertilizer utilization, pesticide utilization, farm machinery, total grain production, and total grain sown area, which were all acquired from socioeconomic statistical yearbooks published by National Statistical Bureau of China.

Geophysical data. The geophysical data are also closely related with the land productivity. In this specific study, the geographic factors mainly include elevation and average slope, the data of which were derived from China's Digital Elevation Model Dataset supplied by the National Geomatics Center of China. Another part of the geophysical data was the proximity variables, including the distance from each grid to the nearest expressway or highway, and other way, which were incorporated to measure the impacts of the infrastructure construction on the land degradation.

Primary Analysis

Changes of *grain production.* We analyzed the changes of grain production and the relationship between total grain production and grain yield (kg/ha) of the North China Plain during the year of 2000–2008. The spatial distribution of grain

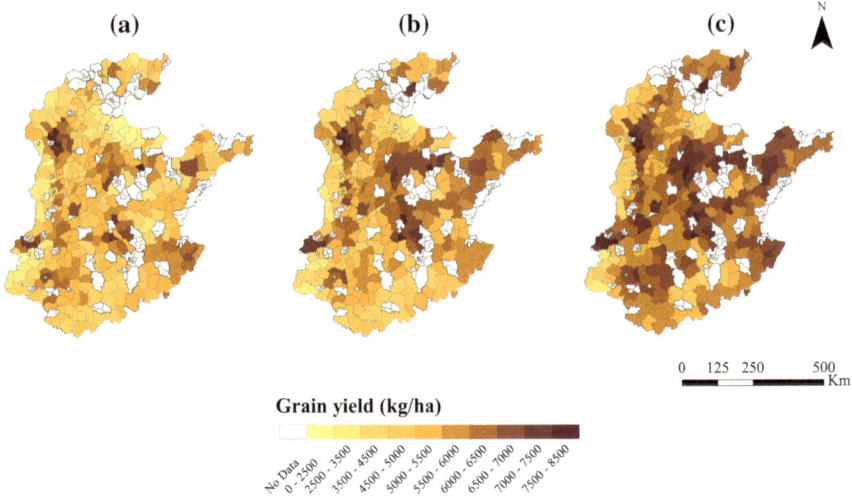

Fig. 2.2 County-level grain yield of the North China Plain, **a** 2000, **b** 2005, and **c** 2008

yield at municipality or county level in the year of 2000, 2005, and 2008 is shown in Fig. 2.2. The grains mainly included wheat and maize, but some rice, soybean, and tuber crops were also included.

During the period of the year of 2000–2008, annual total grain production of the selected counties of the North China Plain increased by more than 40 %, and the average grain yield increased from 4994 to 5986 kg/ha (with an increasing rate of 20 %). The time-series changing trend of the total grain production is similar to that of the average grain yield, and therefore, the increase of agricultural production can be mainly ascribed to the increase in the average grain yield (Fig. 2.3). The time-series change of grain yield from the year of 2000 to 2008 can be divided into two periods. Firstly, the average grain yield decreased after

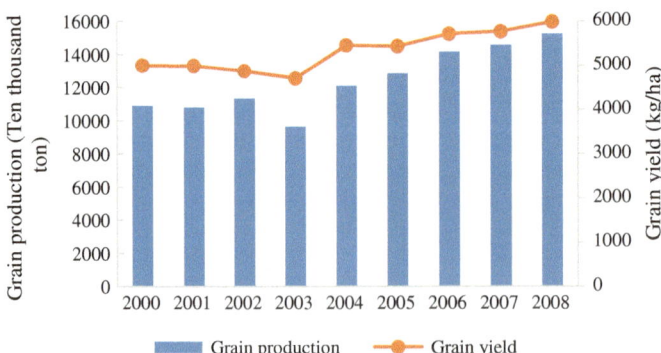

Fig. 2.3 Relationship between grain production and grain yield of the North China Plain, 2000–2008

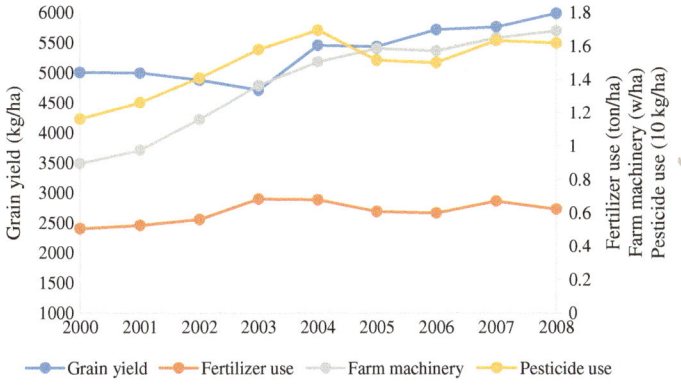

Fig. 2.4 Changes of grain yield, fertilizer use, pesticide use, and farm machinery application of the North China Plain, 2000–2008

the year of 2000 because of a decrease in grain price. Secondly, the price of grain increased again through governmental support after the year of 2003, which made the average yields recover again.

Also, we showed that the application rates of fertilizer and pesticide increased with fluctuations from 525 to 622 kg/ha and from 12.6 to 16.2 kg/ha during the period of the year of 2000–2008, respectively. The application rate of the farm machinery also showed an increasing trend (Fig. 2.4). The increase of the application of farm machinery, fertilizer, and pesticide can be the key determinants of the increase in crop yield; however, further increases in fertilizer and pesticide application are unlikely to be as effective as they previously were in increasing the grain yields because of the gradually diminishing return rate.

Trends in NDVI. The annual average greenness (represented with NDVI) is chosen as the standard proxy of annual average biomass productivity, which generally fluctuates with the rainfall. The NDVI calculated using remote sensing image data, currently is one of the most commonly used indicators to monitor regional or global vegetation and the ecosystem, and is the best instruction factor to reflect vegetation growth conditions and coverage (Zhang et al. 2005). In this study, the cultivated land degradation areas were identified with a sequence of analyses of the NDVI data. The result shows that the annual average NDVI of the North China Plain showed an overall increasing trend during the year of 2001–2004. About 90 % of the study area's annual average NDVI experienced an increasing trend. While only 5 % of the study area's annual average NDVI showed a decreasing trend, which was distributed in the Shandong Province (Fig. 2.5a). In comparison, during the years 2004–2008, about 80 % of the study area's annual average NDVI experienced a decreasing trend, with only 19 % of the study area's annual average NDVI showed an increasing trend (Fig. 2.5b). Besides, the annual average NDVI in some areas showed first an increasing and then a decreasing trend during the year of 2000–2008. Overall, the annual average NDVI showed an increasing trend in most part of the whole study area during the year of 2000–2008 (Fig. 2.5c).

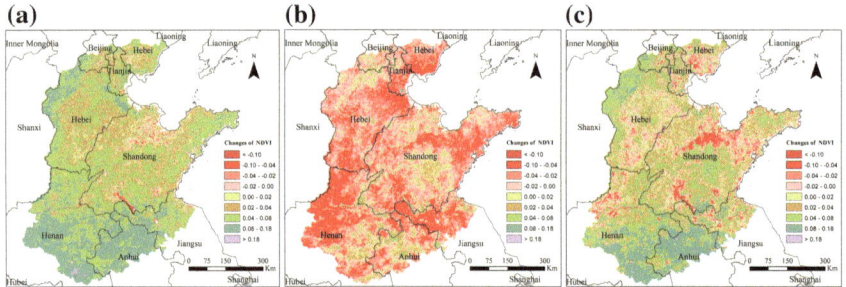

Fig. 2.5 Changes of NDVI of the North China Plain during **a** 2000–2004; **b** 2004–2008, and **c** 2000–2008

Land-use change. During the year of 2000–2008, the land-use change is mainly dominated by the conversion from cultivated land to other land-use types, especially the conversion from cultivated land to built-up land. As cultivated land is the major land-use types in the study area, thus we focus on the cultivated land conversion. In the North China Plain, cultivated land is mainly converted to built-up land (67 %), forest land (9.1 %), and grassland (12.4 %) (Fig. 2.6).

Associations Between Climate Factors and Land Degradation

We investigated the relationship between NDVI (from the year of 2000 to 2008) and climate factors. The relationship between NDVI and climate indicators is widely analyzed in many previous studies, which have confirmed that vegetation changes are closely related to changes in the climate factors at the global or regional scale. In particular, it is widely acknowledged that the temperature and rainfall are the most significant factors influencing the ecosystem characteristics and distribution. We sampled the value of the NDVI, rainfall, and temperature and calculated their annual average values in the whole North China Plain. The mean NDVI, temperature, and rainfall for the study area were plotted along a nine-year time series (Fig. 2.7). Figure 2.7 visually shows that the average NDVI experienced a fluctuated increasing trend, and the average rainfall and temperature slightly showed an increasing trend with some fluctuations.

Specifically, to study the responses of NDVI to the climate change, we analyzed the spatial correlation of NDVI with rainfall and temperature using the correlation analysis at the pixel scale. Figure 2.8 shows the spatial distribution of the correlation of NDVI with rainfall and temperature, respectively. It shows that the rainfall has a strong effect on NDVI, and there was a positive correlation between rainfall and NDVI in most part of the study area, especially in the northern part of the study area where it showed a high correlation. As to the relationship between NDVI and temperature, there was generally a positive relationship between NDVI and temperature, only in the northern part of the study area showed negative relationships.

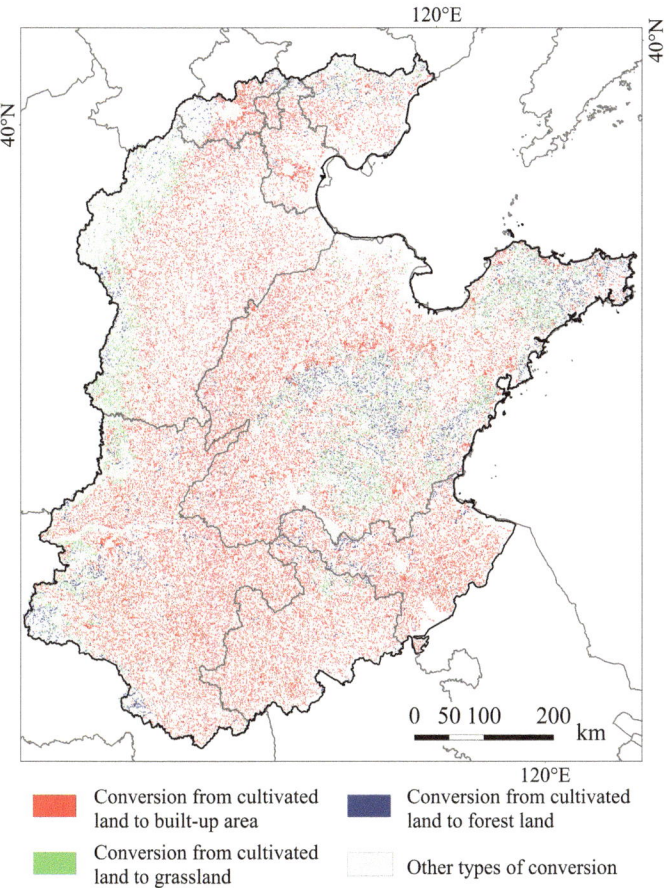

Fig. 2.6   Land conversion from cultivated land to three major land-use types during 2000–2008

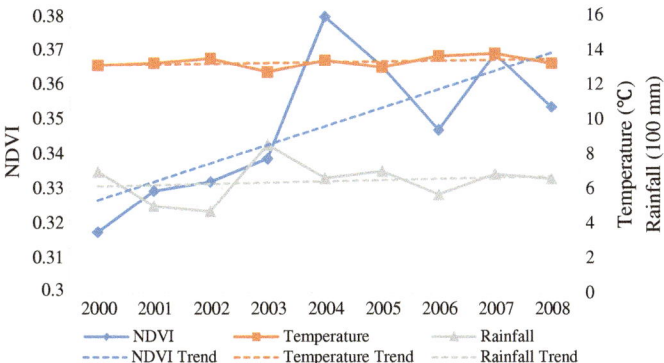

Fig. 2.7   Changing trend of spatially aggregated annual NDVI, rainfall, and temperature

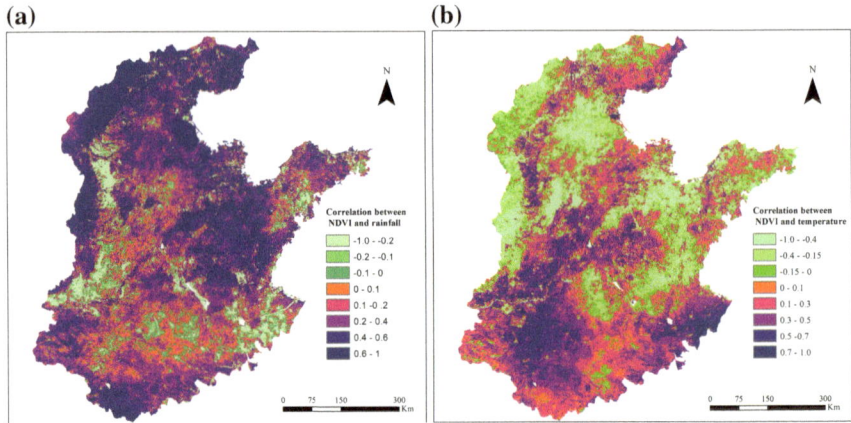

Fig. 2.8 Spatial patterns of correlation **a** between annual average NDVI and rainfall, and **b** between annual average NDVI and temperature at the pixel level in the North China Plain

Assessments of the Driving Forces of Land Degradation

The causes of land degradation are numerous, interrelated, and complex. The relationship between direct and underlying causes of land degradation is complex also, and the impact of underlying factors is context specific (Nkonya et al. 2011). Quite often, the same cause may lead to diverging consequences in different contexts because of its varying interactions with other proximate and underlying causes of land degradation. This implies that targeting only one underlying factor is not sufficient in itself to address land degradation. It is necessary to take into account a number of underlying and proximate factors when designing policies to prevent or mitigate land degradation. For our model specification, it is essential to identify the effects of various combinations and interactions of underlying and proximate causes of land degradation in a robust manner rather than only analyze the individual cause of land degradation. With this regard, we try to identify the key underlying and proximate causes of land degradation. In this part, we conducted econometric analysis of the causes of land degradation in the North China Plain with the binary logistic regression model.

We constructed the model for estimating causes of land degradation or land improvement at the county level, using annualized data, and the model specification is as follows:

$$\text{Logistic } Y_t = \beta_0 + \beta_1 \Delta x_{1t} + \beta_2 \Delta x_{2t} + \beta_3 x_3 + \beta_4 x_4 + \varepsilon_{it}$$

$Y_t = 1$, if $\text{NDVI}_t - \text{NDVI}_{t-1} >= 0$ and $Y_t = 0$ if $\text{NDVI}_t - \text{NDVI}_{t-1} < 0$;

x_{1t} a vector of biophysical causes of land degradation (e.g., climate conditions, topography, soil fertility constraints);

x_{2t} a vector of demographic and socioeconomic causes of land degradation (e.g., population density, per capita GDP, rural farmers' per capita income, and agricultural intensification);

x_3 a vector of variables representing access to infrastructure services (e.g., distance to expressway, distance to high way, and distance to other way);

x_4 a vector of variables representing land-use change (e.g., county-level percentage of land conversion from cultivated land to built-up land, forest land, and grassland during the year of 2000–2008)

Y_t is a binary variable that being used to identify whether land degradation happens, if Y_t equals or is larger than zero, then we assume that there happens land improvement; otherwise, there happens land degradation. Furthermore, we applied a binary panel logistic regression model to simultaneously study the relationship between land degradation conditions and all other variables. Table 2.1 illustrates the results of the regression. The results are to be interpreted with extreme caution due to the complex and multidirectional relationship between NDVI changes and the selected variables. Note that the results might not indicate a causal relationship but only an association between NDVI and the selected biophysical and socioeconomic variables (Nkonya et al. 2011).

Table 2.1 Results for binary panel logistic regression analyses for the biophysical and socioeconomic driving forces of land degradation

Land degradation conditions Y (1 represents land improvement and 0 represents land degradation)	Coef.
△Rainfall (mm)	3.221***
△Temperature (°C)	1.025***
Percentage of land conversion from cultivated land to built-up land (%)	−0.326
Percentage of land conversion from cultivated land to forest land (%)	1.788
Percentage of land conversion from cultivated land to grassland (%)	8.507*
△Fertilizer utilization per unit area (kg/ha)	0.506**
△Population density (ten thousand people/km^2)	−11.282
△Rural farmers' per capita income (ten thousand yuan)	−0.076
△Per capita gross domestic production (ten thousand yuan)	0.0004
△Share of production value of agriculture, animal husbandry, and fishery in GDP (%)	−1.672*
DEM (km)	3.214
Slope (degree)	−0.018
Soil organic matter (%)	0.446***
Distance to highway (km)	0.002*
_cons	0.032

Note *Significant at the 10 % level
**Significant at the 5 % level
***Significant at the 1 % level

Climate change is significantly related with the land degradation conditions which represented by the changes in NDVI. As expected, increments in temperature and rainfall are positively related with the increases in NDVI, and increases in temperature and rainfall will significantly contribute to land improvement. Land-use change is also the major driving force of land degradation. The results showed that the conversion from cultivated land to grassland and forest land will contribute to land improvement, which corresponded to increase in NDVI, while the conversion from cultivated land to built-up land would lead to land degradation. Among the land conversions, only the conversion from cultivated land to grassland significantly affects the land degradation. In comparison, climate change plays a much more important role than land-use change in land degradation. Besides, agricultural intensification represented by fertilizer utilization is significantly and positively related with land improvement, as fertilizer application increases soil carbon (Vlek et al. 2004), which could correspond to an increase in NDVI. Increases in the population density would lead to land degradation, but not significantly. According to the previous studies, the impact of population density on land degradation is ambiguous, while our results suggested that there would be more serious land degradation in areas with higher population density. The population growth will lead to increasing demand for housing and other facilities, which in turn can lead to increase of impervious surface as a result of urban development and infrastructure construction and deforestation. Rapid economic development, which can be represented by changes in the per capita GDP, shows that an increase in both GDP and NDVI had positive impacts on land quality, suggesting the role of ecosystem service could play in economic growth and prosperity. Economic development would stimulate the land improvement, but such impacts were not significant, while the rural economic development, represented by rural farmers' per capita income, showed negative relationship with land improvement, suggesting that the NDVI will decrease as the rural farmer's income increases. It may be due to the fact that rural economic growth develops at the cost of intensive land-use practice, which contributes to environmental degradation further lead to the land degradation. In addition, the increases in the share of the primary industry (agriculture, animal husbandry, and fishery) in GDP significantly corresponded to the decrease in NDVI, as production value of the agriculture increases were mostly resulted from the investment into agricultural production, such as fertilizer, pesticide, and farm machinery, not from natural land-quality improvement. In addition, the geographic and topographic factors also affect the land degradation. As to the topographic factors, the results showed that the higher the elevation and the slower slope, the less serious land degradation. The impacts of elevation was not significant, while it may suggest that in the areas with higher elevation, there would be less impact of anthropogenic factors on the land degradation. The slope had clear significant impacts on land degradation, suggesting steep slopes would lead to land degradation more easily, as steep slope region was more vulnerable to severe water-induced soil erosion. As to the soil fertility represented by SOM, the results showed that the higher the soil fertility was, the less possibility of land degradation. Besides, the distance to highway was significantly positively related with land improvement, which suggested that the larger distances to road network meant there was less human disturbance, less infrastructure development and land-use change, which would exert less impact on the land degradation.

Summary

Land degradation or improvement is an outcome of many highly interlinked direct and underlying causes, including natural, socioeconomic, and related agricultural practices. In China, many types of land degradation are occurring, such as grass-land degradation, deforestation, and cultivated land degradation. In this study, we conducted an empirical study in the North China Plain and analyzed the basic grain production changes, the changing trend of NDVI, and the spatial pattern of relationship between climatic factors and NDVI. Further taking NDVI as a proxy for land degradation and improvement, we applied an econometric model to ana-lyze the causes of land degradation.

According to the analyses, both the grain production and grain yield of the North China Plain showed an increasing trend during the year of 2000–2008; however, estimation based on the remotely sensed NDVI data showed that the North China Plain still experienced land degradation. Degraded areas have been expending in the northern part of the study area, though many parts of the study area showed land improvement. Based on the correlation analysis, the increases in temperature and rainfall corresponded to the increase in NDVI in most parts of the North China Plain, though some parts of the study area showed negative correlation between NDVI and temperature and rainfall. Further, according to the binary panel logistic regression analysis, the results showed there were strong impacts of some key bio-physical and socioeconomic variables on land degradation. Some of these relation-ships were consistent with conclusions in previous case studies, while other showed complex differences. The results for climate factors were not surprising; increase in temperature and rainfall would contribute to land improvement. Different land-use changes exert different impacts on land degradation. Unsurprisingly, increas-ing of forest land and grassland would benefit land quality, while urban expansion would lead to land degradation. Our estimation results about the positive effects of agricultural intensification represented by fertilizer utilization on land degra-dation were also as expected, and the SOM was also a crucial positive factor for land-quality conservation. Rural economic and agricultural development may lead to land degradation. The rural economic and agricultural production growth exerted negative impacts on land quality, which means the development of rural economy and agricultural production led to land degradation. The increases in rural farmers' per capita income and primary production value ratio did not result in the improve-ment of land quality, which may be due to the overexploitation of land with insuf-ficient investment into the land conservation. With this regard, an effective response to land degradation needs increasing incentives of farmers to conserve their culti-vated land and improve their access to the knowledge and inputs that are required for proper conservation. Promotion of such land improvements should be a development policy priority. During the process to promote rural economic development, the governments should also focus on the monitoring and assessment of land quality and make measures to improve land quality, and such improvement measures should be designed together with farmers to meet their prior needs and use appropriate techniques according to the local economic and social conditions. In addition, increasing accessibility to infrastructures may also have negative effects on land quality. Infrastructure development is the basis for regional prosperity, and a booming economy will result in more construction of infrastructure. The expansion of basic infrastructure of transportation such as roads, railways, and airports can further take up land resources and further result in land overexploitation and degradation. The covering of the soil surface with impervious materials as a result of urban development and infrastructure construction is known as soil sealing. Sealing of the soil and land consumption are closely interrelated, when natural, seminatural and cultivated land is covered by impervious surfaces and structures; this will degrade soil functions or cause their loss. To reduce the impacts of infrastructure construction on land quality, the local government should take the assessment of land degradation into consideration during the construction of infrastructures.

Based on the results of the above analysis, to achieve sustainable land management, climate change should be monitored so as to make adaptation measures to mitigate the impacts of climate change on land quality; along with the socioeconomic development, investments and better land management for improving land quality should be definitely encouraged through appropriate policy measures; human activities that change the land surface, such as infrastructure constructions, should be regulated on the basis of the assessment of impacts on land quality, and corresponding land conservation measures should be taken during the construction process.

Predicted Land-Use Conversions in the North China Plain

Introduction

Land-use change, as the direct cause and response of regional environment change, has always been one of the core topics of global change research (Foley et al. 2005). It is difficult to analyze the relationship between land-use and climate change clearly. On the one hand, climate change should exert impacts on the production of cultivated land, forestry, grassland, and so forth. For example, agricultural yields can be directly affected by climate change through changing temperature and precipitation, the distribution of pests, and the frequency of forest fires, and the markets can also be affected by climate change (Haim et al. 2011). Moreover, in recent years, there have been a number of literature-analyzed effects of climate change on agricultural production, with the help of some models (Schlenker et al. 2005; Jiang et al. 2012b). From these pieces of literature, we can learn that hedonic price models are widely used to estimate the relationship between county-level farmland values and climate variables such as temperature and precipitation. These models are then used to simulate the effects of climate change on the value of agricultural production (Deschenes and Greenstone 2007). Even so, some researchers think that it is unreasonable to use mean temperature in the analysis of climate change impacts on agriculture (Schlenker and Roberts 2009). They find that the grain output has a good positive correlation with temperature but then falls quickly as temperature increases above a certain threshold. Other studies support these findings for global timber productivity (Sohngen et al. 2001; Perez-Garcia et al. 2002). In most regions, crops are sensitive to climate change (Izaurralde et al. 2003; Lobell and Field 2008). Usually, global warming will accelerate the crop development, alter the growing season, and enhance the maintenance respiration (Qu et al. 2013), while, as to animal husbandry development, global warming would lead to grassland degradation, which would reduce the amount of livestock grazing and then decrease the output of livestock production. Meanwhile, livestock grazing might promote the CO_2 emissions (Herrero et al. 2009). According to the researches mentioned above, climate change will affect the agriculture, forestry, animal husbandry production, and so forth; then, the demand for products of different land-use types can result in new balance of the supply of and demand for land and further lead to land-use structure changes. On the other hand, most of experts think that the climate system should take account of land use; that is, land-use/land-cover change plays an important role in climate change; for example, afforestation and reforestation can reduce atmospheric CO_2 concentrations, thus, to mitigate the emission of greenhouse gas, and deforestation can result in temperature rising and precipitation decreasing in some regions (Nobre et al. 1991). Then, through the connection of climate change, we can figure out that the land-use structure has relationships with climate change.

Research on land-use structure changes needs to identify the driving factors from a systemic perspective and choose an appropriate model as a tool to reflect the changes to spatial distribution on a certain scale (Deng et al. 2008b). At present, there have been some achievements in applying economic models and empirical statistical methods for analyzing the driving forces of the land-use structure changes, and the simulation about land structure change has a tendency of regionalization and of being microcosmic, while there is much more room for improvement in many links of such simulation researches (Deng et al. 2012). For example, quite a lot of case studies just simulated the changes of one or several types of land use and seldom comprehensively simulated macroscopic structural changes of all land-use types from a systemic perspective (Capozza and Helsley 1989). The change of land-use structure is a dynamic process, which should be based on regional socioeconomic development characteristics, cultural traditions, natural conditions, the previous trend of land-use structure changes, and other factors to acquire valuable simulation results for decision. And it is helpful to make the forecast and evaluation results more science oriented and rational by simulating regional land-use structure changes under different scenarios.

The North China Plain is located in 32–40°30′N, 113–120°30′E, which belongs to the warm temperate zone, with the flat terrain, deep soil layer, and peaks and valleys of rainfall and heat in the same period. It covers seven provinces and cities, including Hebei, Henan, Shandong, Jiangsu, Anhui, Beijing, and Tianjin (containing 387 counties), and has an area of 33.4 million ha, among which the area of land is 21 million ha, and it is the largest plain in China. The region with convenient transportation developed industries, and sufficient labor is one of the most important agricultural regions, and its agricultural production has a great potential for development. The North China Plain is a region that was earlier developed and was greatly influenced by human activities, and it is also one of the economically developed regions in China. The distribution of towns and cities is relatively intensive in the North China Plain; in addition to Beijing and Tianjin, there are also more than 20 cities with a population of more than one million. In the North China Plain, the grain output accounts for 18.4 % of total national output; the output of cotton accounts for 40 % of total national output; and the output of oil-bearing crops also accounts for a large proportion in China.

In 2006, the GDP per capita (GDPPC) in the North China Plain reached 19,224.27 Yuan. According to the experience of industrialization and urbanization in the developed countries and districts, the urbanization development of the North China Plain has entered into the period of cluster development and the later period of industrialization development. Since the North China Plain is located in a vast plain area and the land-use types are relatively unitary, the core problem of land use lies in the contradiction between the cultivated land protection and nonagricultural construction land (Deng et al. 2008b). The North China Plain is one of the important commodity grain bases in China, while its built-up area's expansion has taken up a large amount of high-quality cultivated land resources, which poses threat to regional and national food security and ecological environment.

In the North China Plain, most researches focused on the impact of climate change on cultivated land. There are few researches of the response mechanism on grassland and built-up area, and only a few studies have explored the combined effects of climate change on the agricultural and forestry sectors. The North China Plain is China's traditional and important agricultural production base, and it is also a region where the population grows rapidly; economy and urbanization develop rapidly with obvious land-use conversion, especially the transfer between cultivated land and nonagricultural land use (Jiang et al. 2012a). Therefore, in this study, we selected the North China Plain as case study area to analyze the future land-use structure changes under different scenarios. And based on an econometric model, we analyzed the effect of natural conditions (climate factors such as precipitation, temperature, etc.) and socioeconomic factors on land-use structure changes. This study can provide reference for the decision making of the local land-use planning, urbanization management, and land-use management.

Land-Use Change in the North China Plain

In this study, the information about land-use structure and its changes was derived from the remote sensing images of the North China Plain in years 1988,

Table 2.2 Land-use structure of the North China Plain in various years (*unit* ten thousand hectares)

Land-use type	1988	1995	2000	2005	2010	Change rate during 1988–2010 (%)
Cultivated land	2448.81	2410.86	2402.01	2372.28	2346.51	−4.18
Forestry	467.01	472.01	470.82	471.05	471.48	0.96
Grassland	375.12	363.34	363.48	356.25	355.51	−5.23
Water area	310.09	312.74	315.28	317.84	321.02	3.53
Built-up area	1137.88	1187.64	1197.93	1237.09	1263.32	11.02
Unused land	40.20	32.50	29.58	24.59	21.25	−47.12

*Change rate during (R) 1988–2010 is calculated based on the following formula: $R = (B - A)/A \times 100\%$; A represents the area of each type of land in 1988 and B represents the area of each type of land in 2010

1955, 2000, 2005, and 2010 though the man–computer interactive interpretation (Table 2.2). And the information can be used as the basic data for analyzing the driving mechanism of land-use structure changes.

According to the statistics in Table 2.2, in the North China Plain, cultivated land accounts for the largest share about 50 %, and built-up area, forestry, water area, and grassland overall occupy a certain proportion, which guaranteed the development of local farming, forestry, animal husbandry, side-line production, and fishery. From 1988 to 2010, there had been some changes in the macrostructure of land use; cultivated land and grassland decreased by 4.18 and 5.23 %, respectively. And along with the development of economy and expansion of urbanization, the built-up area increased by 11.02 % from 1988 to 2010. In addition, the change trends of each land-use type during 1988–2010 are also not the same; cultivated and unused land always showed a trend of decrease, while water area and built-up area showed a continual trend of increase. The forestry overall increased with interval decrease from 1995 to 2000, and the grassland overall decreased with a slight interval increase from 1995 to 2000.

Data and Methodology

Data

The data used in this study include the basic geographic information data, socio-economic data, and climate data that can affect land-use structure changes. The basic geographic information data mainly include remote sensing data and biological geographic elements data. And the land-use data and climate information are uniformly rasterized into 1 km × 1 km grid cell. The land-use data mainly come from remote sensing images interpretation during 1988–2010 (Yin et al. 2010), and then, the information of interpreted raster data can be extracted from

Table 2.3 Descriptive statistics of main variables at county level

Variable	Obs	Mean	Std. Dev.	Min	Max
Population	1633	614,909.90	421,749.10	63,245	6,600,841
Annual per capita income (Yuan)	1366	1959.64	2414.40	823	46,739
Total grain output (T)	1392	294,692.9	207,261.60	0	2,231,531
Price index of agricultural products	1935	104.12	3.30	101.30	108.40
Price index of forestry products	1935	105.20	1.98	102.30	108.50
Price index of animal husbandry products	1935	111.62	7.92	100.50	123.90
NPP(net primary productivity) ($gc/m^2 y$)	1933	352.66	85.39	143.74	827.50
Annual average precipitation (mm)	1930	688.14	208.84	257.44	1562.09
Annual average temperature (°C)	1930	13.57	1.46	3.23	16.28
Land quality (dummy variable)	1915	499.58	142.50	54.00	868.07

*The observations are different because of the data availability in several counties (387 counties × 5 periods)

county level. And thus, we derived the geographic elements information of each county, including temperature and precipitation. The concrete operations of data are mainly based on ArcGIS 9.3 and STATA 10.0 platform, and the data eventually were prepared in the form of panel data for analyses (Deng et al. 2012).

In this study, the socioeconomic data and climate data of 378 counties in the North China Plain were collected in 1988, 1995, 2000, 2005, and 2010 (Table 2.3). However, the land-use data are of spatial data type. To make the land-use data match with the statistic data of socioeconomic factors, the land-use data should be aggregated into county level. And in order to increase the sample space, improve the degree of freedom, provide individual information, and make the estimated results more accurate, the data mentioned above were prepared into the format of panel data which integrated the cross-sectional data and the time-series data.

Scenarios

In this study, the basic settings of parameters for future scenarios are as follows: the average annual growth rate of annual per capita income is 7 %; the population in China should increase to 1.36, 1.45, and 1.5 billion in 2010, 2020, and 2050, respectively. The data are derived from the forecast of National Planning and Chinese Academy of Social Sciences.

In this research, through investigating the rule of macroscopic land-use structure changes during the past 20 years in the North China Plain, three scenarios were designed, namely business as usual scenario (BAU), rapid economic growth scenario (REG), and coordinated environmental sustainability scenario (CES). The land-use structure changes of the North China Plain during 2010–2050 are simulated under the three different scenarios. These scenarios are story lines that

Table 2.4 Regional trends in population and annual per capita income in the North China Plain assumed for the BAU, REG, and CES scenarios

Scenario	Year				
	2010	2020	2030	2040	2050
Population	(Hundred million people)				
BAU	2.48	2.51	2.54	2.57	2.6
REG	2.54	2.69	2.64	2.63	2.64
CES	2.42	2.52	2.45	2.35	2.33
Annual per capita income	(Yuan)				
BAU	3375.6	4537.8	5873.6	6974.3	7825.4
REG	3407.9	4769.8	6200.7	7443.3	8930.7
CES	3442.4	4625.4	5921.4	7220.5	8378.5

Note The annual per capita income was calculated and adjusted based on the consumer-price index of year 2000 to eliminate the effect of inflation

represent different future developments regarding population growth, economic growth, and environmental sustainability (Table 2.4). We adopted the population and income assumptions of the three scenarios to develop associated projections of built-up area returns. We modified agricultural returns with projections of agricultural prices produced for each of these scenarios. Finally, we incorporated agricultural and forest yield changes into the scenarios. From 2010 to 2050, population and annual per capita income increase gradually and the population growth rate in the REG is the highest and about 1.3 % higher than that in the CES scenarios. The REG scenario has higher annual per capita income, which is 24 % higher than that in the CES scenario in 2050.

BAU scenario was designed according to the socioeconomic conditions and the local development plan of the North China Plain, which can comprehensively and objectively reflect real conditions of local production activities and economic development.

REG scenario and CES scenario were designed with different conditions of economic development and ecological environmental sustainability to simulate the land-use structure changes at county level. This study intends to simulate the land-use structure changes, respectively, under different scenarios, among which the REG scenario is giving priority to economic development and the CES scenario is giving priority to ecological environmental sustainability and further provides reference for regional-level land-use exploitation, planning, and management.

Most agricultural commodities are traded internationally. And, therefore, it makes sense to focus on how climate change will affect global agricultural prices; thus, this study applied the basic linked system (BLS), a computable general equilibrium model that represents all of the major economic sectors, including agriculture, together with the agroecological zone (AEZ) model, which can assess the effects of climate change on agricultural systems to analyze the impact of climate change. The regional price indexes of all crops in the three scenarios are shown in Table 2.4. From the national point of view, the area of cultivated land will decrease

Table 2.5 National
agricultural price indexes
under the BAU, REG, and
CES scenarios

Scenario	Year					
	2000	2010	2020	2030	2040	2050
BAU	100	108	136.5	180.7	228.4	250.8
REG	100	109.7	147.4	188.7	230.1	267
CES	100	108.6	149	190.6	237.2	274.6

year by year, the future mode of agricultural production will gradually transfer from cultivated land to grassland, and increase of built-up area will not bring small impact on cultivated land, so the national and regional agricultural price index is rising constantly in the future (Table 2.5). The national agricultural price index has obvious rising trend year by year, and it is significantly higher under the REG and CES scenarios than that under the BAU scenario, which means that in order to guarantee the REG and the sustainable development of ecological environment, it will inevitably lead to the changes of land-use structure and further affect the changes of industrial structure and product price.

Methods

As to the methods used to project land-use structure under the BAU, REG, and CES scenarios, in the first subsection, we discuss the land-use simulation model. The key driver of land-use structure changes in the projection model is county-level net profits of cultivated land, grassland, forestry, and built-up area. These variables are modified in the three scenarios according to assumed changes in population and annual per capita income and to predicted changes in regional agricultural prices, NPP, and forest carbon. And in the subsequent subsections, we discuss how these projections are scaled down to the county level and linked to the net profits.

Land-Use Simulation Model

In this study, based on the econometric model that Haim et al. (2011) used to study the relationship between climate change and future land-use change in USA, the land-use simulation model is developed through adjusting the variables that affect the net profit of each land-use type. In the model, the change of the overall land system with various land-use types can be simulated, and it allows transfers of different land-use types. This study applied the historical data of the North China Plain to estimate the land owners' response to economic benefits and further make decisions on land-use allocation. The county-level net profits of cultivated land, forestry, grassland, and built-up area are regarded as the main driving factors of land-use structure changes, and the net profits are calculated synthetically based on population, annual per capita income, agricultural prices, NPP, and so forth. Climate change can indirectly affect the net profits through causing the change of some factors, and according to the principle of maximum benefits, it will affect land-use allocation.

In this model, the area of each land-use type is the dependent variable, and the net profit and quality of each land-use type are independent variables. The net profit variables provide the link between different land-use types, which can be described in detail as follows.

The general equation for the net profit in county c is as

$$\mathrm{NP}_{mc} = \sum_{m} \omega_{mc} p_{mc} q_{mc} - \sum_{m} \omega_{mc} c_{mc} \tag{2.1}$$

where ω_{mc} is the net profits for commodity type m in county c, p_{mc} is the per unit price for commodity type m in county c, q_{mc} is the per acre average yield of commodity type m in county c, and c_{mc} is the per hectare cost of producing commodity type m in county c. Thus, the net profits of different land-use types can be calculated as the income of all the products minus the cost.

In the model, it is assumed that the land owners' expected values of the net profits of land are fixed, and they would change the land-use type in order to get the maximum net profits. The net profits include deterministic and stochastic components; the deterministic components include county-level net profits, land quality, and interaction between the two variables, and the stochastic components are estimated through the nested logistic model according to the distribution hypothesis (Jiang et al. 2012a). In this study, four main types of land use are taken into consideration, which are cultivated land, forestry, grassland, and built-up area. The probabilities of the transfer of different types of land use can be calculated based on the econometric model shown in (2.2), in which the independent variables and estimated parameters are incorporated to calculate the probabilities:

$$P_{ijk}^{t} = P(\varepsilon_{jk}, NP_i^t, LQ_i) \tag{2.2}$$

where P_{ikj}^{t} is the probabilities that the land-use type j be transferred to land-use type k in region i during the research period, ε_{jk} is the vector of estimated parameters, LQ_i is the dummy variable that indicates land quality of region i.

Scenario simulations are performed at county level, taking year 1988 as the baseline, and for simplicity, years 1988, 1995, 2000, 2005, and 2010 are noted as $t = 0, 1, 2, 3, 4,$

$$S_{ijk}^{t+1} = \sum_{k} P_{ijk}^{t} \cdot S_{ikt} \tag{2.3}$$

where S_{ik}^{t} is the area of land-use type j in region i at time t, and we can get the sequence of transfer probabilities with Eq. (2.2), and then future land-use allocation can be calculated. The land-use structure changes from time t to time $t + 1$ indicates the change of supply of land products and services and further indicates the change of net profits and prices of all land-use types. In this study, the prices of land products of cultivated land, forestry, grassland, and built-up land are regarded as exogenous variables.

As the area of each type of land use (S_{ij0}) in baseline year was known, the land-use structure changes till 2050 can be simulated based on the Eq. (2.3). The transfer probability P^t_{ikj} can be derived from Eq. (2.2), and P^t_{ikj} is a function of county-level net profit. Thus, once we get the county-level net profit, the future land-use structure changes can be simulated.

Estimation of Net Profits

Net Profit of Built-Up Area. According to the standard urban land rent theory, population and income growth are the two main determinants of the value of urban land (Capozza and Helsley 1989). Based on the statistical model using county-level panel data, the net profit of built-up area as the dependent variable can be linked with the population and income as independent variables at county level. And according to the theory, population and annual per capita income are positively correlated with the built-up area value. High annual per capita income increases the demand for housing, and high population growth increases the price of habitable land. So, we can choose population and annual per capita income as two indicators to simulate and predicate the future net profit of built-up area.

Net Profit of Cultivated Land. According to (2.1), using the price index of agricultural products released by National Bureau of Statistics, the net profit of cultivated land can be calculated. Considering the condition of marked economy in China, it is reasonable to apply the nation price index to county-level estimation, and it is the same to the county-level price index of forestry products and animal husbandry products.

Net Profits of Forestry and Grassland. Based on the MAPSS model, it was assumed that the changes of forestry products and animal husbandry products were positively correlated with NPP. Through optimal management measures, the NPP of forestry and grassland can reach the maximum value. And the NPP was calculated based on the leaf area index (LAI) and NDVI.

Validation of the Model

In order to validate the accuracy of the model, that is, to verify whether the simulation results can meet with the actual situation of land use, based on the historic data of land use, we do the validation as follows. Based on the model and actual land-use data in 2005, we simulate the land-use structure in 2010, then calculate the deviance between the actual land-use structure in 2010 and the simulated land-use structure, and thus measure the accuracy of the model, as shown in

$$D_{i,t} = \frac{\hat{Y}_{i,t} - Y_{i,t}}{Y_{i,t}} \times 100\% \qquad (2.4)$$

where $D_{i,t}$ is the area deviance of land-use type i at time t, $\hat{Y}_{i,t}$ is the simulated area of land-use type i at time t, and $Y_{i,t}$ is the actual area of land-use type i at time t.

The validation results showed that the deviance of forestry and grassland was ±4 and ±3 %, respectively. While cultivated land and built-up area account for larger shares, the deviance of them is relatively high, but the overall deviance was still controlled within 8 %. Thus, the model can be regarded as robust and can be used to simulate the future land use.

Results

In this study, we established the county-level net return models of cultivated land, forestry, grassland, and built-up area. And various factors that have effects on the net returns were synthetically incorporated into the basic land-use structure change model. Through the analysis of the historical data, land-use structure change model can quantitatively determine the effect degrees of various factors on the land-use structure changes. Then, as to the simulation of future land-use structure, based on the effect degrees of the factors, such as population, annual per capita income, and price index, the future land-use structure changes can be simulated under different scenarios.

The macroscopic change of the land-use structure during 2010–2050 was simulated under different scenarios, with the water area and unused land integrated into other lands since it is difficult to evaluate them. As to the simulation of land-use scenarios, integrating all the factors that can affect the net returns, according to the assumption of the changes of population and annual per capita income, expected agricultural prices, and grassland production, the future land-use structure changes can be simulated from 2010 to 2050.

Under the BAU scenario (Fig. 2.9), cultivated land and built-up area are still the main land-use types, but cultivated land tends to decrease year by year, and built-up area tends to increase year by year, which is the inevitable requirement of economic development. By 2050, the cultivated land still accounts for the largest proportion. The forestry generally shows a growing trend; the grassland overall shows a trend of decrease; the change of both of land-use types is not obvious; and the amount of forestry and grassland is relatively stable.

Under the REG scenario (Fig. 2.10), during 2010–2050, the area of cultivated land decreases, from 2400 ten thousand hectares to 1500 ten thousand hectares, reducing by 37.5 %. However, there has been a rapid growth in built-up area, the area of which increases from 1250 ten thousand hectares to 2000 ten thousand hectares. The growth rate of built-up area is much higher than that under the BAU scenario (Fig. 2.9) and CES scenario (Fig. 2.11). However, the area of forestry shows a trend of fluctuations with slow increase, and the area of grassland decreases year by year with small amplitude.

Under the CES scenario, we give priority to the protection of development of ecological environmental sustainability. According to the simulation results, the area of cultivated land decreases significantly, while the area of built-up area and woodland area increases year by year, and the area of grassland also shows a trend of decrease, but the trend is not obvious. The decreasing trend of cultivated land

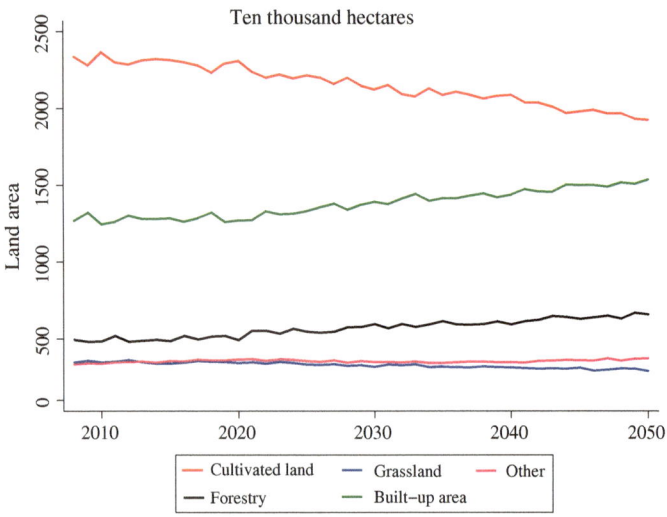

Fig. 2.9 Structure changes of land use under BAU scenario

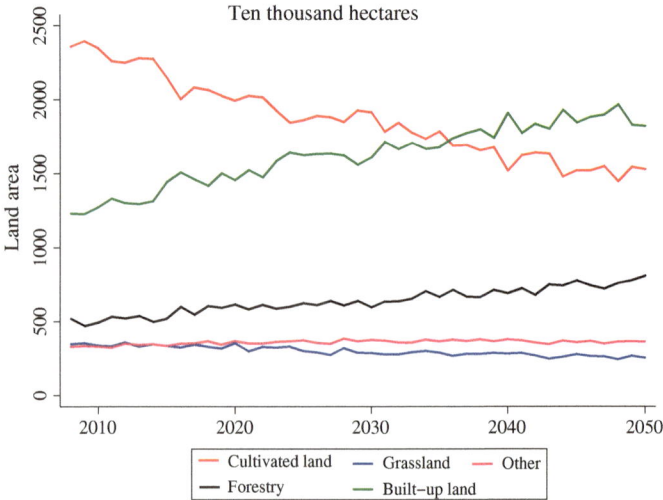

Fig. 2.10 Structure changes of land use under REG scenario

and the expansion trend of built-up area can reach stable balance until 2050, and the area of the forestry continues to increase to maintain the sustainable development of ecological environment.

In general, with the improvement of economic growth rate, the built-up area is increasing year by year, and the decrement rate of cultivated land is also

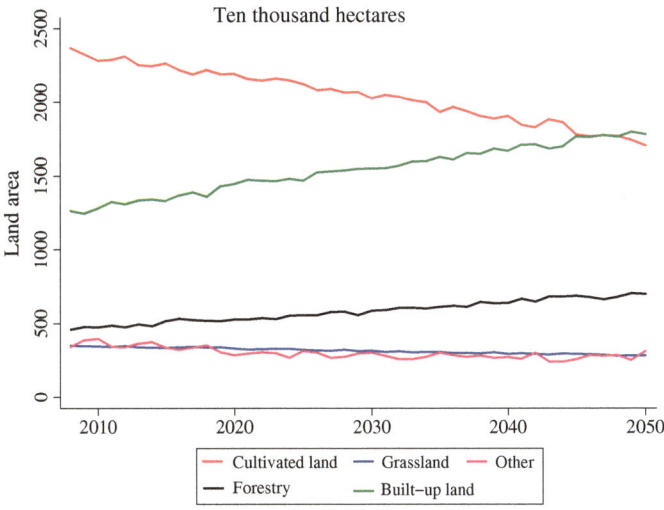

Fig. 2.11 Structure changes of land use under CES scenario

increasing. Under the REG and CES scenarios, the proportion of built-up area in the land system is more than that of the cultivated land by 2050. And under the three different economic growth scenarios, the forestry overall shows a trend of increase, and the growth rate of forestry speeds up as economic growth speeds up. And the grassland overall shows a trend of decrease under the three scenarios.

Summary

The North China Plain is one of the agricultural production bases in China, which is in the important period of increasing population, developing economy, and accelerating urbanization process. In this study, we applied econometric model to analyze the mechanism of county-level land-use structure changes, then simulated the changes of land-use structure from 2010 to 2050 under three scenarios, quantitatively revealed the relationship between the land-use structure changes and the driving factors, and realized the simulation and scenario analysis on structural changes of land use. This relatively sophisticated county-level land-use structure changes simulation method can be applied in other regions, and the conclusions can provide reference for decision making of regional-level land exploitation, planning, and management.

First, comparing the simulation results of future land-use structure changes in the North China Plain of BAU, REG, and CES scenarios, the competition laws among the land-use types under the synthesized effects of various driving factors can be analyzed. Under the BAU scenario, each type of land experience expansion and shrinkage to some extent over time, among which, the cultivated land has an obvious trend of decrease and the built-up area has an obvious trend of increase. And under the REG and CES scenarios, since the economic growth rates are different, the extents of expansion and shrinkage of different land-use types are also different.

Second, the built-up land overall shows a trend of increase under the three scenarios, with its area increasing from 1260 ten thousand hectares in 2010 to 1500 ten thousand hectares, 1910 ten thousand hectares, and 1820 ten thousand hectares in 2050 under the BAU, REG, and CES scenarios, respectively. Under the REG scenario, the significant increase of the built-up land reflects that the REG can lead to the increase in net return of built-up area and it is important to note that the growth rate of built-up area may be overestimated since the endogenous driving factor urbanization is not taken into consideration. There has been a view that, with the expansion of built-up area, the net return of built-up area will decline, thus, to limit the continual expansion of built-up area. In addition, the increase in the net return of built-up area could lead to the increase of urban housing density and further affect land-use structure, while the housing factor is also not taken into consideration in this model. Thus, the absence of these factors may limit the accuracy of the simulation results of land-use structure in these models.

Third, the structural change trends of land use under the three scenarios are relatively consistent, with cultivated land showing a trend of decrease and built-up area showing a trend of increase. It indicates that urbanization in the North China Plain is one of the main driving factors of land-use structure changes. According to the simulation results, cultivated land, as the main land-use type in the North China Plain, accounting for the largest proportion, decreases with the improvement of economic growth. Under the REG scenario, the area of cultivated land will be less than that of built-up land by 2050.

Finally, the simulation results of land-use structure changes under the three scenarios can provide significant reference for making land-use planning and sustainable development strategy of government. On the whole, the cultivated land will decrease and built-up area will increase with the economic development, which may lead to contradiction between cultivated land, food security, and improvement in our living standard. So, more attentions should be paid to the trade-off between the urban expansion and food security, including the land consolidation and rehabilitation and compensation mechanism for protecting cultivated land.

Simulated Shifting Patterns of Agroecological Zones Across China

Introduction

The world is facing a crisis in terms of food security (Rosegrant and Cline 2003). The challenge is from not only the growing global population but also the sustainability of nutritious food supply. In order to meet global demands, food production should increase 60–70 % by 2050 compared with that at the beginning of the twenty-first century (Alexandratos and Bruinsma 2012). Climate change is now widely recognized as one of the most critical influences on sustainability of food supply. It changes the suitability of crop and investment structure in agriculture. Researchers have confirmed that the crop suitability shifts in the context of climate change (Ramirez-Villegas et al. 2013). Lane and Jarvis (2007) predicted the impact of climate change with current and projected future climate data and found that the suitable area of main food crops, including rice, wheat, potato, and some cash crops, such as apple, banana, coffee, and strawberry, would reduce along with climate change. According to the evaluation of Easterling et al. (2007), even in the scenario of Intergovernmental Panel on Climate Change (IPCC) low emissions (B1, with a 2 °C rise in global mean temperatures by 2100), the current farming systems will be destabilized. If the increasing suffering from chronic hunger is frustrating today, the further difficulties, risks, and challenges for achieving food security will make people desperate, especially for those in South Asia and sub-Saharan Africa (Godfray et al. 2010).

China, which experienced rapid growth and increased integration with the global economy in recent years, has significant potential to contribute to global food security not only by alleviating hunger among its own citizens, but also by increasing trade and financial linkages as well as technology and knowledge exchanges with other developing countries (Fan and Brzeska 2010). China has made remarkable progress in reducing poverty, cutting the share of people living on less than $1.25 a day from 84 % of the population in 1981 to 13 % in 2008 and reducing the number of poor people from 835 million to 174 million (Malik 2013). In recent years, the development of China's agricultural production is very stable which significantly contributes to global security (Huang and Rozelle 2009). But there are still some potential challenges for China's food supply. One of the most probable challenges is climate change. Now and in the future, China's food supply will be a key issue for the world food market.

There are various ways such as the frequency and intensity changes of disasters (including flood, drought, hail, and typhoon) and the changes of precipitation, accumulated temperature, and solar radiation that climate change influencing China's agricultural production. Though most of these influences are occasional and small-scaled, there are still some influences that are persistent and massive (Guo et al. 2014). The variation of AEZs may be one of the most notable responses of climate change that can alter the pattern of crop and agricultural management. The AEZ concept was originally developed by the FAO. It has widespread applications in land-use planning, design of appropriate agricultural adaptations, reducing vulnerability, as well as crop water requirements and long-term frost protection measures determination (Wratt et al. 2006). There are many researches on agro-ecological zoning in China and relevant topics (Deng et al. 2006; Yu et al. 2012). Chen classified China into twelve AEZs based on the mode of agricultural production, the productivity of farmland, heat, water, and landform (Chen 2001). Jin and Zhu divided Northeast China into three AEZs and found that the maize yields are reduced significantly along with climate change (Zhi-Qing and Da-Wei 2008).

Yang et al. applied the method proposed by Chen (2001) to classify Northwest China into thirteen AEZs and analyzed the climate-induced changes in crop water balance in each AEZ (Yang et al. 2008). He et al. divided Chinese wheat-producing area into three major regions and ten AEZs depending on the produced grain traits, ecological factors, soil properties, cropping system, and so on (Zhonghu et al. 2002). On the basis of these ten AEZs, Jia et al. studied the variation of precipitation and found the AEZ-based research help to master the precipitation patterns of different agricultural regions. As a determinant of cropping pattern and agriculture management, the pattern of AEZs may change significantly under the background of global climate change. There is no doubt that an assessment of shifting patterns of AEZs in China is helpful to guide the nation's future agricultural development and guarantee global food security (Jia et al. 2012).

The global agroecological zones (GAEZs) were developed in order to provide a prediction for crop potential productivity in a specific environment under limiting factors including climate and soil. They were used as the main analysis units of agricultural production in Global Trade Analysis Project (GTAP) as well as a lot of other researches. For example, Atehnkeng et al. analyzed the distribution and toxigenicity of Aspergillus species isolated from maize kernels from three AEZs in Nigeria (Atehnkeng et al. 2008). Palm et al. analyzed environmental and socioeconomic barriers for plantation activities on local and regional levels and investigated the potential for carbon finance to stimulate the increased rates of forest plantation on wasteland in southern India combined with AEZs (Palm et al. 2011). IIASA and FAO had published the GAEZ version 3.0 aiming to include practical applications such as a significantly updated version, including expanded crop coverage and dryland management techniques. This dataset provides climate change impacts on agroecological suitability and productivity for three time horizons, 2020s, 2050s, and 2080s, for 11 combinations of GCMs and IPCC emission scenarios, which is global significance. But it is a little bit sketchy for guiding national and regional agriculture developments because more detailed and accurate data can always be obtained to support agroecological zoning at such scale. Mugandani et al. reclassified the AEZs of Zimbabwe based on the climatic data from meteorological stations to cover the shortage of GAEZ (Mugandani et al. 2012). Kurukulasuriya and Mendelsohn compared the observed distribution of AEZs from FAO, the calculated distribution of AEZs given climate, and found there were significant local differences (Kurukulasuriya and Mendelsohn 2008). These studies imply that it makes sense to access the impacts of future climate change on shifting patterns of the AEZs by using more detailed and accurate data at regional and national levels.

This study aims to assessing the shifting patterns of AEZs in China driven by future climate change. The major contribution of this study is that it provides future AEZs in China which contains decision-making information for agriculture development in China and global food security. Compared with GAEZ, this study takes full advantage of observed climate data from meteorological stations. And compared with other static assessment results based on historical period data such as Chen (2001), this study provides more predictive information of shifting patterns of AEZs. In addition, the data processing scheme of integrating observed and predicted data of this study has high reference value for similar studies. AEZs in China of four periods: 2011–2020, 2021–2030, 2031–2040, and 2041–2050 were drawn in this study based on the projected climatic changes. It is found that the future climate change will lead to significant change of AEZs in China, while the overall pattern of AEZs in China is stable. The northward expansion of humid AEZs to subhumid AEZs in south China, eastward expansion of arid AEZs to dry and moist semiarid AEZs in north China, and southward expansion of dry semi-arid AEZs to arid AEZs in southwest China are the major characteristics of future AEZs changes in China.

Data and Methodology

Agroecological Zoning Methodology

The AEZ is a zone characterized by specific length of growing period (LGP) and climatic attributes. Several agroecological zoning methods have been previously used for agricultural purposes (Sys et al. 1991). The Center for Sustainability and the Global Environment (SAGE) at the University of Wisconsin derived six global LGPs by aggregating the IIASA/FAO GAEZ data into six categories of approximately 60 days per LGP: (1) LGP1: 0–59 days, (2) LGP2: 60–119 days, (3) LGP3: 120–179 days, (4) LGP4: 180–239 days, (5) LGP5: 240–299 days, and (6) LGP6: more than 300 days (Ramankutty et al. 2004). These six LGPs roughly divide the world along humidity gradients, in a manner that is generally consistent with previous studies in global agroecological zoning (Alexandratos 1995). They are calculated as the number of days with sufficient temperature and precipitation/soil moisture for growing crops. In other words, LGPs are calculated as the number of days in the year when average daily temperature (Ta) and precipitation plus moisture stored in the soil (P) are above their thresholds. In GAEZ, three standard temperature thresholds for temperature growing periods are used: (1) periods with Ta > 0 °C, (2) periods with Ta > 5 °C, which is considered as the period conducive to plant growth and development, and (3) periods with Ta > 10 °C, which is used as a proxy for the period of low risks for late and early frost occurrences. In this study, we choose the second standard temperature threshold (Ta > 5 °C) for calculating LGPs in China. We set the threshold of precipitation plus moisture stored in the soil as half the potential evapotranspiration (ET) (P > 0.5 ET) to keep it consistent with FAO (Doorenbos and Kassam 1979). Therefore, LGP used in this study is defined as follows:

$$LGP = Card\{(Ta_i, P_i) | Ta_i > 5\,°C, P_i > 0.5\,ET_i\} \quad (2.5)$$

where Ta_i (°C) is the average daily temperature of the ith day in the year, P_i (mm day^{-1}) is the precipitation plus moisture stored in the soil of the ith day in the year, ET_i (mmday^{-1}) is ET of the ith day in the year, and Card is the cardinality function counting the number of elements of a set.

In addition to the LGP breakdown, the world is subdivided into three climatic zones, tropical, temperate, and boreal, using criteria based on absolute minimum temperature and growing degree days (GDD), as described by Ramankutty and Foley (1999). Concretely, the climatic zone rules are as follows:

$$
\text{zone} = \begin{cases} \text{tropical,} & \text{if } t_{\min} > 0\,^{\circ}\text{C} \\ \text{temperate,} & \text{else if } t_{\min} > -45\,^{\circ}\text{C and GDD}_5 > 1200 \\ \text{boreal,} & \text{else} \end{cases} \tag{2.6}
$$

where $t_{\min}$ (°C) is the absolute minimum temperature, and GDD_5 is the annual number of accumulated GDD above 5 °C. GDD_5 is calculated by taking the average of the daily maximum and minimum temperatures compared to the base temperature, 5 °C.

$$
\text{GDD}_5 = \sum_{i=1}^{365} \left(\frac{\text{Td}_{i,\min} + \text{Td}_{i,\max}}{2} - \text{Td}_{\text{base}} \right) \tag{2.7}
$$

where $\text{Td}_{i,\min}$ and $\text{Td}_{i,\max}$ are the daily maximum and minimum temperatures, respectively; Td_{base} (=5 °C) is the base temperature. We employ the definition of AEZs used in the GTAP land-use database, which divides the world into 18 AEZs (Table 2.6).

Data and Process

Both observed and projected climate data were used in this research to assess the shifting patterns of AEZs in China driven by climatic changes. This dataset includes climatic data, moisture stored in the soil, and ET. Historically observed climate data including average daily temperature, precipitation, absolute minimum

Table 2.6 Definition of agroecological zones used in Global Trade Analysis Project

AEZs	LGP	Climatic zone	AEZs	LGP	Climatic zone
AEZ1	0–59	Tropical	AEZ4	180–239	Tropical
AEZ7		Temperate	AEZ10		Temperate
AEZ13		Boreal	AEZ16		Boreal
AEZ2	60–119	Tropical	AEZ5	240–299	Tropical
AEZ8		Temperate	AEZ11		Temperate
AEZ14		Boreal	AEZ17		Boreal
AEZ3	120–179	Tropical	AEZ6	≥300	Tropical
AEZ9		Temperate	AEZ12		Temperate
AEZ15		Boreal	AEZ18		Boreal

temperature, and daily maximum and minimum temperatures were collected by the China Meteorological Administration for the year 2006. The spline interpolation method is used to interpolate these data using a 100-m digital elevation model developed by Chinese Academy of Sciences. This interpolative method is selected because it takes into account the climatic dependence on topography by using a trivariate function of latitude, longitude, and elevation (Deng et al. 2008b, 2011) and balances the smoothness of fitted surfaces and the fidelity of the data by minimizing the generalized cross-validation automatically (Tait et al. 2006). Observed data of moisture stored in the soil come from soil categories and soil composite datasets at the scale of 1:1,000,000. This dataset was developed according to the survey results of the second national soil census data, covering 1572 soil profiles in 100-cm-depth layer. The ET data come from the MODIS (Moderate Resolution Imaging Spectroradiometer) Global Evapotranspiration Project (MOD16). This project aims to estimate global terrestrial evapotranspiration by using satellite remote sensing data. The missing values of ET data are complemented using the spline interpolation method.

The projected climate data including average daily temperature, daily precipitation, absolute minimum temperature, and daily maximum and minimum temperatures from 2006 to 2050 comes from the datasets produced from Hadley Center coupled model version 3 (HadCM3) GCM (Global Circulation Model) under greenhouse gas emission scenario of B2. The B2 scenario describes a world in which the emphasis is on local solutions to economic, social, and environmental sustainability. While the scenario is also oriented toward environmental protection and social equity, it focuses on local and regional levels. The data are originally obtained with a spatial resolution of $2.50° \times 3.75$ and downscaled to 1 km $\times$ 1 km using the spline interpolation method. A data processing scheme is designed to reduce systematic errors generated by HadCM3 and improve the assessment accuracy of the shifting patterns of AEZs in China (Fig. 2.12). First, the changes of average daily temperature, daily precipitation, absolute minimum temperature, and daily maximum and minimum temperatures of each year from 2011 to 2050 compared with those of 2006 are calculated using the projected climate data:

$$\Delta K_{i,y} = K_{i,y} - K_{i,2006} \tag{2.8}$$

where $K_{i,2006}$ refers to the projected average daily temperature, daily precipitation, daily maximum and minimum temperatures of the ith day, and absolute minimum temperature, in 2006; $K_{i,y}$ ($y = 2011, 2012, ..., 2050$) refers to the projected average daily temperature, daily precipitation, daily maximum and minimum temperatures of the ith day, and absolute minimum temperature, in the yth year; and $\Delta K_{i,y}$ refers to the changes of average daily temperature, daily precipitation, absolute minimum temperature, and daily maximum and minimum temperatures of each year from 2011 to 2050 compared with those of 2006. Second, the corrected climate data can be generated by adding $\Delta K_{i,y}$ to the observed climate data of 2006.

$$Q_{i,y} = Q_{i,2006} + \Delta K_{i,y} \tag{2.9}$$

Fig. 2.12 Data processing
scheme for assessing
the shifting patterns of
agroecological zones in
China

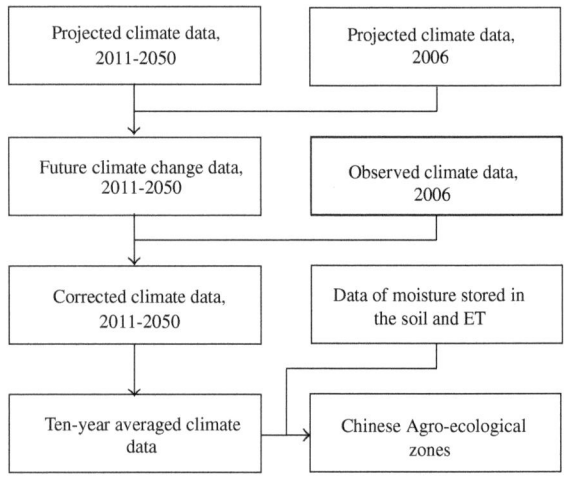

where $Q_{i,2006}$ refers to the observed average daily temperature, daily precipitation, daily maximum and minimum temperatures of the ith day, and absolute minimum temperature, in 2006; and $Q_{i,y}$ refers to the corrected average daily temperature, daily precipitation, daily maximum and minimum temperatures of ith day, and absolute minimum temperature, in the yth year. By these two steps, the systematic errors in projected climate data generated by HadCM3 are reduced. Considering climate change is a long-term shift in weather conditions, we then take the average for ten years of the corrected climate data to assess the shifting patterns of AEZs driven by future climate change.

$$Q_{i,p-(p+9)} = \frac{1}{10} \sum_{y=p}^{p+9} Q_{i,y} \qquad (2.10)$$

where $Q_{i,p-(p+9)}$ ($p = 2011, 2021, 2031, 2041$) refers to the ten-year averaged climate data including average daily temperature, daily precipitation, daily maximum and minimum temperatures of the ith day, and absolute minimum temperature, in the periods of 2011–2020, 2021–2030, 2031–2040, and 2041–2050. Finally, future AEZs in China are obtained with the ten-year averaged climate data and data of moisture stored in the soil and ET.

Results and Discussion

The assessment results show that there are totally fourteen AEZs in China, most of which will experience pattern shifting from 2011 to 2050 (Fig. 2.13). The AEZ4, AEZ5, and AEZ6 characterized by tropical climate will be mainly found in the extreme south of China (Fig. 2.13). The AEZ4 will be stable in pattern and area

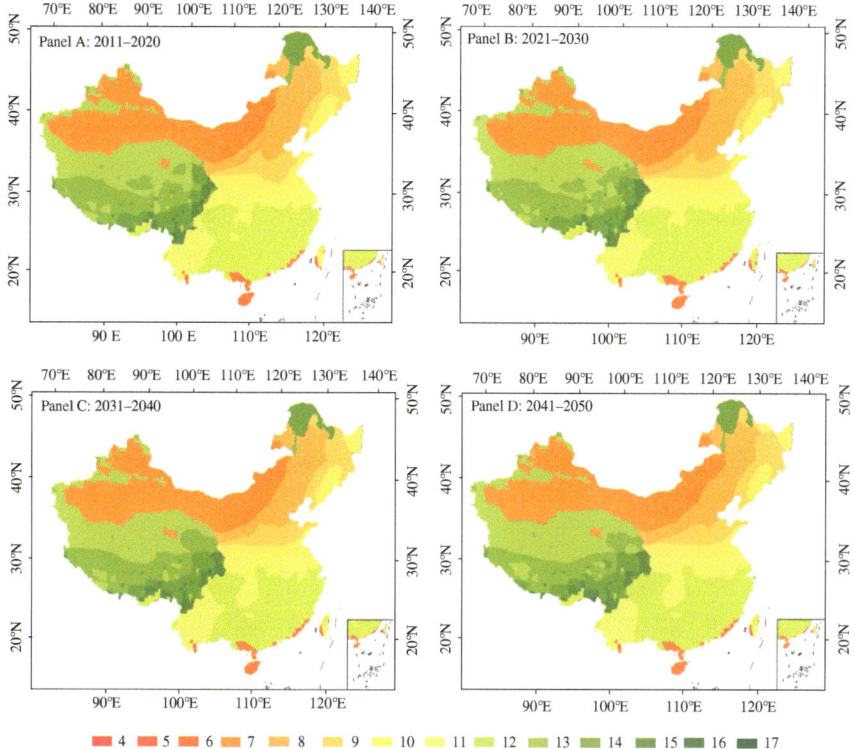

Fig. 2.13 Future agroecological zones in China, 2011–2050

over time and covers 0.16 million ha, and this translates to no more than 0.02 % of the whole country (Table 2.7). The area of AEZ5 will increase from 1.08 million hectares to 1.76 million ha from 2011 to 2050 with an area expansion of 63.73 %. This area increase will mainly happen in the period of 2021–2040. The new area of AEZ5 will be mainly transferred from AEZ6 (Table 2.8). The AEZ6 will experience an area decreasing from 10.44 to 9.57 million ha during the period from 2011 to 2050 (Table 2.7). Besides the conversion to AEZ5, the area decrease of AEZ6 will be also due to the southward expansion of AEZ11 (Table 2.8; Fig. 2.13).

The AEZ11 and AEZ12 will occupy most area of south China (Fig. 2.13). The AEZ12 will cover about 16 % of the whole of China and expand by 3.88 million ha from 2011 to 2050 (Table 2.7). This expansion will be mainly because of the westward expansion being larger than the southward shrink of AEZ12 (Fig. 2.13). And the newly added area of AEZ11 cannot counterbalance the reduced area from AEZ12 expansion (Table 2.8). Consequently, the area of AEZ11 will reduce from 85.50 million ha in 2011 to 82.48 million ha in 2050 (Table 2.7). Moreover, the northward movement of northern side boundary of AEZ11 implies that there will be LGP reduction that happened in central China. The area of

Table 2.7 Areas of future agroecological zones in China

AEZs	Area (*unit* million hectares)			
	2011–2020	2021–2030	2031–2040	2041–2050
AEZ4	0.16	0.16	0.16	0.16
AEZ5	1.08	1.08	1.71	1.76
AEZ6	10.44	11.08	9.62	9.57
AEZ7	200.23	199.37	203.73	207.76
AEZ8	90.05	89.85	85.72	83.41
AEZ9	60.26	58.38	60.29	60.53
AEZ10	53.52	55.80	48.39	47.34
AEZ11	85.50	81.78	88.18	82.48
AEZ12	151.90	155.90	149.59	155.78
AEZ13	141.05	136.86	143.84	156.05
AEZ14	86.57	90.65	88.98	74.27
AEZ15	55.60	53.83	51.94	52.64
AEZ16	15.96	17.58	19.71	20.10
AEZ17	0.49	0.49	0.96	0.97

AEZ10 will decline from 53.52 to 47.34 million ha from 2011 to 2050 mainly due to the southward expansion of AEZ11 in central China and the expansion of AEZ9 in northwest China (Table 2.8). The center of gravity of AEZ9 will shift to the east due to the overall northward movement of AEZ10, but its area which is about 60.00 million ha will keep stable (Table 2.7).

Most of the area of northern China will belong to AEZ7 and AEZ8 (Fig. 2.13). The AEZ7 will cover 200.23 million ha in the period of 2011–2020 and expand to 207.76 million ha in the period of 2041–2050 (Table 2.7). This area increase is mainly because of the conversion from AEZ8 to AEZ7 (Table 2.8). And there will be significant area reduction of AEZ7 due to the expansion of AEZ13 in northwest China (Table 2.8). The area of AEZ8 will decrease from 90.05 to 83.41 million ha from 2011 to 2050 mainly due to the eastward expansion of AEZ7 (Table 2.7 and Fig. 2.13). The boundary change of AEZ7 and AEZ8 implies that there will be LGP reduction in northern China and temperature decrease in northwest China. Another significant change in northeast China will be the area reduction of AEZ15 (Table 2.7 and Fig. 2.13). About one-fifth of the AEZ15 in northeast China will convert to AEZ8, AEZ9, and AEZ10 due to temperature rise (Table 2.8).

The AEZ13 will cover about 15 % of the whole China and more than a half of southwest China (Table 2.7 and Fig. 2.13). Its area will expand by 15.00 million ha from 2011 to 2050 mainly due to the conversion from AEZ14 (Table 2.8). This indicates that the LGP in southwest China will be generally increased driven by future climate change. And AEZ14 will become more continuous along with climate change (Fig. 2.13). The total area of AEZ15 will be reduced by 2.95 million ha mainly due to its area decrease in northeast China (Table 2.7 and Fig. 2.13). The AEZ16 and AEZ17 will mainly spread over southwest China

Table 2.8 Transition matrix of agroecological zones in China (*unit* million hectares)

Transition from	Transition to													
	AEZ4	AEZ5	AEZ6	AEZ7	AEZ8	AEZ9	AEZ10	AEZ11	AEZ12	AEZ13	AEZ14	AEZ15	AEZ16	AEZ17
AEZ4	–													
AEZ5		–	0.25[c]											
AEZ6		0.63[b] 0.30[c]	–					0.42[b]	0.40[b]					
AEZ7				–						2.44[a] 3.29[b] 2.14[c]				
AEZ8				6.63[b] 5.08[c]	–	0.72[b]				0.31[b]	0.20[a] 0.87[b]			
AEZ9					3.29[b] 2.77[c]	–	3.37[a] 1.17[c]					0.21[b]		
AEZ10						1.09[a] 4.37[b] 2.30[c]	–	2.82[b] 1.02[c]				0.24[b]		
AEZ11			0.24[a]					–	4.68[a] 0.24[b] 7.85[c]			0.44[b] 0.24[c]	0.28[b]	0.47[b] 0.29[c]
AEZ12			0.40[a]					0.57[a] 8.81[b] 1.66[c]	–				0.26[b]	

(continued)

Table 2.8 (continued)

	Transition to													
	AEZ4	AEZ5	AEZ6	AEZ7	AEZ8	AEZ9	AEZ10	AEZ11	AEZ12	AEZ13	AEZ14	AEZ15	AEZ16	AEZ17
AEZ13				1.58[a] 1.02[b] 1.10[c]						–	5.53[a] 9.81[b]	2.61[a] 0.25[b] 3.12[c]		
AEZ14					0.91[b]					3.10[a] 14.43[b] 13.74[c]	–	1.41[a] 2.00[b] 0.98[c]		
AEZ15					0.20[b]	0.40[a] 0.32[b] 1.88[c]	1.10[c]	0.27[a]		0.02[b] 0.55[c]	2.86[a] 3.54[b] 0.01[c]	–	2.26[a] 3.78[b] 0.72[c]	
AEZ16								0.35[a] 0.40[b]	0.29[a]		1.45[b]	2.83[b]	–	0.33[c]
AEZ17												0.61[c]		–

Notes [a]From period of 2011–2020 to period of 2021–2030
[b]From period of 2021–2030 to period of 2031–2040
[c]From period of 2031–2040 to period of 2041–2050

(Fig. 2.13). The AEZ16 will cover 15.96 million ha in the period of 2011–2020 and expand to 20.10 million ha in the period of 2041–2050 (Table 2.7). These newly expanded areas of AEZ16 will mainly come from AEZ15 (Table 2.8). The area of AEZ17 in China will be almost doubled due to future climate change though it will still cover an area of not more than 1 million ha (Table 2.7).

On the whole, the pattern of AEZs in China will change significantly due to future climate change. The changed area will reach 32.00 million ha accounting for 3.36 % of the country's total area (Table 2.8). Certainly, the changes of AEZs in China are not monotonous. The direction inversion of boundary movement of AEZ12 in southwest China from the period of 2011–2040 to the period 2041–2050 is one of the most prime examples of such changes (Fig. 2.13). But the overall change trend of AEZs in China including northward expansion in south China, eastward expansion in north China, and southward expansion in southwest China is persistent. These results are obtained using datasets produced from HadCM3 under greenhouse gas emission scenario of B2. It is sure that some new results will be found by applying dataset from other models and scenarios. And we will address them in future studies.

Summary

The shifting patterns of AEZs in China driven by climatic changes were assessed. The projected climate data and historically observed climate data as well as moisture stored in the soil and ET data are used in this study. By using the agroecological zoning methodology proposed by IIASA and FAO, we draw AEZs in China of each of the four periods: 2011–2020, 2021–2030, 2031–2040, and 2041–2050. The main conclusions from this study could be summarized as follows. (i) The overall pattern of AEZs in China will be stable in the future. The AEZ4, AEZ5, and AEZ6 will distribute mainly in the extreme south of China. The AEZ12, AEZ11, AEZ10, AEZ9, AEZ8, and AEZ7 will sequentially spread from north to south. The AEZ13 and AEZ14 will occupy most of the southwest of China. And besides AEZ15 partly distributing in the northeast of China, most of other AEZs (including AEZ15, AEZ16, and AEZ17) will distribute mainly in the southwest of China. (ii) The future climate change will lead to significant local changes of AEZs in China. The changes of AEZs will be not monotonous over time, but the overall change trend of AEZs is persistent. The future change of AEZs in China is characterized by area expansion of AEZ7, AEZ12, AEZ13, and AEZ16 and area reduction of AEZ8, AEZ10 AEZ11, AEZ14, and AEZ15. The shifting patterns of AEZs in China are characterized by northward expansion of humid AEZs to sub-humid AEZs in south China, eastward expansion of arid AEZs to dry and moist semiarid AEZs in north China, and southward expansion of dry semiarid AEZs to arid AEZs in southwest China. (iii) The data processing scheme proposed in this study is helpful in reducing systematic errors in projected climate data which has high reference value for similar studies. Its feasibility has been proved by our empirical analysis. The application of observed data from meteorological stations will improve the assessment accuracy of impacts of future climate change on shifting patterns of the AEZs in principle.

References

Alexandratos N (1995) World agriculture: towards 2010: an FAO study, Food & Agriculture Organization, Rome

Alexandratos N, Bruinsma J (2012) World agriculture towards 2030/2050: the 2012 revision. ESA Work. Pap 3

Atehnkeng J, Ojiambo PS, Donner M, Ikotun T, Sikora RA, Cotty PJ, Bandyopadhyay R (2008) Distribution and toxigenicity of *Aspergillus* species isolated from maize kernels from three agro-ecological zones in Nigeria. Int J Food Microbiol 122(1):74–84

Bai Z, Dent D (2009) Recent land degradation and improvement in China. AMBIO J Hum Environ 38(3):150–156

Bajocco S, De Angelis A, Perini L, Ferrara A, Salvati L (2012) The impact of land use/land cover changes on land degradation dynamics: a Mediterranean case study. Environ Manag 49(5):980–989

Capozza DR, Helsley RW (1989) The fundamentals of land prices and urban growth. J Urban Econ 26(3):295–306

Chen B (2001) The integrated agricultural resources production capability and population supporting capacity in China. Meteorological Press, Beijing (in Chinese)

Chen J, Jönsson P, Tamura M, Gu Z, Matsushita B, Eklundh L (2004) A simple method for reconstructing a high-quality NDVI time-series data set based on the Savitzky-Golay filter. Remote Sens Environ 91(3):332–344

Deng X, Huang J, Rozelle S, Uchida E (2006) Cultivated land conversion and potential agricultural productivity in China. Land Use Policy 23(4):372–384

Deng X, Huang J, Rozelle S, Uchida E (2008a) Growth, population and industrialization, and urban land expansion of China. J Urban Econ 63(1):96–115

Deng X, Su H, Zhan J (2008b) Integration of multiple data sources to simulate the dynamics of land systems. Sensors 8(2):620–634

Deng X, Huang J, Huang Q, Rozelle S, Gibson J (2011) Do roads lead to grassland degradation or restoration? A case study in Inner Mongolia, China. Environ Dev Econ 16(06):751–773

Deng X, Yin F, Lin Y, Jin Q, Qu R (2012) Equilibrium analyses on structural changes of land uses in Jiangxi Province. J Food Agric Environ 10(1):846–852

Deschenes O, Greenstone M (2007) The economic impacts of climate change: evidence from agricultural output and random fluctuations in weather. Am Econ Rev 354–385

Doorenbos J, Kassam A (1979) Yield response to water. Irrig Drainage Pap 33:257

Easterling W, Aggarwal P, Batima P, Brander K, Bruinsma J, Erda L, Howden M, Tubiello F, Antle J, Baethgen W (2007) Food, fibre, and forest products 3

Elhag M, Walker S (2009) Impact of climate variability on vegetative cover in the Butana area of Sudan. Afr J Ecol 47(s1):11–16

Fan S, Brzeska J (2010) The role of emerging countries in global food security. IFPRI Policy Brief 15

FAO (1996) Agro-ecological zoning guidelines. FAO Soils Bulletin, vol 73. Food and agriculture organization of the United Nations, Rome, Italy

Foley JA, DeFries R, Asner GP, Barford C, Bonan G, Carpenter SR, Chapin FS, Coe MT, Daily GC, Gibbs HK (2005) Global consequences of land use. Science 309(5734):570–574

Godfray HCJ, Beddington JR, Crute IR, Haddad L, Lawrence D, Muir JF, Pretty J, Robinson S, Thomas SM, Toulmin C (2010) Food security: the challenge of feeding 9 billion people. Science 327(5967):812–818

Guo D, Gao Y, Bethke I, Gong D, Johannessen OM, Wang H (2014) Mechanism on how the spring Arctic sea ice impacts the East Asian summer monsoon. Theor Appl Climatol 115(1–2):107–119

Haim D, Alig RJ, Plantinga AJ, Sohngen B (2011) Climate change and future land use in the United States: an economic approach. Clim Change Econ 2(01):27–51

Herrero M, Thornton PK, Gerber P, Reid RS (2009) Livestock, livelihoods and the environment: understanding the trade-offs. Curr Opin Environ Sustain 1(2):111–120

Huang J, Rozelle S (2009) Agricultural development and nutrition: the policies behind China's success. World Food Programme

Izaurralde RC, Rosenberg NJ, Brown RA, Thomson AM (2003) Integrated assessment of Hadley Center (HadCM2) climate-change impacts on agricultural productivity and irrigation water supply in the conterminous United States: part II. Regional agricultural production in 2030 and 2095. Agric For Meteorol 117(1):97–122

Jia SJ, Han SQ, Wang HJ (2012) The research on the pattern of precipitation in Hebei agricultural regions. Adv Mater Res 518:4062–4067

Jiang QO, Deng X, Zhan J, He S (2011) Estimation of land production and its response to cultivated land conversion in North China Plain. Chin Geogr Sci 21(6):685–694

Jiang L, Deng X, Seto KC (2012a) Multi-level modeling of urban expansion and cultivated land conversion for urban hotspot counties in China. Landscape Urban Plan 108(2):131–139

Jiang QO, Deng X, Yan H, Liu D, Qu R (2012b) Identification of food security in the mountainous guyuan prefecture of China by exploring changes of food production. J Food Agric Environ 10(1):210–216

Kurukulasuriya P, Mendelsohn RO (2008) How will climate change shift agro-ecological zones and impact African agriculture? World Bank Policy Res Working Pap Ser Vol

Lane A, Jarvis A (2007) Changes in climate will modify the geography of crop suitability: agricultural biodiversity can help with adaptation. SAT ejournal 4(1):1–12

Li H, Liu Z, Zheng L, Lei Y (2011) Resilience analysis for agricultural systems of north China plain based on a dynamic system model. Scientia Agricola 68(1):8–17

Liu J, Zhang Z, Xu X, Kuang W, Zhou W, Zhang S, Li R, Yan C, Yu D, Wu S (2010) Spatial patterns and driving forces of land use change in China during the early 21st century. J Geogr Sci 20(4):483–494

Lobell DB, Field CB (2008) Estimation of the carbon dioxide (CO_2) fertilization effect using growth rate anomalies of CO_2 and crop yields since 1961. Glob Change Biol 14(1):39–45

Malik K (2013) Human development report 2013. The rise of the south: human progress in a diverse world. In: The rise of the south: human progress in a diverse world (15 March 2013). UNDP-HDRO human development reports

MEA (2005) Ecosystems and human well-being: desertification synthesis. World Resources Institute

Mugandani R, Wuta M, Makarau A, Chipindu B (2012) Re-classification of agro-ecological regions of Zimbabwe in conformity with climate variability and change. Afr Crop Sci J 20(2):361–369

Nkonya E, Gerber N, von Braun J, De Pinto A (2011) Economics of land degradation. IFPRI Issue Brief 68

Nobre CA, Sellers PJ, Shukla J (1991) Amazonian deforestation and regional climate change. J Clim 4(10):957–988

Palm M, Ostwald M, Murthy IK, Chaturvedi RK, Ravindranath N (2011) Barriers to plantation activities in different agro-ecological zones of Southern India. Reg Environ Change 11(2):423–435

Perez-Garcia J, Joyce LA, McGuire AD, Xiao X (2002) Impacts of climate change on the global forest sector. Clim Change 54(4):439–461

Qu R, Cui X, Yan H, Ma E, Zhan J (2013) Impacts of land cover change on the near-surface temperature in the North China Plain. Adv Meteorol 2013

Ramankutty N, Foley JA (1999) Estimating historical changes in global land cover: croplands from 1700 to 1992. Glob Biogeochem Cycles 13(4):997–1027

Ramankutty N, Hertel T, Lee HL, Rose S (2004) Global land use and land cover data for integrated assessment modeling. In: Book chapter for the Snowmass conference, Snowmass, CO, July

Ramirez-Villegas J, Jarvis A, Läderach P (2013) Empirical approaches for assessing impacts of climate change on agriculture: the EcoCrop model and a case study with grain sorghum. Agric For Meteorol 170:67–78

Rosegrant MW, Cline SA (2003) Global food security: challenges and policies. Science 302(5652):1917–1919

Schlenker W, Hanemann WM, Fisher AC (2005) Will US agriculture really benefit from global warming? Accounting for irrigation in the hedonic approach. Am Econ Rev 395–406

Schlenker W, Roberts MJ (2009) Nonlinear temperature effects indicate severe damages to US crop yields under climate change. Proc Nat Acad Sci 106(37):15594–15598

Shi W, Tao F, Liu J (2013) Changes in quantity and quality of cropland and the implications for grain production in the Huang-Huai-Hai Plain of China. Food Secur 5(1):69–82

Sohngen B, Mendelsohn R, Sedjo R (2001) A global model of climate change impacts on timber markets. J Agric Res Econ 326–343

Sys C, Van Ranst E, Debaveye J (1991) Land evaluation. Part. II. Methods in land evaluation. Agric Publ 7:247

Tait A, Henderson R, Turner R, Zheng X (2006) Thin plate smoothing spline interpolation of daily rainfall for New Zealand using a climatological rainfall surface. Int J Climatol 26(14):2097–2115

Tucker CJ (1979) Red and photographic infrared linear combinations for monitoring vegetation. Remote Sens Environ 8(2):127–150

Vlek PG, Rodríguez-Kuhl G, Sommer R (2004) Energy use and CO_2 production in tropical agriculture and means and strategies for reduction or mitigation. Environ Dev Sustain 6(1–2):213–233

Wessels K, Prince S, Frost P, Van Zyl D (2004) Assessing the effects of human-induced land degradation in the former homelands of northern South Africa with a 1 km AVHRR NDVI time-series. Remote Sens Environ 91(1):47–67

WMO (2005) Cliamte and land degradation. World Meteorological Organization

Wratt D, Tait A, Griffiths G, Espie P, Jessen M, Keys J, Ladd M, Lew D, Lowther W, Mitchell N (2006) Climate for crops: integrating climate data with information about soils and crop requirements to reduce risks in agricultural decision-making. Meteorol Appl 13(04):305–315

Yang Y, Feng Z, Huang HQ, Lin Y (2008) Climate-induced changes in crop water balance during 1960–2001 in Northwest China. Agric Ecosyst Environ 127(1):107–118

Yin RS, Xiang Q, Xu JT, Deng X (2010) Modeling the driving forces of the land use and land cover changes along the Upper Yangtze River of China. Environ Manage 45(3):454–465

Yin F, Deng X, Jin Q, Yuan Y, Zhao C (2014) The impacts of climate change and human activities on grassland productivity in Qinghai Province, China. Front Earth Sci 8(1):93–103

Yu Y, Huang Y, Zhang W (2012) Changes in rice yields in China since 1980 associated with cultivar improvement, climate and crop management. Field Crops Res 136:65–75

Zhang X, Ge Q, Zheng J (2005) Impacts and lags of global warming on vegetation in Beijing for the last 50 years based on remotely sensed data and phonological information. Chin J Ecol 24(2):123–130

Zhi-Qing J, Da-Wei Z (2008) Impacts of changes in climate and its variability on food production in Northeast China. Acta Agronomica Sinica 34(9):1588–1597

Zhonghu H, Zuoji L, Loongjun W (2002) Classification of Chinese wheat regions based on quality. Scientia Agricultural Sinica (China)

Chapter 3
Impact Assessments of Land-Use Change on Valued Ecosystem Services

Wei Song, Xiangzheng Deng, Bing Liu, Zhaohua Li, Gui Jin and Xin Wen

Abstract Land-use change is a major factor driving ecosystem service change. Measuring the ecosystem service variation in response to land-use change is an effective way to assess the environmental costs and benefits of different approaches to policy-based planning. In this chapter, we examined changes in valued ecosystem services induced by the land encroachment of urbanization in North China Plain (NCP). China is on the track for rapid urbanization since the implementation of a market-oriented economic reform in 1978. The unprecedented urbanization in China has resulted in substantial cultivated land loss and rapid expansion of urban areas. The cultivated land loss due to urbanization not only threatens food security in China, but also leads to the ecological system degradation, to which close attention should be paid. Therefore, we examined the effects of conversion from cultivated to urban areas on ecosystem service in the NCP on the basis of an NPP-based ecosystem service model and a buffer

W. Song (✉) · X. Deng
Institute of Geographic Sciences and Natural Resources Research,
Chinese Academy of Sciences, Beijing 100101, China
e-mail: songw@igsnrr.ac.cn

X. Deng
Center for Chinese Agricultural Policy, Chinese Academy of Sciences,
Beijing 100101, China

B. Liu
College of Geomatics, Shandong University of Science and Technology,
Qingdao 266590, Shandong, China

Z. Li · G. Jin
Resources and Environmental Science,
Hubei University, Wuhan 430062, Hubei, China

X. Wen
Centre D'Applications et de Recherches En Télédétection,
Université de Sherbrooke, Sherbrooke, QC J1K 2R1, Canada

© Springer-Verlag Berlin Heidelberg 2015
J. Zhan (ed.), *Impacts of Land-use Change on Ecosystem Services*,
Springer Geography, DOI 10.1007/978-3-662-48008-3_3

comparison method. Cultivated land loss due to urbanization in the NCP led to a total loss of ecosystem service value (ESV) of 34.66 % during 1988–2008. Urban expansion significantly decreased the ecosystem service function of water conservation (100 %), nutrient cycling (−31.91 %), gas regulation (−7.18 %), and organic production (−7.18 %), while it improved the soil conservation function (2.40 %). Land-use change accounted for 57.40 % of the changes in ecosystem service and had a major influence on changes in nutrient cycling and water conservation. However, climate change mainly determined the changes in functions of gas regulation, organic production, and soil conservation. Further, we analyzed the changes in valued ecosystem services resulting from ecological restoration programs in Shandong province. China launched a series of ecological restoration policies to mitigate its severe environmental challenges in the late 1990s. From the beginning, the effects and influences of the ecological restoration policies have been hotly debated. In the present study, we assessed the effects of two vital ecological restoration policies (Grain-for-Green and Grain-for-Blue) on valued ecosystem services in Shandong province. A new method based on the net primary productivity (NPP) and soil erosion was developed to assess the ESV. In the areas implementing the Grain-for-Green and Grain-for-Blue policies, the ESV increased by 24.01 and 43.10 % during 2000–2008, respectively. However, comparing to the average increase of ESV (46.00 %) in the whole of Shandong province in the same period, Grain-for-Green and Grain-for-Blue did not significantly improve overall ecosystem services. The ecological restoration policy led to significant trade-offs in ecosystem services. Grain-for-Green improved the ecosystem service function of nutrient cycling, organic material provision, and regulation of gases but decreased that of water conservation. Grain-for-Blue increased the water conservation function but led to a reduction in the function of soil conservation and nutrient cycling.

Keywords Ecosystem service values · NPP · Urbanization · Ecological restoration program · North China Plain

Changes in Valued Ecosystem Services Induced by the Land Encroachment of Urbanization in North China Plain

Introduction

Humankind is entering an urban era (Seto et al. 2012), and the average urbanization rate of the world is projected to be 67.2 % in 2050 (Heilig 2012). Urban areas will therefore become the major living environment for most of the world's population in the future (Haase et al. 2014). Currently, urbanization in most developed countries is almost complete; that is, almost 80 % of Europeans have already lived in urban areas, and the urbanization rate in the USA has reached 81.28 %

(Ntelekos et al. 2010). However, most of the developing countries are on the track of rapid urbanization (Song 2014).

In 1978, the Chinese government launched a market-oriented economic reform, namely the Open Door Policy. Since then, urbanization in China has accelerated at an unprecedented speed. The urbanization rate in China has increased to 53.7 % in 2013 from 17.9 % in 1978. This trend is projected to continue in the next few decades, and over 65 % of the Chinese population will live in urban areas in 2050 (Zang et al. 2011).

Urbanization inevitability leads to the expansion of urban areas (Song and Pijanowski 2014). For example, the total global urban area quadrupled during 1970–2000 (Seto et al. 2011). Urban areas in developing countries are projected to increase from to 300,000 km^2 in 2000 to 770,000 km^2 in 2030, and 1,200,000 km^2 in 2050 (Angel et al. 2011). In the 1990s, urban areas of 145 of cities in China expanded by 39.8 % (Tan et al. 2005). Because of the close location of cultivated land to urban areas, urban expansion usually results in the loss of a large amount of cultivated land. During 1986–2003, urban expansion in China occupied more than 33,400 km^2 of cultivated land, accounting for 21 % of total cultivated land loss.

Many negative effects of urbanization have been well documented, such as resource removal, the decrease of native biodiversity, the urban heat island effect, and air and water pollution. Ecological system degradation is a significant problem. Urbanization influences the ecosystem service by converting agricultural land to built-up areas. Many researchers have taken notice of this and assessed the changes in ecosystem service in response to urbanization. For example, Long et al. 2014 assessed that the ecosystem service value (ESV) of Tianjin Binhai New Areas decreased by 25.9 % during 1985–2010 due to the conversion from ecological land to construction land. Su et al. (2014) found that changes in ESVs were negatively correlated with urbanization indicators in Shanghai, China. Research of Lin et al. (2013) showed that urbanization in the island city of Xiamen, China, resulted in the decrease in ESV and significantly changed in the landscape.

Although great efforts have been made in assessing the effects of urbanization on ecosystem service, there are still many knowledge gaps in this field. First, the proxy method is usually adopted to assess the ESV, which generates great uncertainties. The proxy method pre-assigns each land-use type an invariable ESV regardless of the spatial heterogeneity. In fact, even in a small region, ESV of the same kind of land-use types usually varies according to the physiographic conditions. For example, the ESV of cultivated land in flat areas is usually different from that in hilly areas. Second, many ESV models can dynamically assess the spatial heterogeneity of ESV but fail to distinguish the effects of climate change and land-use change on ESV (Harmáčková and Vačkář 2015), such as the Integrated Valuation of Ecosystem Services and Tradeoffs (InVEST) which is a suite of software models used to map and value the goods and services from nature that sustain and fulfill human life. Last, the effects of cultivated land loss on ESV induced by urbanization were paid less attention.

Considering these knowledge gaps, we developed a net primary productivity (NPP)-based ecosystem service model (NESM) and a buffer comparison method to quantitatively assess the effects of cultivated land conversion on ESV in the process of urbanization. Specifically, the purposes of this study are to (1) examine the urban expansion and consequent cultivated land loss in the North China Plain (NCP) during 1988–2008; (2) assess the changes in ESV that resulted from the conversion from cultivated land to urban areas; and (3) separate the effects of land-use change from climate change on ESV.

Data Source

Two raster land-use maps in 1988 and 2008 are utilized to analyze the urban expansion and cultivated land loss in NCP. The maps are provided by Data Center for Resources and Environmental Sciences, Chinese Academy of Sciences (RESDC) (Liu et al. 2014). The overall accuracy of the land-use maps is over 95 %.

The NPP data utilized in this study are from NASA's EOS/moderate resolution imaging spectroradiometer (MODIS). MODIS is a key instrument aboard the Terra (EOS AM) and Aqua (EOS PM) satellites. The Normalized Difference Vegetation Index (NDVI) data are from vegetation of Satellite Pour l' Observation de la Terre (SPOT) which is a commercial high-resolution optical imaging Earth observation satellite system operating from space. The climate data are collected from the China Meteorological Data Sharing Service System (http://cdc.cma. gov.cn/home.do). The soil map including attributes of nutrient and soil texture is from the second soil survey in the 1980s in China. The actual evapotranspiration data are from the Data Sharing Infrastructure of Earth System Science, China. The spatial resolution of these maps is 1 km except for the actual evapotranspiration map (about 850 m). The temporal resolution of the vegetation and climate inputs are all annual values in 1988 and 2008.

Methodology

NPP-Based Ecosystem Service Model

Ecosystem services in NCP were assessed by NESM and developed by Song et al. (2015) NESM can quantify and map the distribution of ecosystem services under alternative scenarios. NESM's multi-service design provides an effective tool for exploring the likely outcomes of alternative management and climate scenarios and for evaluating trade-offs among services.

In total, five kinds of ecosystem services were designed in NESM: organic material production, nutrient cycling, soil conservation, water conservation, and

Table 3.1 Descriptions of the net primary productivity based ecosystem service model

Ecosystem service	Data requirement
1. Organic material production	NPP, the conversion coefficient from biochar to organic material, price of standard coal
2. Nutrient cycling Nitrogen Phosphorus Potassium	NPP, distribution rate of nutrient elements in organic material, conversion coefficients of nutrient elements to corresponding chemical fertilizer, price of chemical fertilizer, land-use map
3. Soil conservation	
Soil fertility	Digital elevation model, precipitation, soil texture, soil organic carbon, soil nutrients, vegetation coverage, price of chemical fertilizer, land-use map
Soil sedimentation	Digital elevation model, precipitation, soil texture, soil organic carbon, soil density, soil thickness, vegetation coverage, economic benefit of forest planting, land-use map
Surface soil	Digital elevation model, precipitation, soil texture, soil organic carbon, soil density, vegetation coverage, cost of reservoir construction, land-use map
4. Water conservation	
Soil	Precipitation, ratio of runoff generated from precipitation, the coefficient of reducing runoff compared to bare land, cost of reservoir construction, land-use map
Water area	Precipitation, actual evapotranspiration, cost of reservoir construction, land-use map
5. Gas regulation	
Oxygen production	NPP, parameter of absorbing carbon dioxide when producing 1-g dry matter
Carbon dioxide absorption	NPP, parameter of producing oxygen when producing 1-g dry matter

gas regulation. Detailed parameters of NESM are shown in Table 3.1. For the convenient calculation process and simple data requirements, NESM can quickly update and map an ecosystem service.

Many parameters (such as NPP and vegetation coverage) in the model can be either calculated by the users or directly downloaded from the Internet database (e.g., products of MODIS and SPOT). Other parameters (e.g., the conversion coefficient from biochar to organic material, ratio of runoff generated from precipitation) could be acquired from the previous researches. Soil erosion amount and actual evapotranspiration are two vital parameters of the model. NESM can automatically calculate the soil erosion amount using the Universal Soil Loss Equation according to the data input in Table 3.1. However, the calculation of actual evapotranspiration was not designed in the model for the complicated calculation process. Users need to generate this data input themselves.

Most of the parameters have obvious spatial variation. However, because of the lack of reliable data, we utilized a unified parameter in the whole NCP for several parameters such as the distribution rate of nutrient elements in organic material, the ratio of runoff generated from precipitation, and the coefficient of reducing runoff compared to bare land.

Buffer Comparison Method

Using NESM and required data input, we can map the ESV under different land use and climate scenarios. However, whether the changes in ESV were a result of land-use change or climate change is not known. To solve the problem, we developed a buffer comparison method.

As shown in Fig. 3.1, pixel *e* was converted from cultivated land to urban areas during 1988–2008. The changes in ESV in pixel *e* are the combined effects of land use and climate change. However, pixels *c, d, f, h,* and *i* did not undergo land-use change from 1988 to 2008. Climate change itself determines the changes in ESV. In other words, changes in ESV in pixels *c, d, f, h,* and *i* are the consequences of climate change.

In 1988, pixels *c, e,* and *f* are all cultivated land (Fig. 3.1). Since climate is highly autocorrelated in adjacent pixels, we assumed that the physical geography conditions and climate change in pixels *c, e,* and *f* are the same for the close location. The change percentages in ESV resulting from climate in pixel *e* should be equal to that of pixel *c* and *f*. Thus, we assessed the effects of climate change on ESV in pixels *e,* i.e., the average change percentage in ESV of pixels *c* and *f*. Subtracting the average change percentage in pixel *c* and *f* (ESV changes due to climate change) from the actual change percentage in pixel *e* (combined influences of land-use and climate change), we can calculate that the ESV changes resulted from land-use change in pixel *e*.

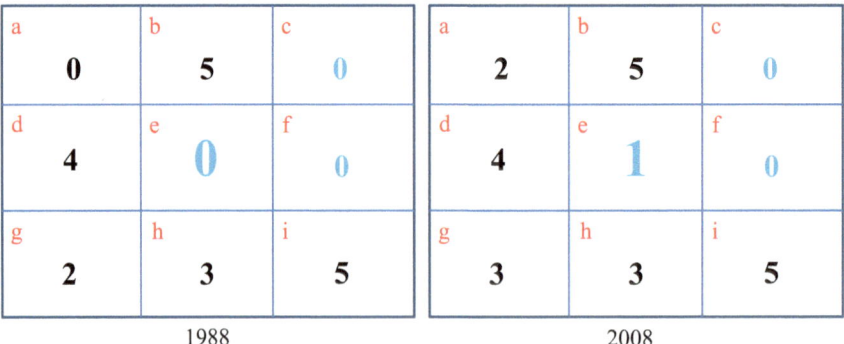

1988   2008

Numbers 0-5 represent cultivated land, urban areas, forestry areas, water areas and unused land, respectively.

Fig. 3.1 The mechanism distinguishing the effects of land-use change from climate change on ESV

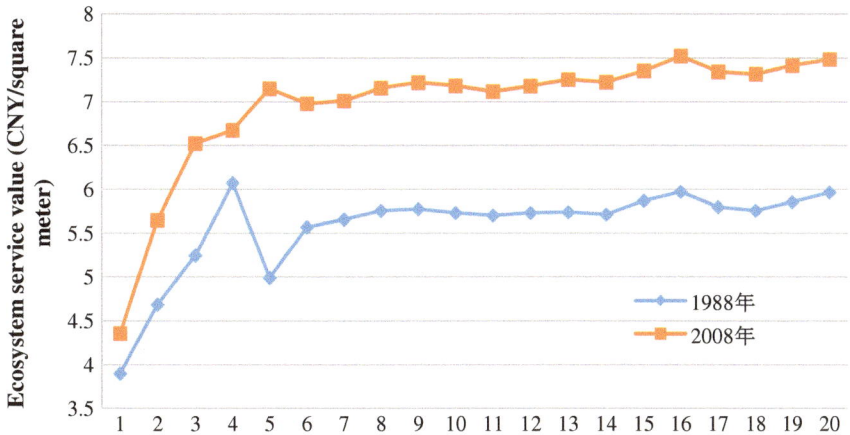

Fig. 3.2 Sensitivity of ESV changes in response to buffer distances in the North China Plain for both 1988 and 2008

How to determine the buffer distance is the next problem. If the buffer distance is too close, we may not find the same unchanged land-use type (e.g., pixel c and f in Fig. 3.1) in 1988. In addition, the errors in land-use classification could generate interference on the result. If the buffer distance is too far, the physical geography conditions and climate change in these pixels could be significantly different from that in pixel e (Fig. 3.1). To solve the problem, we assessed the sensitivity of the ESV change to the buffer distances (Fig. 3.2). It was found that the ESV of cultivated land has a significant inflection point at the distance of 5 km. It means that the physical geography conditions and climate change within 5 km are similar to that of the converted areas, while significant differences exist beyond 5 km. Thus, we decided to create a 5 km buffer zone to find the unchanged cultivated land during 1988–2008, which could provide a reference for the effects of climate change on ESV.

Contributions of Land-Use and Climate Change on ESV

Based on the buffer comparison method, we further developed an equation to distinguish the contributions of land-use and climate change on ESV.

$$C_{\mathrm{lucc}} = \frac{|CP_{\mathrm{act}} - CP_{\mathrm{bz}}|}{|CP_{\mathrm{act}} - CP_{\mathrm{bz}}| + |CP_{\mathrm{bz}}|} \times 100\,\% \tag{3.1}$$

$$C_{\mathrm{cc}} = 100\,\% - C_{\mathrm{lucc}} \tag{3.2}$$

where C_{lucc} is the contribution of land-use change on ESV; CP_{act} is the actual change percentage in ESV in converted areas, i.e., areas converted from cultivated land to urban areas during 1988–2008; CP_{bz} is the change percentage in ESV in buffer zone; and C_{cc} is the contribution of climate change on ESV.

Results

Cultivated Land Loss Due to Urbanization in NCP

Urban areas in NCP increased to 12,648 km^2 in 2008 from 6036 km^2 in 1988, with a change percentage of 108.54 %. The annual expansion ratio of urban areas in NCP reached 5.48 % a year. Urban areas expanded faster in western NCP, i.e., regions around Beijing, central Hebei, central Henan, and northwestern Shandong, with urbanization ratios over 300 % (Fig. 3.3). Urban expansion ratios are lower in southeastern NCP, i.e., regions in northern Anhui and Jiangsu provinces. Most of the urban expansion ratios in these areas were less than 50 % during 1988–2008 (Fig. 3.3).

Ratio of cultivated land loss due to urban expansion reached 76.48 % in NCP during 1988–2008. In addition, about 19.89 % of expanded urban areas were converted from other construction land (i.e., rural settlement, industrial, and mining land). The lost proportions of forestry areas, grassland, and water areas due to urban expansion are as low as 0.75, 1.19, and 1.36 %, respectively. Cultivated land

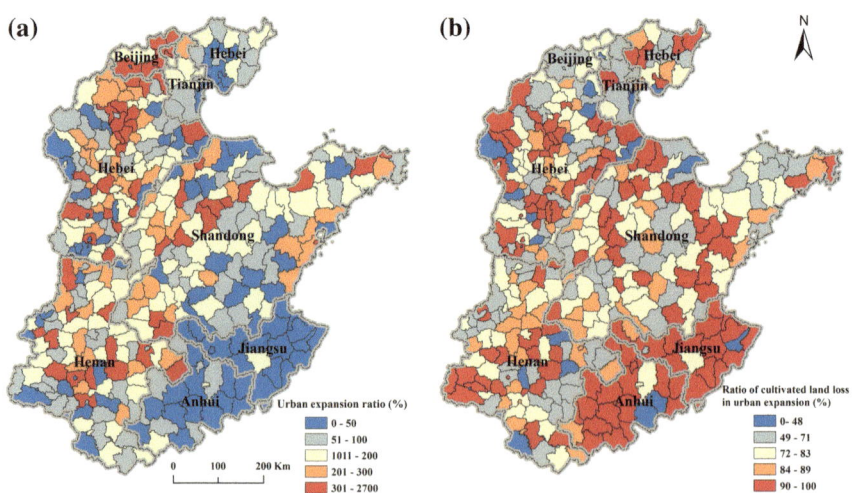

Fig. 3.3 Urban expansion rate (**a**) and cultivated land loss ratio in urban expansion (**b**) in the North China Plain during 1988–2008

contributed to most of the expanded urban areas. The spatial pattern of cultivated land loss ratio in urban expansion is opposite to that of urban expansion ratio. The ratio of cultivated land loss in urban expansion is higher in southeastern NCP, ranging from 90 to 100 % (Fig. 3.3). Urban expansion consumes less cultivated land in western NCP, with a ratio of cultivated land loss lower than 60 %.

Actual Changes in ESV Due to Conversions from Cultivated Land to Urban Areas

The value of total ecosystem service of converted cultivated land in 1988 is about 3911.17 USD/ha (e.g., in 2008 USD) (Table 3.2). Value percentage of nutrient cycling, water conservation, gas regulation, provision of organic production, and soil conservation are 0.32, 22.77, 15.69, 51.17, and 10.05 %, respectively. After the conversion from cultivated land to urban areas, the value of total ecosystem service decreased by 8.94 % (349.33 USD/ha) during 1988–2008. Three kinds of ecosystem services presented a decreasing trend, while two ecosystem services increased.

The ecosystem service of water conservation vanished (decreased by 100 %) when cultivated land was converted to urban areas. The ESV of soil conservation and nutrient cycling also decreased by 5.50 and 3.18 %, respectively. However, ESV of gas regulation and organic production increased by 21.54 and 21.54 %, respectively. The changes in ESV are the combined results of climate and land-use change.

Effects of Land-Use Change on ESV

According to the buffer comparison method, we assessed the effects of land-use change on ESV when converting cultivated land to urban areas. Land-use change

Table 3.2   Actual changes in ecosystem service value in NCP during 1988–2008

	Ecosystem service value in (USD/hm^2)		Value change (USD/hm^2)	Change percentage (%)
	1988	2008		
Total ecosystem service	3911.17	3561.84	−349.33	−8.94
Nutrient cycling	12.33	10.96	1.37	−3.18
Water conservation	890.46	0.00	−890.46	−100
Gas regulation	613.73	746.62	132.88	21.54
Organic production	2001.48	2431.64	431.53	21.54
Soil conservation	393.17	372.62	−20.55	−5. 5

Table 3.3 Effects of climate and land-use change on ESV in NCP during 1988–2008

	Actual change (%)	Effects of climate change (%)	Effects of land-use change (%)
Total ecosystem service	−8.94	25.72	−34.66
Nutrient cycling	−3.18	28.73	−31.91
Water conservation	−100.00	24.03	−100.00
Gas regulation	21.54	28.73	−7.18
Organic production	21.54	28.73	−7.18
Soil conservation	−5.50	−7.90	2.40

led to the overall decrease of ESV except for soil conservation in the conversion (Table 3.3). Urban expansion resulted in the complete loss of water conservation function. Ecosystem service functions of nutrient cycling, gas regulation, and organic production also decreased by 31.91, 7.18, and 7.18 %, respectively. However, urban expansion improved the function of soil conservation with an increase of ESV of 2.40 %.

Land-use change significantly reduced the total ESV in central NCP (decreased by 43–100 %), while it increased the ESV in western NCP (ranging from 15 to 128 %) (Fig. 3.4). Except for several regions in northwest, the water conservation function all decreased by 100 % in NCP. The soil conservation function was improved in 65.28 % of the counties in NCP due to urban expansion. The increase of the soil conservation function is particularly significant in eastern NCP, i.e., Shandong Peninsula. However, it decreased remarkably in southwestern NCP (Fig. 3.4).

Organic material production and the gas regulation function decreased in 70.91 % of counties in NCP, while the nutrient cycling function decreased in over 98 % of counties in NCP. The decrease of nutrient cycling is more significant than that of organic production and gas regulation. The decrease of nutrient cycling is larger in central NCP than that in western NCP. The regions that the gas regulation and organic production increased was scattered across the entire NCP. However, the number of counties experiencing increase in gas regulation and organic production gradually decreased from the western NCP to eastern NCP.

Contributions of Land-Use and Climate Change on ESV

In the conversion from cultivated land to urban areas, land-use change accounted for 57.40 % of the changes in ESV, while climate change resulted in 42.60 % of the changes in ESV. For different ecosystem services, land-use change accounted, on average, for 52.62, 80.63, 19.99, 19.99, and 23.30 % of changes in ESV for nutrient cycling, water conservation, gas regulation, organic production, and soil

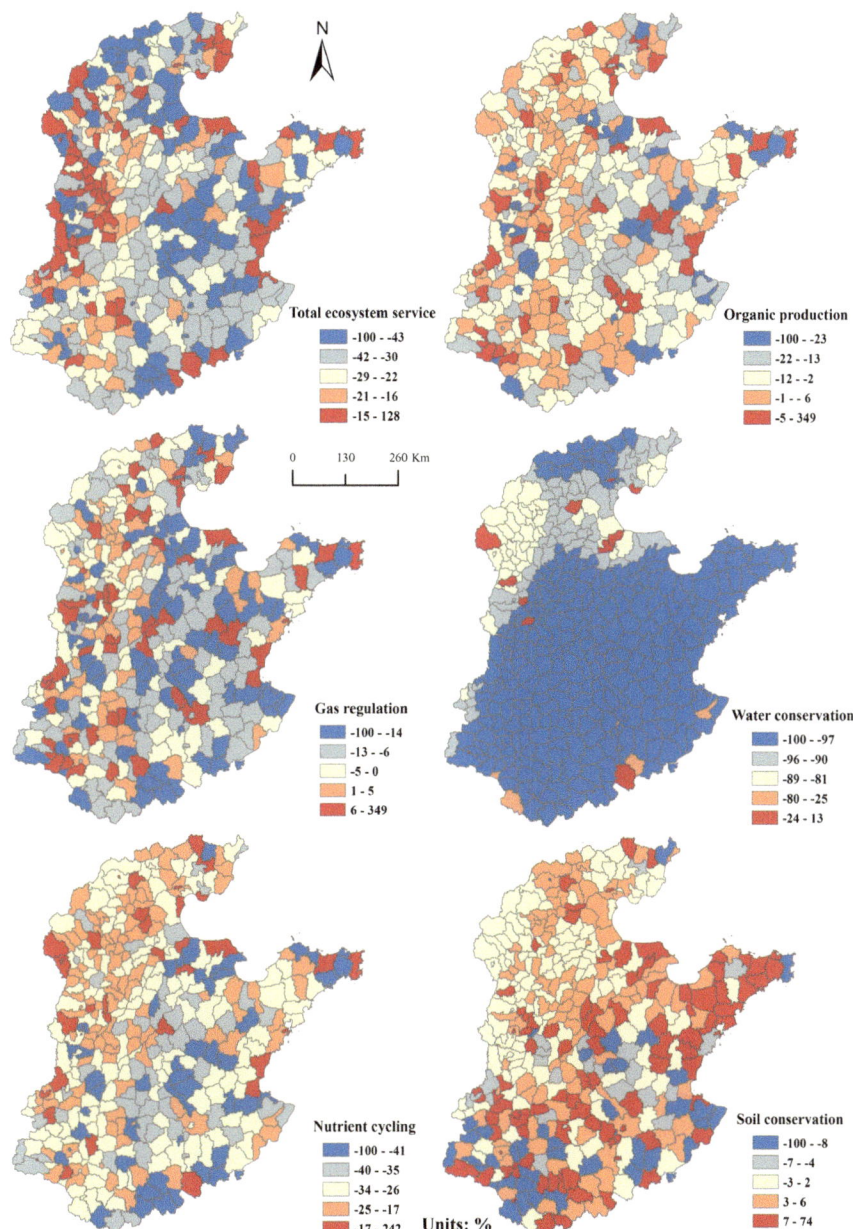

Fig. 3.4 Percentage change in ESV induced by land-use change in the North China Plain

conservation, respectively (Fig. 3.5). Land-use change has a major influence on water conservation and nutrient cycling, while the main effects of climate change are on gas regulation, organic production, and soil conservation.

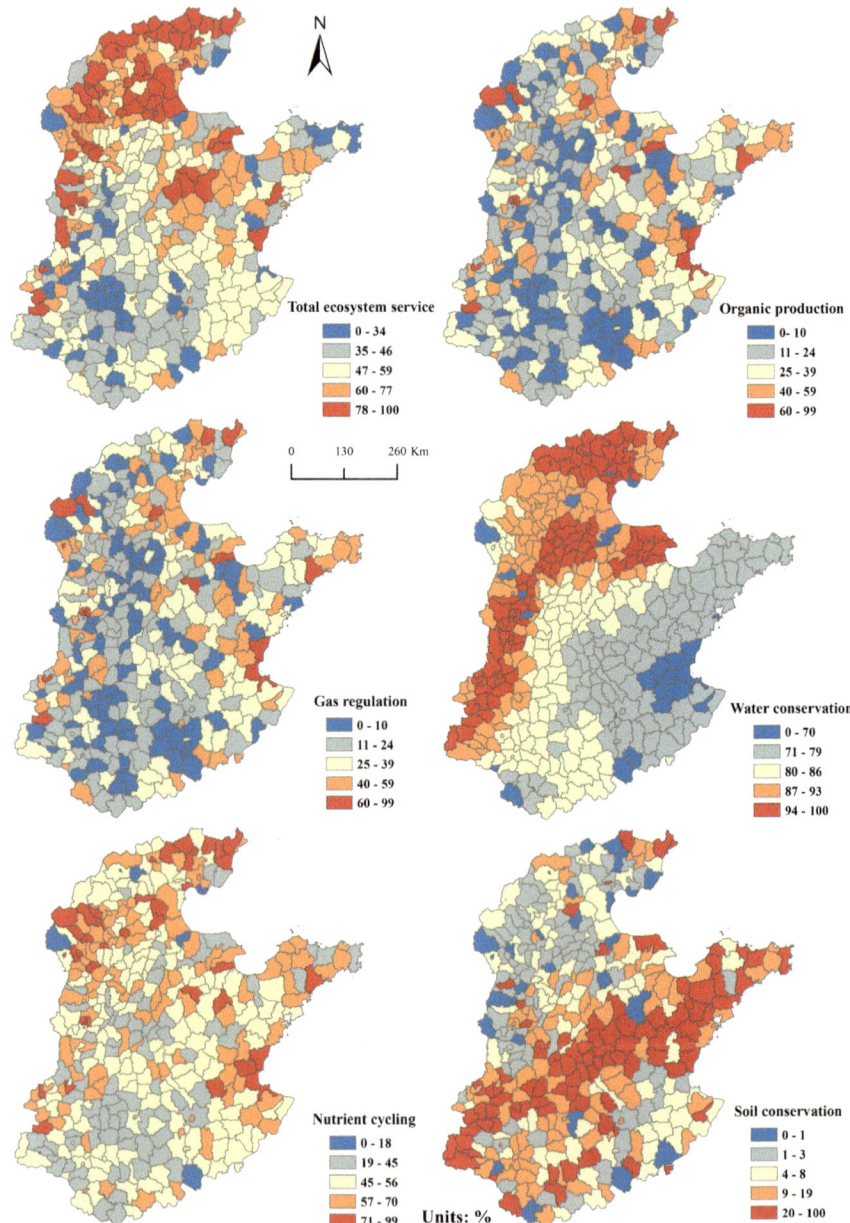

Fig. 3.5 Contribution of land-use change on changes in ESV in NCP

The effects of land-use change on ESV are more significant in northwestern NCP than any other region in NCP (Fig. 3.5). Land-use change in these areas accounted for over 60 % of the changes in ESV. However, in northern NCP, the effects of land-use change on ESV are generally lower than 45 %. The influence of land-use change on water conservation gradually decreased from northwestern NCP to southeastern NCP. The effects of land-use change on soil conservation is particular significant in a northeast–southwest strip (Fig. 3.5). The effects of land-use change on organic production, gas regulation, and nutrient cycling are similar, i.e., they gradually decreased from the east to the west in NCP.

Causes of the Changes in Ecosystem Service in the Process of Urbanization

There are many trade-offs in ecosystem service in urban expansion, i.e., an increase in soil conservation, while there is a decrease of water conservation, nutrient cycling, gas regulation, and organic production. Urban areas are mostly soil sealed, which means that soil erosion does not easily occur. However, cultivated land usually leads to soil erosion under certain conditions, such as in sloping areas, low vegetation cover, and heavy rainfall.

For the soil sealing in urban areas, urban land almost has no water conservation function. When cultivated land was converted to urban areas, the water conservation function almost vanished. The nutrient content of nitrogen in cultivated land is similar to that of an urban ecosystem. However, the nutrient content of phosphorus and potassium in urban ecosystem is far below than that in cultivated land. Therefore, in the conversion from cultivated land to urban areas, the nutrient cycling value of phosphorus and potassium significantly decreased.

Theoretically, if expanded urban areas are all soil sealed, the functions of organic material production and gas regulation will significantly decrease for the loss of vegetation. In this study, we found that land-use change resulted in a decrease of 7.18 and 7.18 % for the organic material production and gas regulation. It can be explained from two perspectives. First, we utilized the land-use change data with a spatial resolution of 1 km. There will be many mixed pixels of urban areas and cultivated land. Although many pixels in land use were classified as urban areas, vegetation could still emerge in these pixels forming functions of organic material provision and gas regulation. Second, there may be many mistaken classification errors in urban areas, i.e., much of the cultivated land close to cities was reclassified as urban areas.

Comparisons Between the Buffer Comparison Methods with Other Previous Methods

Separating the effects of land-use change from climate change on ESV has always been difficult for the complicated mutual influences between them. In the previous researches, two approaches were usually adopted to do this work, i.e., econometrics analysis and scenario analysis. Econometrics analysis identifies the effects of land-use and climate change on ESV by regression or correlation analysis

(Heubes et al. 2012). For example, Su and Fu (2013) found that precipitation in the Chinese Loess Plateau significantly influenced the water yield, while the land-use conversion from cropland to grass/woodland significantly influences the sediment control. Econometrics analysis can judge whether land-use or climate change significantly influences ESV, while it fails to assess the degree of the influence.

In scenario analysis, three scenarios were generally developed, i.e., pure climate scenario (no land-use change), pure land-use change scenario (no climate change), and combined scenario (changes in both climate and land use) (Geng et al. 2014). The influence of land-use and climate change on ESV can be clearly presented by assessing the changes in ESV under different scenarios. The contribution of climate and land-use change on ESV can also be assessed in scenarios analysis. However, scenario analysis still faces great challenges in scenario creation. Many pure scenarios of climate and land-use change are difficult to be created for the complicated mutual effects between them. For example, soil erosion is determined by a great deal of factors, e.g., precipitation, land use, slope, and vegetation coverage. Precipitation, land use, and slope can be easily controlled in scenario creation. However, the vegetation coverage, which is usually directly calculated by remote sensing, is a combined result of land-use and climate change. It is difficult to create a pure climate vegetation coverage without the influences of land-use change and vice versa. Therefore, the application of scenario analysis method is greatly limited.

The buffer comparison method developed in this study can directly assess the contribution of land-use and climate change on ESV. It could be viewed as an improved scenario analysis method. Buffer comparison method factually developed two scenarios, i.e., pure climate scenario and combined scenario. The influence of land-use change on ESV was indirectly deduced by the two scenarios. Since there is no land-use change in buffer zones, pure climate scenario can be easily created and applied in all ESV assessment. The application range of this method is widely enhanced.

Uncertainty of the Assessment

The MEA classifies the functions of ecosystem services into provisioning (e.g., provision of food and fiber), regulation (e.g., regulation of climate through carbon storage), cultural (e.g., recreational values), and supporting services (e.g., nutrient cycling and soil formation). Because of the lack of available data and method in mapping cultural services, we only assessed five ecosystem functions in provisioning, regulation, and supporting services. The cultural function changes in the process of urbanization were not assessed, which added much uncertainty in the assessment.

The spatial resolution of the data we utilized to assess ESV and urban expansion is 1 km. The data could be a little coarse for analyzing urban expansion and generate many mixed pixels of urban areas and other land-use types. If the mixed pixels were classified as urban areas, the NPP and NDVI in urban areas could be overestimated. The ESV closely related NPP and NDVI (e.g., organic material production, nutrient cycling) could be consequently overestimated.

Summary

We examined the urban expansion and the cultivated loss in NCP during 1988–2008. It was found that urban areas in NCP increased by 108.54 % during 1988–2008. Urban areas expanded faster in northwestern NCP, i.e., regions around Beijing, but expanded slower in southeastern NCP, i.e., northern Anhui and Jiangsu provinces. Urban expansion in NCP leads to significant cultivated land loss. About 76.48 % of the expanded urban areas were converted from cultivated land. Furthermore, cultivated land loss is more serious in the regions with a low urban expansion speed.

Using NESM and a buffer comparison method, we assessed the changes in ESV due to land-use change when converting cultivated land to urban areas. The ecosystem service function of water conservation, nutrient cycling, gas regulation, organic production, and soil conservation changed −31.91, −100.00, −7.18, −7.18, and 2.40 % due to land-use change, respectively. Urban expansion leads to a total loss of ESV of 34.66 %, while it slightly improves the soil conservation function.

Land-use change has a major effect on changes in ESV in urban expansion. As a whole, land-use change accounted for 57.40 % of the changes in ESV. However, the effect of land-use change on ESV varied according to different ecosystem service functions. Land-use change has a major influence in ESV of nutrient cycling (52.62 %) and water conservation (80.63 %), while it has a minor influence on changes in ESV of gas regulation (19.99 %), organic production (19.99 %), and soil conservation (23.30 %).

Changes in Valued Ecosystem Services Resulting from Ecological Restoration Programs in Shandong Province

Introduction

Over the past 50 years, 60 % of worldwide ecosystem services have degraded due to social and economic development (Reid et al. 2005). In China, ecological degradation is also extensive; for example, 23 % of the area in China has suffered ecological degradation (Lü et al. 2012). The economic loss from ecological degradation accounted for over 13 % of the national gross domestic product (Shi et al. 2011). Awakening to the severe effects of ecological degradation, the Chinese government implemented two ecological restoration policies at the end of the 1990s, that is, the Grain-for-Green and Grain-for-Blue policies.

The Grain-for-Green policy, the largest land retirement/afforestation program in China, was launched in 1999 to mitigate land degradation (soil erosion) by returning steeply sloping cultivated land to forest area or grassland (Song and Liu 2014). The Grain-for-Blue policy was launched in 1998, aiming to return cultivated land to water areas, that is, relinquishing the cultivated land at the periphery of water areas (Nelson et al. 2010).

Policy can affect decision making and change the ways people utilize and manage ecosystems (Sutherland et al. 2014). Evaluating the consequences of a policy is a critical lesson from the Millennium Ecosystem Assessment (Perrings et al. 2011). In recent years, the policy research interest in ecosystems has focused on the policy tools, policy impacts, policy incentives, policy option, policy assessment approach, and policy making. For example, Brady et al. (2012) modeled the impacts of agricultural policy on ecosystem services using an agent-based approach; Bronner et al. (2013) assessed the impacts of US stream compensatory mitigation policy on ecosystem functions and services; Geneletti (2013) assessed the impacts of alternative land-use zoning policies on future ecosystem services; Simpson (2014) estimated the effects of conservation policy on ecosystems.

As an effective approach to recognize the multiple benefits provided by ecosystems, the work of economic valuation of ecosystem services has been widely conducted since 1990 (Zhao et al. 2004). These works cover the method developed in valuating ecosystem services and the estimation of ESV in different regions and ecosystems. One of the most notable assessments of ESV was conducted by Costanza et al. (1997) who estimated 17 ESV provided by 16 dominant global biomes using a market valuation method. Other researchers also estimated ESV of tropical forests and protected areas, endangered species management, and different biological resources (Peters et al. 1989; Gardiner et al. 2013; Silvestri et al. 2013).

Although Grain-for-Green and Grain-for-Blue have been implemented over 10 years in China, few works have been conducted to assess the ecological impacts of the two ecological restoration policies particularly from the perspective of changes in ESV. In this study, we evaluated the effects of Grain-for-Green and Grain-for-Blue on ESV by a newly developed approach. Specifically, the aims of this study are to (a) examine land-use change in Shandong province during 2000–2008, (b) assess the changes in ESV, and (c) estimate the effects of Grain-for-Green and Grain-for-Blue on ESV.

Study Area, Data, and Methods

Study Area

Shandong province, located on the eastern edge of the NCP (114°19′–122°43′E, 34°22′–38°15′N) and the lower reaches of the Yellow River, is a coastal province

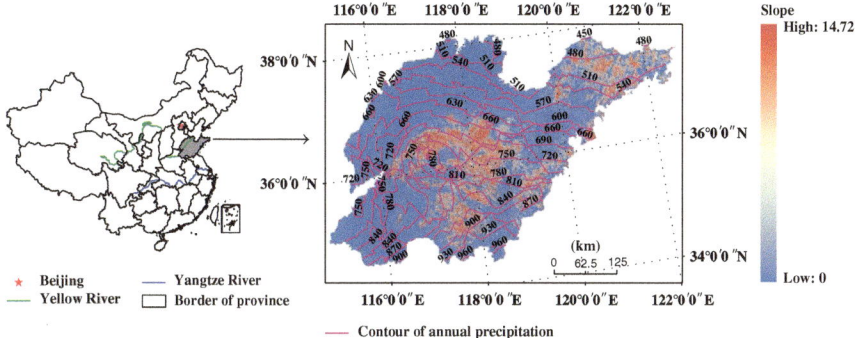

Fig. 3.6 Slope and precipitation of Shandong province, China

in China (Fig. 3.6). It covers a total area of over 151,100 km^2 with plain terrain accounting for 55 %, mountainous area for 15.5 %, and hilly area for 13.2 % (Fig. 3.6). The northwest, west, and southwest of the Shandong are all part of the NCP. The central region of the province is mountainous, with Mount Tai being the most prominent. Shandong province lies in the warm-temperate zone with a continental monsoon climate.

Data Sources

The land-use maps of Shandong province in 2000 and 2008 were obtained from the Institute of Geographic Sciences and Natural Resources Research of the Chinese Academy of Sciences. The maps were interpreted from Landsat Thematic Mapper (TM) satellite images by the human–machine interactive approach. The average accuracy of the maps is over 95 % (Liu et al. 2010). The land use was divided into six primary types and 25 subclasses (Liu et al. 2010).

The NPP data in Shandong are the products of EOS/MODIS of NASA. The NDVI data are sourced from SPOT-vegetation with a temporal step of 10 days. The data formats of NPP and NDVI are all 1 km grid. Precipitation and temperature in Shandong province were gained from the China Meteorological Data Sharing Service System (http://cdc.cma.gov.cn/home.do) and interpolated in spatial. The soil nutrients data such as the contents of N, P, K, and organic matter were taken from the soil map generated from the second soil survey of China. The actual evapotranspiration data, which were calculated by IBIS model with a temporal step of 8 days, were sourced from the China Data Sharing Infrastructure of Earth Systems Science.

Method

Value Quantification of Ecosystem Services

The method for assessing ESV can be summarized into two categories, i.e., the primary data-based method and proxy-based method (Su et al. 2014). The primary data-based method directly assesses the ESV according to the primary data from the study area. For the complicated calculation process, many models have been developed to assess ESV, e.g., InVEST, ARIES, SoLVES, MIMES. The proxy-based method assesses ESV by assigning each biome a value. One of the most representative proxy-based methods was developed by Costanza et al. (1997). The merit of the primary data-based method is high accuracy, while the demerit is the complexity. Since running ESV models usually needs numerous parameters, the primary data-based method is usually not convenient. The merit of the proxy-based method is the convenience, while the accuracy of this method is usually challenged.

In this study, we developed a novel method to assess ESV. The method is one of the primary data-based methods. However, since the calculation of ESV in this method is based on remote sensing data such as NPP, the calculation process is very convenient. Considering data accessibility and technique feasibility, we estimated five kinds of ESV, i.e., values for provision of organic material, nutrient cycling, soil conservation, water conservation, and regulation of gases.

Assessment of the Value of Provision of Organic Material

The value of organic material of an ecosystem was assessed by NPP. The equation is as follows:

$$V_{\mathrm{om}} = \sum \mathrm{NPP}(x) \times P_{\mathrm{om}}, \tag{3.3}$$

$$P_{\mathrm{om}} = \mathrm{NPP}(x) \times 2.2 \times 0.67 \times P_{\mathrm{sc}}, \tag{3.4}$$

where V_{om} is the value of provision of organic material; $\mathrm{NPP}(x)$ is the organic material produced in x pixel; and P_{sc} is the price of standard coal in 2000.

Assessment of the Value of Nutrient Cycling

We utilized the saved inputs due to nutrient cycling in agricultural production to assess the value of nutrient cycling. The formula is as follows:

$$V_{\mathrm{nc}} = \sum V_{\mathrm{nc}_i}(x) = \sum \mathrm{NPP}(x) \times R_{i1} \times R_{i2} \times P_i, \tag{3.5}$$

where V_{nc} is the value of nutrient cycling in an ecosystem; i is the nutrient elements of nitrogen (N), phosphorus (P), or potassium (K); NPP(x) is the organic material produced in x pixel; $V_{nc_i}(x)$ is the accumulated value of N, P, and K; R_{i1} is the distribution rate of nutrient elements of N, P, and K in different ecosystems; R_{i2} is the converted coefficients of N, P, and K to corresponding chemical fertilizer; and P_i is the chemical fertilizer price of N, P, and K in 2000.

Assessment of the Value of Water Conservation

The function of water conservation of an ecosystem is similar to that of a reservoir. Therefore, we utilized the average construction cost of a reservoir to assess the value of water conservation. The formula is as follows:

$$V_{wc} = \sum V(x) \times P_w,$$ (3.6)

where V_{wc} is the value of water conservation; $V(x)$ is the water conservation amount in x pixel; and P_w is the average cost of reservoir construction in 2000.

According to the difference in underlying surface, we utilized two different approaches to assess the water conservation amount of an ecosystem. When the underlying surface is soil, the equation is as follows:

$$V_s(x) = \sum P_{mean}(x) \times K_w \times R_w,$$ (3.7)

where $V_s(x)$ is the annual water conservation amount in x pixel; $P_{mean}(x)$ is the monthly precipitation in x pixel; K_w is the ratio of runoff generated from precipitation; and R_w is the coefficient of reducing runoff by comparison with bare land without vegetation. R_w of cultivated land, woodland, and grassland is valued as 0.4, 0.29, and 0.24, respectively, while R_w of other land-use types is valued as 0.

When the underlying surface of an ecosystem is an area of water, the equation is as follows:

$$V_{wc}(x) = \sum (P_{mean}(x) - ET_a(x)),$$ (3.8)

where $V_{wc}(x)$ is the annual water conservation amount in x pixel; $P_{mean}(x)$ is the monthly precipitation in x pixel; and $ET_a(x)$ is the monthly actual evaporation in x pixel.

Assessment of the Value of Soil Conservation

Soil erosion usually leads to three different kinds of value loss in an ecosystem, i.e., a reduction in soil fertility, river channel sedimentation, and loss of top soil.

Therefore, the assessment of the value of soil conservation is composed of three sections:

$$V_{ac} = V_{ef} + V_{en} + V_{es}, \tag{3.9}$$

where V_{ac} is the value of soil conservation; V_{ef} is the value of soil fertility conservation; V_{en} is the value of reducing soil sedimentation in a river channel; and V_{es} is the value of the loss of top soil.

1. Assessment of the soil conservation amount
   The soil erosion amount without any vegetation is viewed as the potential soil erosion amount. The soil erosion amount that actually occurred is looked on as the actual soil erosion amount. The difference between the amount of potential soil erosion and actual erosion is the soil conservation amount. The soil erosion was assessed by Universal Soil Loss Equation:

$$A_p = R \times K \times L \times S, \tag{3.10}$$

$$A_v = R \times K \times L \times S \times C \times P, \tag{3.11}$$

$$A_c = A_p - A_v, \tag{3.12}$$

where A_p is the amount of potential soil erosion; A_v is the amount of actual soil erosion; A_c is the soil conservation amount; K is the soil erodible factor; L is the slope length; S is the slope factor; C is the vegetation and crop management factor; and P is the soil conservation measure factor. Detailed parameters of R, K, L, S, C, and P were referenced from Wischmeier (Wischmeier et al. 1971), Renard et al. (1997), Flanagan et al. (1989), Kelvin et al. (Kuok et al. 2013a, b), Renard and Foster (1983).

2. Assessment of the value of soil fertility conservation
   N, P, and K are the three most important nutrients in soil. Soil erosion could lead to the nutrients' losses and a decrease in soil fertility. Thus, we estimated the value of soil fertility conservation by assessing the nutrient element value of the soil conservation amount. The equation is as follows:

$$V_{ef} = \sum A_c(x) \times C_i \times P_i, \tag{3.13}$$

where V_{ef} is the value of soil fertility conservation; $A_c(x)$ is the soil conservation amount in x pixel; and C_i is the price of N, P, and K fertilizer in 2000.

3. Assessment of the value of the loss of top soil
   The value of the loss of top soil was assessed by calculating the value of conserved land where soil erosion has been avoided due to the protection of vegetation coverage in an ecosystem. The formula is as follows:

$$V_{es} = \frac{\sum A_c(x) \times P_f}{D_{soil} \times T_{soil}}, \qquad (3.14)$$

where ESV is the value of the loss of top soil; $A_c(x)$ is the soil conservation amount in x pixel; D_{soil} is the soil density; P_f is the economic benefits of forest planting which is valued as 26,400 CNY/km^2 year; and T_{soil} is the average soil thickness.

4. Assessment of the value of reducing soil sedimentation in a river channel

   In China, about 24 % of sediments from soil erosion are deposited in reservoirs, lakes, and rivers according to the research of Ouyang et al. (1998). Thus, the value of reducing soil sedimentation in a river channel was assessed as follows:

$$V_{en} = \frac{\sum A_c(x) \times 0.24 \times P_w}{D_{soil}}, \qquad (3.15)$$

where V_{en} is the value of reducing the soil sedimentation in a river channel; P_w is the construction cost of a reservoir per unit; $A_c(x)$ is the quantity of soil conserved; and D_{soil} is the soil density.

Assessment of the Value of Regulation of Gases

The value of regulation of gases was assessed on the basis of the functions of CO_2 absorption and O_2 generation. The equation is as follows:

$$V_{gr} = \sum 1.62 \times NPP(x) \times P_{CO_2} + \sum 1.2 \times NPP(x) \times P_{O_2}, \qquad (3.16)$$

where $NPP(x)$ is the organic material in x pixel; according to the photosynthesis and breathing reaction equation, it can be deduced that producing 1-g dry matter absorbs 1.62 g CO_2 and releases 1.2 g O_2; P_{CO_2} is the price of carbon tax, valued as 7.39×10^{-4} CNY/g C; and P_{O_2} is the price of producing O_2, valued as 8.8×10^{-4} CNY/g C.

Results

Land-Use Change

In 2000, cultivated land in Shandong province comprised 67.62 % of the total area, followed by built-up areas (12.62 %), grassland (8.72 %), forestry area (6.44 %), water area (3.44 %), and unused land (1.17 %).

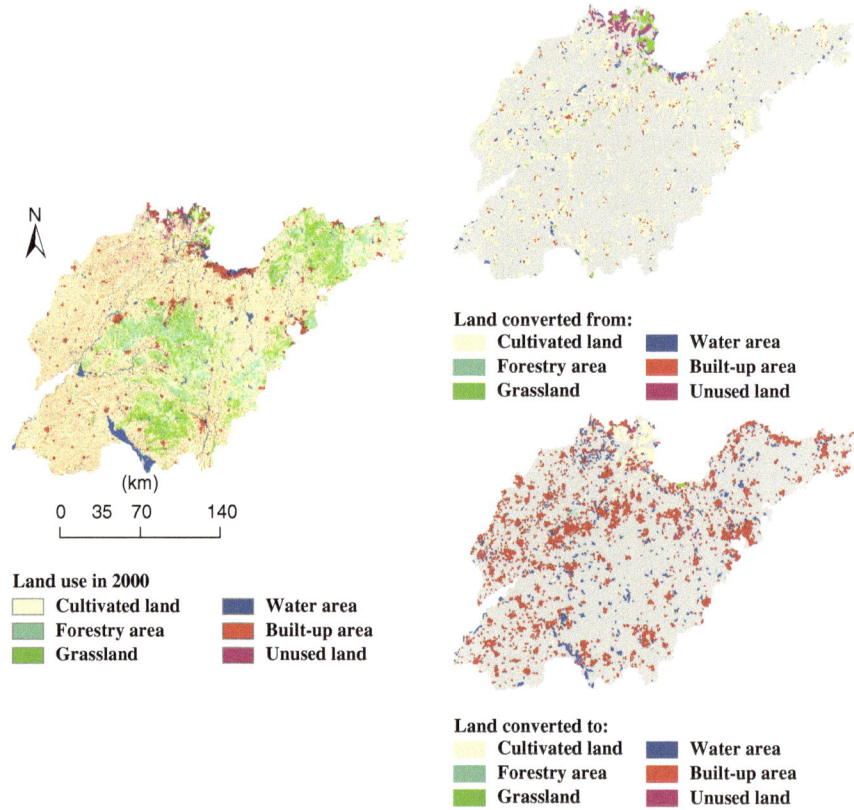

Fig. 3.7 Land-use change in Shandong province during 2000–2008

Shandong province experienced drastic land-use change during 2000–2008. The features of land-use change in Shandong were the expansion of built-up areas, reclamation of unused land, and drastic conversions in water areas. Due to rapid urbanization, urban built-up areas expanded by 56.13 %, with the built-up area increasing from 301,939 to 471,418 ha. However, unused land shrank by 16.54 % due to reclamation (Fig. 3.7). Both drastic increase and decrease occurred in the water areas. For example, reservoirs/ponds and streams/rivers significantly increased by 42.30 and 5.67 %, respectively, while beach and shore, lakes, and bottomland decreased by 53.46, 23.66, and 21.99 %, respectively.

During 2000–2008, the conversion from cultivated land to built-up area reached 235,256 ha, contributing to 45.00 % of the total conversions, followed by the conversion from cultivated land to water areas (42,236 ha) and from grassland to cultivated land areas (33,112 ha).

Changes in Value of Ecosystem Services

The ESV of Shandong was as high as 267.53 billion CNY in 2000. The value of regulation of gases contributed to 38.69 % of the total ESV, followed by the values of soil conservation (34.83 %), water conservation (16.69 %), nutrient cycling (5.05 %), and organic material provision (4.73 %). The ESV per unit in Shandong ranged from 0 to 71.07 CNY/m^2, with an average value of 1.76 CNY/m^2 (Fig. 3.8). The value of organic material provision per unit ranged from 0 to 0.37 CNY/m^2, while the values of regulating gases, nutrient cycling, and soil conservation ranged from 0 to 2.90 CNY/m^2, from 0 to 0.52 CNY/m^2 and from 0 to 69.36 CNY/m^2, respectively. The average values of regulating gases, organic material provision, water conservation, nutrient cycling, and soil conservation are 0.68, 0.08, 0.29, 0.09, and 0.61 CNY/m^2, respectively, in 2000.

In spatial, total ESV is high in central and eastern Shandong while low in northern and western Shandong (Fig. 3.8). The spatial feature of soil conservation value is similar to that of total ESV. The spatial distribution of organic material

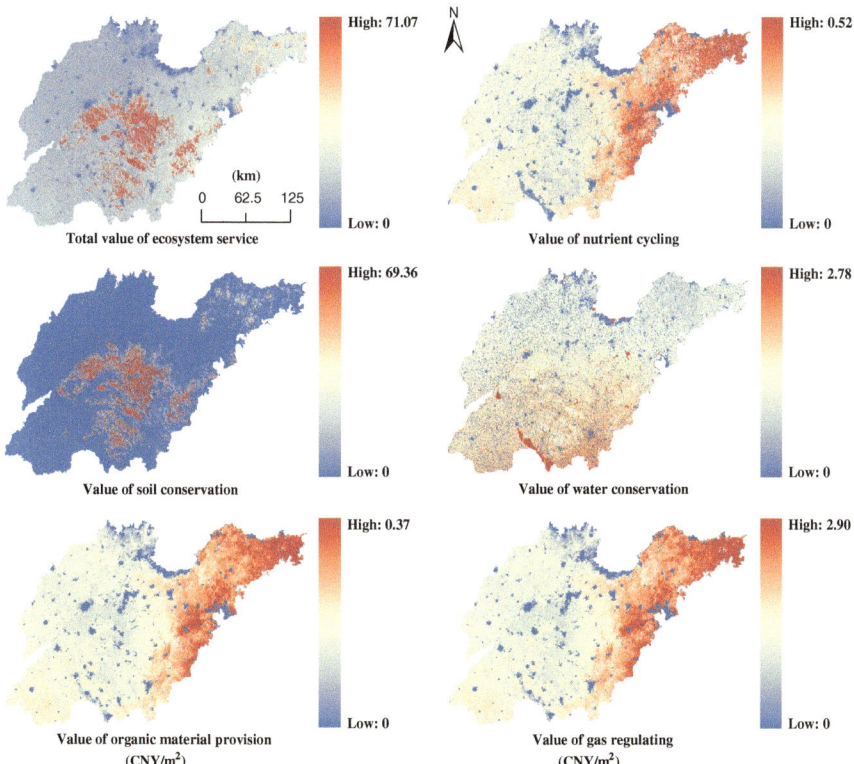

Fig. 3.8 Value of ecosystem services in Shandong in 2000

provision value, gases regulating value, and nutrient cycling value all gradually decreased from the northeast to the southwest. However, the value of water conservation gradually decreased from the southwest to the northeast.

In 2008, the ESV of Shandong province reached 390.59 billion CNY, with an increase of 46.00 % compared to 2000. The soil conservation value increased by 71.64 billion CNY contributing to 58.21 % of the total increase. Values of regulating gases, water conservation, nutrient cycling, and organic material provision increased by 37.40, 4.23, 5.00, and 4.80 billion CNY, respectively, contributing 30.39, 3.44, 4.06, and 3.90 % of the total increase, respectively. In 2008, the average ESV per unit of regulating of gases, water conservation, soil conservation, nutrient cycling, and organic material provision reached 0.93, 0.32, 1.09, 0.12, and 0.12 CNY/m^2, respectively. The total average ESV per unit is 2.58 CNY/m^2.

In spatial, total ESV and value of soil conservation significantly increased in central and eastern Shandong province. In these areas, mountains (Fig. 3.9) are the

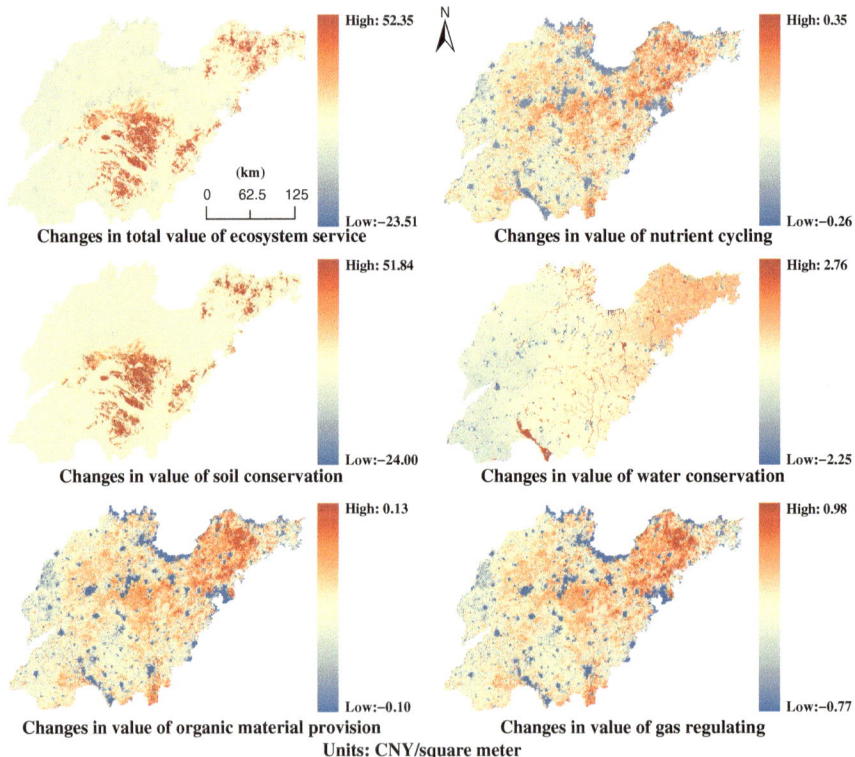

Fig. 3.9 Changes in value of ecosystem services in Shandong during 2000–2008

dominant topography. The changes in value of organic material provision, regulating gases, and nutrient cycling all significantly increased in eastern Shandong while decreased in the coastal zone of Shandong. Water conservation value significantly increased in water areas.

Changes in Value of Ecosystem Services in Response to Land-Use Change

One of the most important consequences of land-use change is land-use conversion. During 2000–2008, 40.57×10^4 ha of land in total in Shandong experienced conversions. The ESV of these converted lands in 2000 was 3.78 billion CNY, while that in 2008 was 3.81 billion CNY, showing an increase of 0.74 % (Fig. 3.10).

However, conversions from other land-use types to built-up areas significantly decreased the ESV in Shandong. The conversion from cultivated land to built-up areas reduced the ESV by 54.75 million CNY, contributing 89.49 % of the ESV loss from land-use conversion. Conversion from water areas, forestry areas, grassland, and unused to built-up areas increased by 2418.54×10^4 CNY, 1322.89×10^4 CNY, 1283.96×10^4 CNY, and 932.15×10^4 CNY, respectively (Fig. 3.10). The conversion from water areas to unused land also led to a decrease of 469.42×10^4 CNY. Other kinds of land-use conversion do not lead to a decrease in ESV.

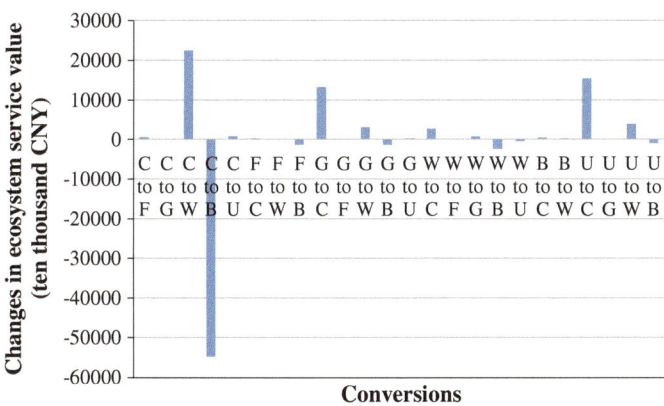

Fig. 3.10 Changes in ESV due to land-use conversions. *Notes C, F, G, W, B,* and *U* are cultivated land, forestry area, grassland, water area, built-up area, and unused land, respectively; **C–F** means the conversion from cultivated land to forestry and similarly for the others

Table 3.4 Changes in ESV in response to ecological restoration policies

Policy	Total value ($\times 10^6$ CNY)			Value per unit (CNY/m^2)			Change percentage (%)
	2000	2008	Change	2000	2008	Change	
Grain-for-Green	21.36	26.55	5.19	1.23	1.53	0.30	24.30
Grain-for-Blue	479.68	686.45	206.77	1.14	1.63	0.49	43.11
Total	501.04	713.00	211.96	1.141	1.62	0.48	42.30

Impacts of Ecological Restoration on Ecosystem Services

The vital consequences of the two ecological restoration policies on land use are the conversions from cultivated land to forestry area/grassland and to water areas. Theoretically, if the sites implementing the two ecological restoration policies in spatial are known, changes in ESV in these areas during 2000–2008 could be viewed as the effects of the two ecological restoration policies.

However, information is lacking on the location and the degree of implementation for the two ecological restoration policies in Shandong. It should be noted that the starting time of our research was 2000 when the two ecological restoration policies had just been launched. Furthermore, during 2000–2008, no other policies were implemented to drive the conversion from cultivated land to grassland, forestry area, or water areas. Most of the conservations from cultivated land to grassland/forestry area or water area during 2000–2008 could be similarly looked upon as the consequence of the Grain-for-Green and Grain-for-Blue policies.

Accordingly, we assessed the impacts of Grain-for-Green and Grain-for-Blue on changes in ESV. Grain-for-Green in total added an ESV of 5.19×10^6 CNY, while Grain-for-Blue led to an increase of ESV of 206.77×10^6 CNY (Table 3.4). The two ecological restoration policies in total led to an increase of 42.30 % in ESV. The ESV per unit increased from 1.23 CNY/m^2 in 2000 to 1.53 CNY/m^2 in 2008 due to the implementation of Grain-for-Green. The ESV per unit increased from 1.14 CNY/m^2 in 2000 to 1.63 CNY/m^2 in 2008 due to the implementation of Grain-for-Blue. The effects of Grain-for-Blue on ecosystem services seem better than those of Grain-for-Green.

The two ecological restoration policies produced different results for the ESV. For the Grain-for-Green policy, the value of nutrient cycling, regulating gases, organic material provision, and soil conservation increased by 64.12, 38.98, 40.00, and 18.25 %, respectively, while the value of water conservation decreased by 21.48 % (Table 3.5). For the Grain-for-Blue policy, the value of water conservation significantly increased by 103.12 %, followed by the increase in regulating gases (23.60 %) and organic material provision (25.72 %). However, the values of soil conservation and nutrient cycling decreased by 100.00 and 56.20 %.

Table 3.5 Changes in different ecosystem service values in response to ecological restoration policies

Policy	Ecosystem service value	2000(10^4 CNY)	2008(10^4 CNY)	Change percentage (%)
Grain-for-Green	Regulating gases	1169.29	1625.10	38.98
	Water conservation	554.92	435.71	−21.48
	Soil conservation	101.31	119.80	18.25
	Nutrient cycling	166.66	273.53	64.12
	Organic material provision	143.36	200.71	40.00
	Total	2135.54	2654.85	24.32
Grain-for-Blue	Regulating gases	20,550.32	25,399.44	23.60
	Water conservation	19,239.73	39,079.81	103.12
	Soil conservation	3300.42	0.00	−100.00
	Nutrient cycling	2400.89	1051.63	−56.20
	Organic material provision	2476.81	3113.85	25.72
	Total	47,968.17	68,644.73	43.10

Summary

The main objective of Grain-for-Green was to restore forests and grasslands in an effort to control soil erosion in China from 1998. In this study, we found that the values of soil conservation increased by 18.25 % due to Grain-for-Green during 2000–2008. However, the value of soil conservation in the whole of Shandong province increased by 76.87 % during the same period. Grain-for-Green in Shandong does not significantly reduce soil erosion. The reason could be that Grain-for-Green in Shandong was mainly implemented in low sloping areas where soil erosion is not significant. The implementation of Grain-for-Green in these areas has limited positive effects on controlling soil erosion.

In comparison with the soil conservation value, Grain-for-Green did improve the ecosystem service function of nutrient cycling, organic material provision, and regulating gases. During 2000–2008, the values of nutrient cycling, organic material provision, and regulating gases increased by 64.12, 40.00, and 38.98 %, respectively, in the areas implementing Grain-for-Green, but increased by 37.02, 37.90, and 36.13 %, respectively, in the whole of Shandong. The ESV of the areas implementing the Grain-for-Blue policy increased by 43.10 % during 2000–2008. The increase in water conservation value accounted for most of the added ESV. In the same period, the value of water conservation in the whole of Shandong increased by 16.69 %, while it increased by 103.12 % in Grain-for-Blue areas. Grain-for-Blue significantly improved the ecosystem service function of water conservation due to the improvement of the water regulating function in water areas.

We assessed five kinds of ESVs in Shandong province. The ecological restoration policies led to significant trade-offs in ESV. Although Grain-for-Green improved the ecosystem service function of nutrient cycling, organic material provision, and regulating gases, it also decreased the ecosystem service function of water conservation. Grain-for-Blue increased the ecological service function of water conservation but led to a decrease in soil conservation (100 %) and the nutrient cycling function (56.20 %).

Ecosystems comprise several different ecosystem service functions. A single ecological restoration policy cannot improve all the ecosystem service functions. Thus, we argue that integral ecological restoration policy should be adopted in ecological restoration. For example, when implementing Grain-for-Green in sloping areas, a water conservation project, such as rainfall collection, should be adopted simultaneously. Rainfall collection can improve not only the irrigation condition of farmland but also the water conservation function. When implementing the Grain-for-Blue policy, a soil conservation project (such as afforestation) should also be considered. Furthermore, close attention must be given to the suitability of the policy. For example, Grain-for-Green in plain areas is of no benefit to the overall improvement of the ecosystem service function and should be strictly limited to sloping areas.

Most of the maps (e.g., NPP, NDVI) utilized to assess the changes in ESV in this study are 1 km grid maps. The resolution of these maps could be too low at the provincial scale. In addition, owing to a lack of suitable method and data, we did not assess the ESV of regulating waste which is one of the most important ecosystem services of water areas. Thus, the ESV of water areas assessed in this study could be underestimated. For the same reason, we also did not assess the ESV of the culture function which could also generate some uncertainty in the results for ESV.

References

Angel S, Parent J, Civco DL, Blei A, Potere D (2011) The dimensions of global urban expansion: estimates and projections for all countries, 2000–2050. Prog Plann 75(2):53–107

Brady M, Sahrbacher C, Kellermann K, Happe K (2012) An agent-based approach to modeling impacts of agricultural policy on land use, biodiversity and ecosystem services. Landscape Ecol 27(9):1363–1381

Bronner CE, Bartlett AM, Whiteway SL, Lambert DC, Bennett SJ, Rabideau AJ (2013) An assessment of U.S. stream compensatory mitigation policy: necessary changes to protect ecosystem functions and services. JAWRA J Am Water Resour Assoc 49(2):449–462

Costanza R, dArge R, deGroot R, Farber S, Grasso M, Hannon B, Limburg K, Naeem S, Oneill RV, Paruelo J, Raskin RG, Sutton P, vandenBelt M (1997) The value of the world's ecosystem services and natural capital. Nature 387(6630):253–260

Flanagan D, Foster G, Neibling W, Burt J (1989) Simplified equations for filter strip design. Trans ASAE 32(6):2001–2007

Gardiner MM, Burkman CE, Prajzner SP (2013) The value of urban vacant land to support arthropod biodiversity and ecosystem services. Environ Entomol 42(6):1123–1136

Geneletti D (2013) Assessing the impact of alternative land-use zoning policies on future ecosystem services. Environ Impact Assess Rev 40:25–35

Geng X, Wang X, Yan H, Zhang Q, Jin G (2014) Land use/land cover change induced impacts on water supply service in the upper reach of Heihe river basin. Sustainability 7(1):366–383

Haase D, Frantzeskaki N, Elmqvist T (2014) Ecosystem services in urban landscapes: practical applications and governance implications. Ambio 43(4):407–412

Harmáčková ZV, Vačkář D (2015) Modelling regulating ecosystem services trade-offs across landscape scenarios in Třeboňsko Wetlands Biosphere Reserve, Czech Republic. Ecol Model 295:207–215

Heilig GK (2012) World urbanization prospects: the 2011 revision. United Nations, Department of Economic and Social Affairs (DESA), Population Division, Population Estimates and Projections Section, New York

Heubes J, Heubach K, Schmidt M, Wittig R, Zizka G, Nuppenau E-A, Hahn K (2012) Impact of future climate and land use change on non-timber forest product provision in Benin, West Africa: linking niche-based modeling with ecosystem service values. Econ Bot 66(4):383–397

Kuok KK, Mah DY, Chiu P (2013a) Evaluation of C and P factors in universal soil loss equation on trapping sediment: case study of Santubong River. J Water Resour Prot 2013

Kuok KKK, Mah DYS, Chiu PC (2013b) Evaluation of C and P factors in universal soil loss equation on trapping sediment: case study of Santubong River. J Water Resour Prot 05(12):6

Lin T, Xue X, Shi L, Gao L (2013) Urban spatial expansion and its impacts on island ecosystem services and landscape pattern: a case study of the island city of Xiamen, Southeast China. Ocean Coast Manage 81:90–96

Liu J, Zhang Z, Xu X, Kuang W, Zhou W, Zhang S, Li R, Yan C, Yu D, Wu S (2010) Spatial patterns and driving forces of land use change in China during the early 21st century. J Geogr Sci 20(4):483–494

Liu J, Kuang W, Zhang Z, Xu X, Qin Y, Ning J, Zhou W, Zhang S, Li R, Yan C, Wu S, Shi X, Jiang N, Yu D, Pan X, Chi W (2014) Spatiotemporal characteristics, patterns, and causes of land-use changes in China since the late 1980s. J Geogr Sci 24(2):195–210

Long H, Liu Y, Hou X, Li T, Li Y (2014) Effects of land use transitions due to rapid urbanization on ecosystem services: implications for urban planning in the new developing area of China. Habitat Int 44:536–544

Lü Y, Fu B, Feng X, Zeng Y, Liu Y, Chang R, Sun G, Wu B (2012) A policy-driven large scale ecological restoration: quantifying ecosystem services changes in the Loess Plateau of China. PLoS ONE 7(2):e31782

Nelson E, Sander H, Hawthorne P, Conte M, Ennaanay D, Wolny S, Manson S, Polasky S (2010) Projecting global land-use change and its effect on ecosystem service provision and biodiversity with simple models. PLoS ONE 5(12):e14327

Ntelekos AA, Oppenheimer M, Smith JA, Miller AJ (2010) Urbanization, climate change and flood policy in the United States. Clim Change 103(3–4):597–616

Ouyang Z, Wang X, Miao H (1998) A primary study on Chinese terrestrial ecosystem services and their ecological-economic values. Acta Ecol Sinica 19(5):607–613

Perrings C, Duraiappah A, Larigauderie A, Mooney H (2011) The biodiversity and ecosystem services science-policy interface. Science 331(6021):1139–1140

Peters CM, Gentry AH, Mendelsohn RO (1989) Valuation of an Amazonian rainforest. Nature 339(6227):655–656

Reid W, Mooney H, Cropper A, Capistrano D, Carpenter S, Chopra K, Dasgupta P, Dietz T, Duraiappah A, Hassan R (2005) Ecosystems and human well-being: synthesis, MA (millennium ecosystem assessment). Island Press, Washington, DC

Renard KG, Foster GR (1983) Soil conservation: principles of erosion by water. Dryland Agric Agron Monogr 23:156–176

Renard KG, Foster GR, Weesies GA, McCool D, Yoder D (1997) Predicting soil erosion by water: a guide to conservation planning with the revised universal soil loss equation (RUSLE). Agric Handb (Washington) 703

Seto KC, Fragkias M, Güneralp B, Reilly MK (2011) A meta-analysis of global urban land expansion. PLoS ONE 6(8):e23777

Seto KC, Reenberg A, Boone CG, Fragkias M, Haase D, Langanke T, Marcotullio P, Munroe DK, Olah B, Simon D (2012) Urban land teleconnections and sustainability. Proc Natl Acad Sci 109(20):7687–7692

Shi M, Ma G, Shi Y (2011) How much real cost has China paid for its economic growth? Sustain Sci 6(2):135–149

Silvestri S, Zaibet L, Said MY, Kifugo SC (2013) Valuing ecosystem services for conservation and development purposes: a case study from Kenya. Environ Sci Policy 31:23–33

Simpson RD (2014) Ecosystem services as substitute inputs: basic results and important implications for conservation policy. Ecol Econ 98:102–108

Song W (2014) Decoupling cultivated land loss by construction occupation from economic growth in Beijing. Habitat Int 43:198–205

Song W, Liu M (2014) Assessment of decoupling between rural settlement area and rural population in China. Land Use Policy 39:331–341

Song W, Pijanowski BC (2014) The effects of China's cultivated land balance program on potential land productivity at a national scale. Appl Geogr 46:158–170

Song W, Deng X, Yuan Y, Wang Z, Li Z (2015) Impacts of land-use change on valued ecosystem service in rapidly urbanized North China Plain. Ecol Model (in press)

Su C, Fu B (2013) Evolution of ecosystem services in the Chinese Loess Plateau under climatic and land use changes. Glob Planet Change 101:119–128

Su S, Li D, Hu YN, Xiao R, Zhang Y (2014) Spatially non-stationary response of ecosystem service value changes to urbanization in Shanghai, China. Ecol Ind 45:332–339

Sutherland WJ, Gardner T, Bogich TL, Bradbury RB, Clothier B, Jonsson M, Kapos V, Lane SN, Möller I, Schroeder M, Spalding M, Spencer T, White PCL, Dicks LV (2014) Solution scanning as a key policy tool: identifying management interventions to help maintain and enhance regulating ecosystem services. Ecol Soc 19(2)

Tan M, Li X, Lu C (2005) Urban land expansion and arable land loss of the major cities in China in the 1990s. Sci China Ser D Earth Sci 48(9):1492–1500

Wischmeier W, Johnson C, Cross B (1971) Soil erodibility nomograph for farmland and construction sites

Zang S, Wu C, Liu H, Na X (2011) Impact of urbanization on natural ecosystem service values: a comparative study. Environ Monit Assess 179(1–4):575–588

Zhao B, Kreuter U, Li B, Ma Z, Chen J, Nakagoshi N (2004) An ecosystem service value assessment of land-use change on Chongming Island, China. Land Use Policy 21(2):139–148

Chapter 4
Impact Assessments of Land-Use Change and Climate Change on Ecosystem Services of Grassland

Jinyan Zhan, Feng Wu, Fang Yin, Tao Zhang and Rongrong Zhang

Abstract The grassland is an important land-use type that plays an important role in the ecosystem service supply in China. It is of great significance to the grassland management to determine the changing trend of grassland productivity and its response to land-use change and climate change. In this chapter, we first examined changes in grassland productivity due to climate and land-use change in the Three-River Headwaters region (TRHR) of Qinghai Province. In the macrocontext of climatic change, we analyzed the possible changing trends of the net primary productivity (NPP) of local grasslands under four representative concentration pathways (RCPs) scenarios (i.e., RCP2.6, RCP4.5, RCP6.0, and RCP8.5) during 2010–2030 with the model estimation, and the grass yield and theoretical grazing capacity under each scenario were further qualitatively and quantitatively analyzed. The results indicate that the grassland productivity in the TRHR will be unstable under all the four scenarios. The grassland productivity will be greatly influenced by the fluctuations of precipitation, and the temperature fluctuations will also play an important role during some periods. The local grassland productivity will decrease to some degree during 2010–2020 and then will fluctuate and increase slowly during 2020–2030. The theoretical grazing capacity was analyzed in this study and calculated on the basis of the grass yield. The result indicates that the theoretical grazing capacity ranges from 4 to 5 million sheep under the

J. Zhan (✉) · F. Wu · T. Zhang
State Key Laboratory of Water Environment Simulation, School of Environment,
Beijing Normal University, Beijing 100875, China
e-mail: zhanjy@bnu.edu.cn

F. Yin
Leibniz Institute of Agricultural Development in Transition Economies (IAMO),
Theodor-Lieser-Str. 2, 06120 Halle (Saale), Germany

R. Zhang
Faculty of Resources and Environmental Science, Hubei University, Wuhan 430062, China

© Springer-Verlag Berlin Heidelberg 2015
J. Zhan (ed.), *Impacts of Land-use Change on Ecosystem Services*,
Springer Geography, DOI 10.1007/978-3-662-48008-3_4

four scenarios and it can provide quantitative information reference for decision making on how to determine the reasonable grazing capacity, promote the sustainable development of grasslands, and so forth. Further, we estimated changes in economic returns of livestock production in the TRHR of China. The land surface in TRHR, a typical ecological fragile zone of China, is quite sensitive to the climate change which will destabilize certain ecosystem service valuable to the entire nation and neighboring countries. We analyzed the impacts of climate change and agents' adaptive behaviors on the regional land-use change with the agent-based model (ABM). First, the main agents were extracted according to the production resource endowments and socioeconomic background. Then, the agents' land-use behaviors were analyzed and parameterized. Thereafter, the ABM model was built to simulate the impacts of climate change on the regional land-use change and agents' economic benefits. The results showed that the land-use change was mainly characterized by the increase of grassland and decrease of unused land area. Besides, the agents would get more wealth under the scenario without climate change in the long term, even though the total income is lower than that under the scenario with climate change. In addition, the sensitivity analysis indicated that the model is sensitive to the climatic conditions, market price of agricultural and animal husbandry products, government subsidies, and cost control. Finally, we predicted changes in grassland productivity in China. The results showed that, firstly, the relationship between grassland productivity and climate change, geographical conditions, and human activities was analyzed with the panel data of the whole China during 1980–2010. The result indicated that the temperature and precipitation were very important to grassland productivity at the national scale; secondly, the grassland in China was divided into seven grassland ecological–economic zones according to the ecosystem service function and climate characteristics. The relationship between grassland productivity and climate change was further analyzed at the regional scale. The result indicated that the temperature is more beneficial to the increase of the grassland productivity in the Qinghai–Tibet Plateau and the Southwest Karst shrubland region; thirdly, the increase of the temperature and precipitation can increase the grassland productivity and consequently relieve the pressure according to the climate factors of simulation with the community climate system model v4.0 (CCSM). However, the simulation result indicates that the human pressure on grasslands is still severe under the four RCPs scenarios and the grassland area would reduce sharply due to the conversion from the grassland to the cultivated land. What is more, there is still a great challenge to the increase of total grassland productivity in China.

Keywords Grassland productivity · Livestock production · Climate change · Land-use change · Agent-based model · Three-River Headwaters region · Qinghai

Changes in Grassland Productivity Due to Climate and Land-Use Change in Qinghai Province

Introduction

The net primary productivity (NPP) of vegetation reflects the productivity of the vegetation under the natural conditions (Liu et al. 1999). The climatic change is one of the key driving forces of the interannual change in NPP of vegetation (Piao et al. 2006). The climate is undergoing the change which is mainly characterized by the global warming. The land surface temperature has increased significantly since the 1980s, especially in the northern region of China. The grassland is one of the most important land-use types in China, which has essential functions in the development of the animal husbandry (Wu et al. 2013). The grassland is greatly influenced by the climatic change, and the spatiotemporal change in NPP of grasslands and the influencing mechanism of the climatic change on it have been one of the research focuses at home and abroad (Cao et al. 2003).

The Three-River Headwaters region (TRHR) is the headstream of the Yellow River, Yangtze River, and Lancang River, which is one of the most ecologically sensitive areas in China. Besides, it is also the largest animal husbandry production base in Qinghai Province, with about 21.3 thousand km^2 of native pasture and native grassland. Many researchers have analyzed the change in NPP in this area from different perspectives, and there have been many research works on the pattern and spatiotemporal characteristics of NPP of ecosystems. However, there have been few comprehensive studies on the spatiotemporal change in NPP of grasslands and the consequent effects in the TRHR. On the one hand, owing to the distinctive natural ecological conditions in this region, the development of local animal husbandry always depends on the increase of the livestock amount, which increases the income of local people and meanwhile leads to the long-term overgrazing. The serious grassland degradation has greatly restrained the development of the local animal husbandry and change of landscape (Zhan et al. 2004). On the other hand, the global warming has led to the decrease of the average annual precipitation, and the grass yield per unit area also decreases slightly year by year, which has threatened the development of local animal husbandry. Therefore, a description of the climate and relevant economic activities in the study area is detailed and significant (González et al. 2012). In order to solve the problems brought by the grassland degradation and promote the sustainable development of the local animal husbandry in the TRHR, it is necessary to carry out scientific prediction of the local grass yield, determine the reasonable grazing capacity, and guide the production of local animal husbandry.

The research on NPP of the grassland is the basis for the study of the grass yield and prediction of grazing capacity, and there have been many investigations and research works on the estimation of NPP of the grassland in China in recent years. The methods to estimate NPP of grassland vary greatly due to the difference in the natural environment of the study area, data availability, and so forth. There are mainly four kinds of models to estimate NPP of grassland, that is, the light-use efficiency model, ecosystem process model, remote sensing-process coupling model, and climatic statistic model. There are both advantages and disadvantages in these models. For example, the light-use efficiency model based on the mechanism of vegetation photosynthesis is easy to be constructed and has high calculation efficiency, but there are some faults in the factors taken into account: parameter selection, calculation result, and so forth. The ecosystem process model simulates the physiological processes of vegetation and applies the technologies such as the remote sensing, which makes it possible to carry out multiscale dynamic monitoring of the spatiotemporal change of NPP. However, this model is very complex and requires high-quality data, which restrains its practicability to some degree, especially in the regional estimation. The remote sensing-process coupling model integrates the advantages of both the models mentioned above, but the accuracy of its calculation result is greatly influenced by other factors. The climatic statistic model introduces the regression models constructed with the simple climatic factors such as temperature and precipitation and has a low data requirement. This kind of models is more practical, but it is still limited by the low accuracy of the result. There is great complexity, uncertainties, and inaccuracy in the extraction of vegetation indices and soil parameters with the remote sensing data, all of which make it very difficult and very inaccurate to calculate these data with the light-use efficiency model, ecosystem process model, and ecological remote sensing-process coupling model. Besides, it is a fact that the climatic conditions have great impacts on the livestock production in the study area. Therefore, the climatic statistic model was finally used to estimate the future grassland productivity in the TRHR.

The most widely used climate models mainly include the Miami Model, Thornthwaite Memorial Model, Chikugo Model, and the comprehensive model. The climate model is an effective tool in the study of climate (Deng et al. 2013). The comprehensive model is more suitable for the estimation of NPP of vegetation in the arid area than the Chikugo Model. Besides, in comparison with other three models, the comprehensive model has a solid theoretical foundation, takes more into account of the physiological processes of vegetation, and consequently can obtain a better estimation result in Zhejiang Province and Inner Mongolia.

In order to overall forecast the changing trend of the grassland NPP and theoretical grazing capacity in the study area in the context of climate change, four representative concentration pathways (RCPs) (which represent the emission trajectories under the natural and social conditions and the corresponding scenarios) were selected to analyze the changing trend of grassland NPP in the TRHR. This study is of both theoretical and practical significance. In theory, this study extends the field of application of the estimation of NPP and explored the theoretical grazing capacity in the TRHR in the future, which provides certain references for the relevant research works in other similar regions. In practice, this study qualitatively and quantitatively analyzed the changing trend of the grass yield and grazing capacity in the study area, which can provide some guidance for the local grassland utility and management and the development of animal husbandry and promote the harmonious and sustainable development of the local man–land relationship.

Study Area

The TRHR is located in the southern part of Qinghai Province of China, between 31°39′–36°12′N and 89°45′–102°23′E and with an area of 363 thousand km² which accounts for 43 % of the total area of Qinghai Province. The TRHR with the altitude ranges from 3500 to 4800 m is the headstream of the Yellow River, Yangtze River, and Lancang River and has a dense network of rivers. The administrative regions cover 16 counties, including Yushu, Xinghai, Tongde, Zeku, Matuo, Maqin, Dari, Gande, Jiuzhi, Banma, Chengduo, Zaduo, Zhiduo, Qumalai, Nangqian, and Henan, except for Tanggula Mountain Town which is under the charge of Golmud City.

The grassland area is 203 thousand km² in the TRHR, accounting for 65.4 % of the total area of this region (Fig. 4.1). The vegetation diversity of the TRHR is the richest among the regions at the same altitude all over the world. The grassland type changes from alpine meadow to high-cold steppe and alpine desert, with the productivity also gradually decreasing. The grassland resource is very rich in this region; however, the grass yield per unit area has decreased year by year due to the climatic change and overgrazing in recent years, which has threatened the development of the local animal husbandry.

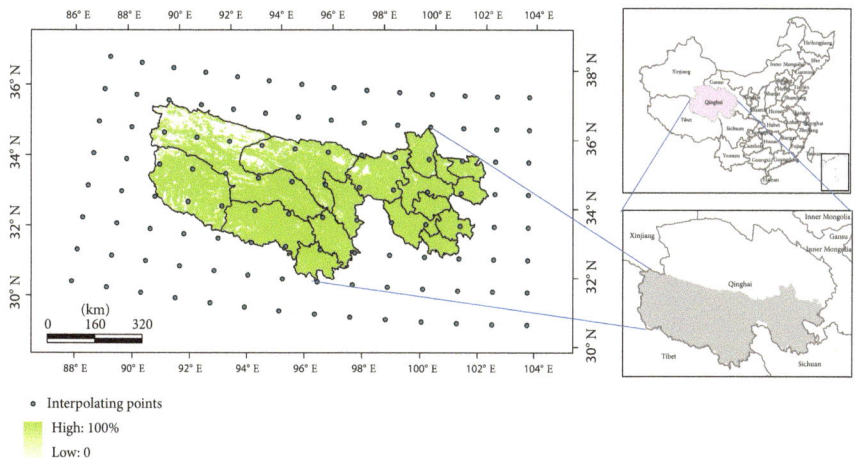

Fig. 4.1 Location of the TRHR and distribution of grassland

Methodology and Data

Models

Comprehensive Model

The comprehensive model was developed on the basis of two well-known balance equations, that is, the water balance equation and heat balance equation. Zhou deduced the regional evapotranspiration model that links the water balance equation and heat balance equation from the physical process during the energy and moisture influence of the vaporization and then constructed the natural vegetation NPP model based on the physiological characteristics (Zhou and Zhang 1996), that is, the comprehensive model. The comprehensive model can calculate the potential NPP of natural vegetation on the basis of the precipitation and net radiation received by the land surface in the study area. This model is of great significance to the reasonable use of climatic resource and fulfillment of the climatic potential productivity (Zhou and Zhang 1996). The formula of this model is as follows:

$$\text{NPP} = \text{RDI} \times \frac{P_r R_n \left(P_r^2 + R_n^2 + P_r R_n\right)}{(P_r + R_n)\left(P_r^2 + R_n^2\right)} e^{\sqrt{9.87 + 6.25 \times RDI}}$$

$$\text{RDI} = \frac{R_n}{L \cdot P_r}$$

where R_n is the annual net radiation, P_r is the annual precipitation, L is the annual latent heat of vaporization, and RDI is the radiation aridity.

Model of the Hay Yield of Grassland

There are mainly three indicators of the grassland productivity, that is, the hay yield, theoretical grazing capacity, and animal products. The hay yield, that is, the total dry matter yield of a certain area during a certain period, reflects the primary productivity of grassland and is a basic indicator of the grassland productivity. In this study, the hay yield of grassland during 2010–2030 was calculated based on NPP of grassland with the following formula:

$$B_g = \frac{\text{NPP}}{S_{\text{bn}}\left(1 + S_{\text{ug}}\right)}$$

where, B_g is the annual total hay yield per unit area (g m^{-2} a^{-1}), NPP is the annual total NPP of grassland (gC m^{-2} a^{-1}), S_{bn} is the coefficient of the conversion coefficient of the grassland biomass and NPP (g/gC), which is 0.45, and S_{ug} is the proportionality coefficient of the over ground biomass and underground biomass, which varies among different vegetation types. S_{ug} of the alpine meadow, high-cold steppe and alpine desert is 7.91, 4.25, and 7.89, respectively. According to the location of the study area, S_{ug} of the alpine meadow was used to calculate the grass yield.

Model of the Theoretical Grazing Capacity of Grassland

The theoretical grazing capacity of grassland during 2010–2030 was calculated on the basis of the grass yield. Since the grazing capacity of grassland is customarily represented by the unit of livestock in China, that is, the number of adult livestock that can be supported by per unit of land area every year, and the number of sheep is generally used as the unit, the grazing capacity of grassland is also represented by the number of sheep per unit of land area.

There have been many methods to calculate the theoretical grazing capacity of grassland. The estimation method of "limiting livestock based on grassland carrying capacity" can better reflect the restriction of the practical situation in the grazing districts on the livestock production, and hence, the following formula was used:

$$CA = \frac{G \cdot Cuse}{U_G \cdot DOY}$$

where CA is the theoretical annual grazing capacity of grassland (Unit: number of sheep per unit of land area), G is the annual hay yield of grassland per square meter (Unit: kg/m^2), and Cuse is the utilization efficiency of grass by the livestock varying among different grassland types. In this study, Cuse of the alpine meadow, high-cold steppe and alpine desert, shrubbery, and swamp meadow is 60, 50, 40, and 55 %, respectively. U_G is the hay quantity needed by per unit of sheep every day (Unit: kg/d), which was set to be 2.0 kg according to the relevant criterion. DOY (Unit: d) is 365. Since the grassland is the main vegetation type in the TRHR, Cuse of the high-cold steppe was used to estimate the theoretical grazing capacity of grassland during 2010–2030.

Data Source

The data of precipitation and near-surface air temperature in the study area were simulated with the models of Coupled Model Intercomparison Project Phase 5 (CMIP5). There are three steps in the data processing. (a) The data were first selected and downloaded, including the model (CCSM4), modeling realm (atmosphere), ensemble (r6i1p1 and r5i1p1), and climatic variables (precipitation and near-surface air temperature). (b) The data of study area were then extracted and calculated. The annual average value was calculated based on the monthly data, and the annual precipitation was calculated as the sum of the monthly precipitation and then extracted 112 points covering the study area. (c) The point data with the spatial resolution 0.9° × 1.25° were interpolated in 1 km × 1 km raster using the Kriging method and were projected with the Albers 1940 coordinate system.

Results

Changing Trend of NPP of Grassland in the TRHR

There is significant spatial heterogeneity of the NPP in the TRHR, decreasing from the southeast to the northwest on the whole (Fig. 4.2). The results indicate that the NPP of grassland mainly increases in the east and southeast part, while it decreases significantly in the northwest, southwest, and middle part. There is no significant change of the NPP of grassland in most of other parts. The changing trends of NPP during every ten years indicate that the NPP changes significantly under the RCP2.6 scenario and RCP4.5 scenario, increasing in the east and southeast part to some degree and decreasing in the south part to some extent. The NPP changes slightly under the RCP6.0 scenario and RCP8.5 scenario. Under the

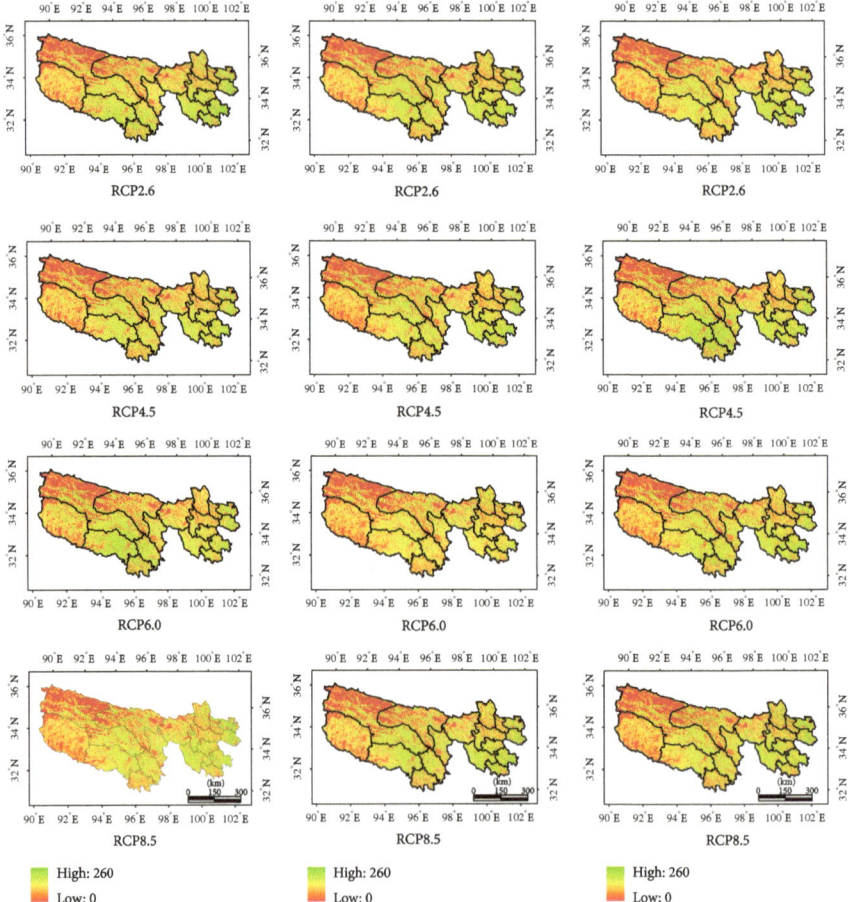

Fig. 4.2 The NPP map of the TRHR in 2010, 2020, and 2030 under the four RCPs scenarios

RCP4.5 scenario, the NPP decreases obviously in the middle and south part during 2010–2020 and increases slightly during 2020–2030, indicating that there is serious desertification of the local grassland. Besides, the increase of NPP by 2030 suggests that there is some improvement in the conditions of the local grassland.

In this study, the influence of temperature and precipitation on the change in NPP was analyzed. The result indicates that the NPP of grassland will range from 100 to 130 g m^{-2} a^{-1} during 2010–2030. The results under different scenarios are shown as follows (Fig. 4.3).

The result under the RCP2.6 scenario indicates that the temperature and precipitation would present a decreasing trend during 2015–2020 and 2025–2030 and show an opposed trend during 2010–2015 and 2020–2025 (Fig. 4.3). The precipitation will fluctuate more greatly than the temperature on the whole. By contrast, the NPP will change in an opposite way during these periods, but with smaller amplitude of fluctuation. Therefore, there is a significant negative relationship between the NPP and temperature, while there is only a weak relationship between the NPP and precipitation under this scenario.

The result under the RCP4.5 scenario indicates that the NPP and precipitation show a similar changing trend, that is, a concave-down parabolic trajectory on the whole. The precipitation will fluctuate most greatly during 2010–2020, while the NPP first decreases with the precipitation and then increases rapidly after reaching

Fig. 4.3 Changing trends of the NPP of grassland, temperature, and precipitation (the average number in every year) in the TRHR during 2010–2030

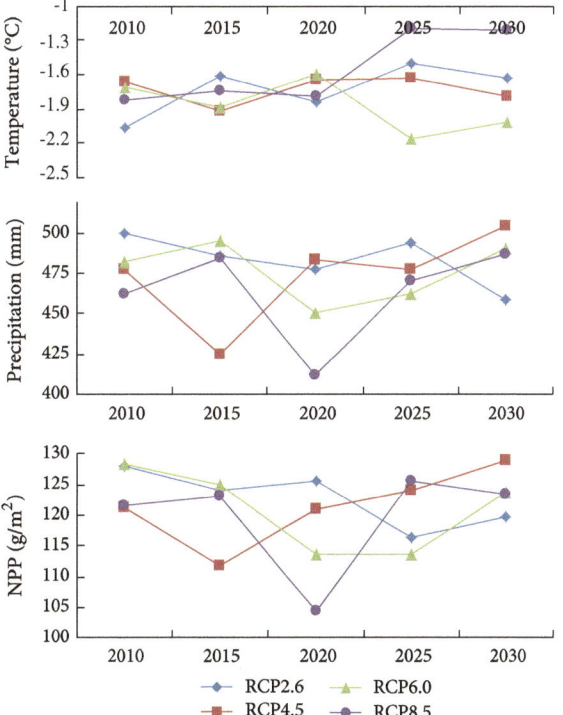

a relatively low level. The NPP will decrease by 8.4 % from 2010 to 2015, but it will increase by 8.2 % from 2015 to 2020. Then, the NPP will increase slowly while fluctuating slightly during 2020–2030. Therefore, there is a significant negative relationship between the NPP and precipitation under this scenario, while the relationship between the NPP and temperature is very weak.

The result under the RCP6.0 scenario indicates that the NPP will first increase and then decrease during 2015–2025, while temperature will show an opposite changing trend during this period, as during other period, they will change in a similar way. The precipitation will show an increasing trend during 2010–2015 and 2020–2030. The NPP and precipitation will both decline obviously during 2015–2020 and reach the bottom around 2020. The NPP will decrease by 10 % in 2020 when compared with 2015, which indicates that the change in NPP is greatly influenced by the change of precipitation during this period and they are strongly correlated. The result suggests that the changing trends of the NPP are consistent with those of the precipitation on the whole, but the fluctuation range of the NPP is small, indicating that there is some lag in the response of the NPP to the change of precipitation under this scenario. According to the analysis above, the NPP responds more sensitively to the change of precipitation than to the change of temperature.

Under the RCP8.5 scenario, the temperature changes slightly during 2010–2025 and 2025–2030 and shows an increasing trend during 2020–2025. Besides, the NPP also fluctuates slightly during 2010–2025 and 2025–2030, indicating that the temperature plays a dominant role in influencing the NPP. The NPP and precipitation both fluctuate significantly during 2015–2020, and there is an obvious low ebb around 2020. The NPP decreases by 15.3 % in 2020 in comparison with 2015, indicating that there is a strong correlation between the change in NPP and the change of precipitation. During 2020–2025, the NPP, temperature, and precipitation all show an obvious increasing trend. The NPP increases by 20.4 % from 2020 to 2025 and reaches a significant peak around 2025, indicating that there is significant relationship between the change in NPP and the changes of both the precipitation and temperature. Therefore, the temperature plays a key role in influencing the change in NPP during 2010–2020 and 2025–2030, while the precipitation plays a dominant role during 2015–2020. Besides, during 2020–2025, both the temperature and precipitation greatly influence the NPP.

Changing Trend of Grass Yield of Grassland

The result under the RCP2.6 scenario indicates that the changing trend and fluctuation range of the grass yield are both pretty consistent with those of the NPP mentioned above; that is, both increase during 2015–2020 and 2025–2030 and decrease during 2010–2015 and 2020–2025 (Fig. 4.4). On the whole, the grass yield is generally above 6.3 million tons under this scenario except for the period around 2025, and the fluctuation range is not great and the average yield level is very stable.

Fig. 4.4 Changing trends of grass yield of grassland in the TRHR during 2010–2030 under the four scenarios

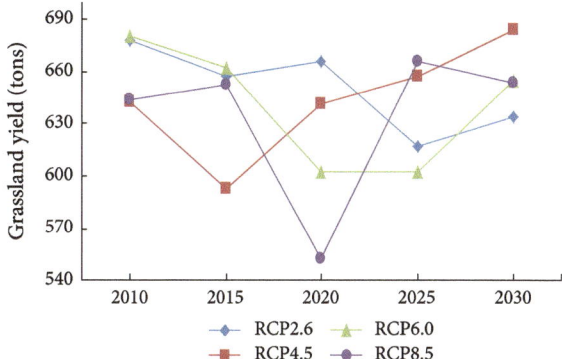

The result under the RCP4.5 scenario indicates that the grassland yield will fluctuate greatly but will still increase slightly on the whole during 2010–2015. The grass yield will keep a stable increasing trend during 2015–2030. In comparison with the changing trend of NPP mentioned above, the changing trend of the grass yield is consistent with that of the NPP during 2015–2030, but they are not closely related during 2010–2015.

The result under the RCP6.0 scenario indicates that the grass yield will decline during 2010–2020 and then tends to increase slowly after reaching a low level around 2020, indicating that the grass yield of the TRHR will fluctuate greatly under this scenario.

The result under the RCP8.5 scenario suggests that the grass yield of the TRHR will fluctuate slightly during 2010–2015, then declines significantly, and thereafter keeps an increasing trend, but the increment will gradually decline and there may even be some slight decrease. According to the analysis above, it is predictable that the grass yield will fluctuate obviously around 2020 and decline to a very low level and will only fluctuate slightly during other periods under this scenario.

To sum up, there is a positive relationship between the grass yield and NPP of grassland in the TRHR. The change of the NPP of grassland has an impact on the grass yield, but its effects vary among different RCPs scenarios. The grass yield is very stable under the RCP2.6 scenario, generally above 6.3 million tons every year. Under the RCP4.5 and RCP6.0 scenarios, the changing trends of the grass yield and NPP of grassland are generally similar during most periods except for 2015–2020, during which their changing trends are contrary. Under the RCP8.5 scenario, the grass yield fluctuates most greatly, and the precipitation, grass yield, and NPP of grassland will all descend to the bottom around 2020, indicating that the grass yield is most greatly influenced by the precipitation under this scenario.

Analysis of the Grazing Capacity of Grassland

The theoretical grazing capacity during 2010–2030 was analyzed in this study. The grazing capacity of grassland in the TRHR was calculated on the basis of the grass yield. The result indicates that the theoretical grazing capacity ranges from 4 to 5 million sheep under the four scenarios (Fig. 4.5).

The result under the RCP2.6 scenario indicates that the theoretical grazing capacity in the TRHR will show a significant decreasing trend and reach the minimum in 2017, and it will then increase rapidly during 2018–2021 but will thereafter keep a decline trend on the whole (Fig. 4.5). Besides, the interannual fluctuation range is very great under this scenario. According to the changing trends of the temperature and precipitation mentioned above, the grazing capacity responds very slowly to the change of temperature within a certain scope and there is no significant relationship between them, while the grazing capacity shows a changing trend similar to that of the precipitation. The result indicates that, under the condition of no great fluctuation in the temperature, the grazing capacity mainly depends on the precipitation, which is consistent with the actual condition that the local animal husbandry is mainly restricted by the water resource.

The result under the RCP4.5 scenario indicates that the grazing capacity fluctuates greatly during 2010–2015, but the fluctuation range will gradually decrease with the time. It shows an increasing trend during 2015–2025, especially during 2019–2025, and there will be a stable and continuous increase. The grazing capacity will first decrease sharply and then increase rapidly during 2025–2030. On the whole, the grazing capacity fluctuates very greatly and the stability is very low under this scenario. It suggests that the grazing capacity of the local grassland increases with the precipitation within a certain scope, beyond which the temperature will play a more important role. In comparison with the changing trends of the temperature and precipitation mentioned above, it can be seen that the influence of the temperature on the grazing capacity is always very significant under this scenario, while that of the precipitation is only significant during 2015–2025.

The result under the RCP6.0 scenario indicates that the grazing capacity shows a decreasing trend on the whole during 2010–2020, during which there

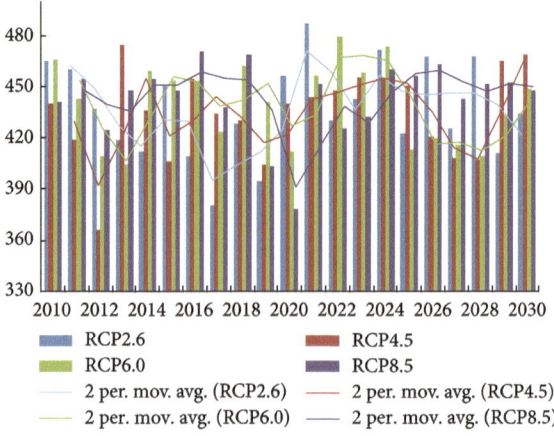

Fig. 4.5 Theoretical grazing capacity in the TRHR during 2010–2030 (ten thousand sheep)

is great fluctuation. The grazing capacity will first increase and then decrease during 2020–2030, and it shows a decreasing trend on the whole under this scenario. In comparison with the changing trends of the temperature and precipitation mentioned above, the changing trend of the grazing capacity is more consistent with that of the precipitation. However, during 2020–2030, the change of the grazing capacity is negatively related with the change of temperature, and it responds very slowly to the change of precipitation, and even not obviously. It indicates that the precipitation has more important impacts on the grazing capacity when the temperature is within a certain range; but on condition that the temperature decreases by a certain degree, the precipitation will only play a secondary role.

The result under the RCP8.0 scenario indicates that the local grazing capacity will fluctuate slightly during 2010–2015, but without significant change on the whole. It will continually decrease during 2016–2020 and reach the bottom around 2020 and then will keep increasing and finally fluctuate around 4.5 million sheep. According to the changing trends of the temperature and precipitation, the changing trend of the grazing capacity is more consistent with that of the precipitation, indicating that the precipitation plays a more important role in influencing the grazing capacity than the temperature does.

In summary, the precipitation plays a dominant role in influencing the grazing capacity under the RCP2.6 scenario, and the water resource is the main limiting factor of the development of the local animal husbandry. The precipitation has limited impacts on the development of the local animal husbandry under the RCP4.5 scenario. The theoretical grazing capacity increases with the precipitation within a certain scope, beyond which the temperature will play a more important role. The precipitation and temperature both have some influence on the grazing capacity under the RCP6.0 scenario. The precipitation plays a more important role when the temperature reaches a certain scope and vice versa. The precipitation plays a more important role in influencing the grazing capacity under the RCP8.5 scenario. On the whole, the theoretical grazing capacity in the TRHR ranges from 4 to 5 million sheep.

Summary

This study estimated the NPP of grassland in the TRHR under four RCPs scenarios based on the comprehensive model and estimated the local grass yield and theoretical grazing capacity in the future. Besides, the future changing trends of the NPP, grass yield, and grazing capacity were analyzed under four scenarios. In this paper, we draw the following conclusions.

There are very complex influences of the precipitation and temperature on the grassland productivity, and the effects of the precipitation and temperature on the NPP, grass yield, and grazing capacity are very complex and unstable under different scenarios. For example, the theoretical grazing capacity in 2029 is 4.1072 million sheep under the RCP2.6 scenario, while it is 4.6527 million sheep under the RCP4.5 scenario, which also differs greatly under another two scenarios.

The grassland productivity in the TRHR is unstable on the whole. The grass yield is greatly influenced by the fluctuation of the precipitation and the temperature which also plays a more important role and subsequently influences the grazing capacity. This conclusion is consistent with that of the previous research on the changing trend of vegetation NPP in the past 50 years in the Yellow River headwater area, which was carried out by Yao et al. (2011), indicating that the precipitation plays a dominant role in influencing the grassland productivity in the TRHR.

The grassland productivity in the TRHR will decrease slightly during 2010–2020, especially around 2020 when there will be a minimum, while the grazing capacity will first increase and then decrease during this period under all the scenarios except the RCP8.5 scenario. According to the analysis of the changing trend of the grazing capacity, there is a dramatic change in the grazing capacity in the TRHR due to the influence of the climatic factors. Therefore, it is necessary to reinforce the control on the grazing capacity, eliminate some livestock species in time, and replace the dominant grass species with the grass species that can better adapt to the climatic change. Besides, it is necessary to prepare for the various responses to the climatic change and formulate the artificial intervention mechanism as early as possible so as to reasonably guide the development of the local animal husbandry.

This study forecasted and analyzed the grassland NPP, hay yield of grasslands, and theoretical grazing capacity with the comprehensive model on the basis of the simulation of temperature and precipitation under the four scenarios. The research result is only obtained on the basis of the hydrothermal conditions, while in fact various factors, such as the soil, terrain, and solar radiation, all have some impacts on the grassland NPP. Therefore, there is still some limitations in the result of this study, and it is necessary to carry out more in-depth research works on the modification of the simulation result with the comprehensive model through including more other factors.

Estimated Changes in Economic Returns of Livestock Production in the TRHR of China

Introduction

Both the global and regional climate change have greatly undermined the terrestrial landscapes, ecological processes, and ecosystem services (Jentsch and Beierkuhnlein 2008), which have subsequently threatened not only the human societies but also the natural environment itself by the recondite feedback effects

(Drinkwater et al. 2010). There have been dramatic changes in the climatic conditions of China, which will continue to last for a long time in the future. The atmospheric temperature of China has increased by 0.5–0.8 °C in the twentieth century, with the upper limit exceeding the global average level. Meanwhile, the precipitation has continuously decreased in northern part of China while increased greatly in the southern and southeast part of China. Additionally, the extreme climate and subsequent natural disasters have also showed a longitudinal distribution pattern. A number of researchers have reported the impacts of climate change on various factors of the ecological and social systems (Yue et al. 2011), such as terrestrial surface (Briner et al. 2012), the market price (Fleischer and Sternberg 2006), and externality of public goods and even the political negotiations among various stakeholders (Rustad 2008).

Climate change associated with human adaptive activities, especially the social agents' decisions on land use, has exerted synergistic impacts on the land surface at different scales (Rounsevell and Reay 2009). There are generally two approaches through which the climate change influences the dynamic land surface properties, that is, the natural processes and human adaptive behaviors. The natural processes, for example, the change of thermodynamic properties and hydrological changes, will alter the terrestrial land surface at the long-term scale, and the inherent properties of different land-cover types limit the land-use activities of human beings, while the human adaptive behaviors may mitigate the adverse effects of climate change through changing the land-use practices. For instance, the herdsmen prefer pasturing sheep on the grassland to other land-cover types, and therefore, they may transform the cultivated land and forest land into grassland so as to adapt to the climate change and obtain more ecosystem services. However, there has been very limited knowledge on the impacts of climate change on the land-use dynamics through disturbing microagents' behaviors so far. Besides, climate change has also influenced the behaviors of government agents, who have taken measures to mitigate the impacts of the climate change and may subsequently influence the behaviors of other agents. For example, the long-term strategic development policy in the western region of China has greatly promoted the ecological restoration and protection of the regional ecosystem; the subsequent national-wide policies of Grain for Green (GFG) and Overgrazing Forbidden have made remarkable progresses in the land planning and ecological construction through compensating farmers and herdsmen for the change of their unreasonable land-use practices; what is more important is that the establishment of the TRHR nature reserve has imposed restrictions on the local human land-use practices and consequently protected the local ecological environment.

It is necessary to analyze the driving mechanism of social agents' behaviors and clarify their behaviors or decision algorithms with a proper model in order to more accurately study the temporal and spatial transitions of land use under the impacts of climate change and social agents' behaviors. The agent-based model (ABM) is a mainstream model to specify the behaviors of agents. Although there have been many ABM studies on the regional and local simulation, especially in the urban areas and transition zones between cultivated land and grassland, it is still necessary to carry out more in-depth ABM studies in areas without significant land-use/land-cover change (LUCC) and easy-distinguished agents. Besides, previous researches on the identification and parameterization of the agents' behaviors have mainly relied on the mathematical or econometric theories, which are different from the thinking ways of human beings from the ethological perspective in the real life. What is more, agents will actually choose the land-use practices for a targeted land parcel according to their own or neighbors' successful land-use experiences, which has served as the theoretical basis of the case-based reasoning (CBR) in this study.

The grassland is the dominant land-use type in this area, accounting for 71.5 % of the TRHR and having a total decrement of 0.03 % in the past two decades, while the unused land accounted for 16.5 % of the study area and increased by 0.10 % in the same period. The high and mediate coverage grasslands have experienced continuous shrinkage, while the low coverage grassland area has had continually expanded. The spatial pattern of land-use change in the past decades was characterized by the grassland degradation in the middle and east regions and the shrinkage of the water body in the north and west regions (Xu et al. 2008). Besides, the economy in the study area has experienced unprecedented development during the past 20 years. The gross domestic product (GDP) in 2010 is 15 times of that in 1990. The animal husbandry is the dominant sector in the primary industry of this region, appropriately accounting for 79.5 % of the total output of the primary industry.

This study has aimed to analyze the impacts of climate change and variability and corresponding agents' adaptive land-use decisions on the land surface in the TRHR with the ABM. This paper will firstly extract the main agents based on the analysis of production resource endowments and socioeconomic situation in the study area and then parameterizes the behaviors of various agents by utilizing county-level statistical data and household survey data. Thereafter, the ABM model was constructed and used to simulate the land-use change in the next 50 years. The economic benefits of agents are also discussed in this paper.

Data and Methodology

Data Collection

The data used in this study mainly include the land-use data, socioeconomic and demographic information, the natural environment, and climatic conditions, all of which were derived from the following sources: (1) Environmental Science Data Center of Chinese Academy of Sciences, which offers the datasets of land use, digital elevation model (DEM), topography, the distance to water and road, and climate and soil properties; (2) statistical yearbooks, such as Qinghai Province Statistical Yearbook, China Agricultural Yearbook, and China Animal Husbandry

Yearbook, which mainly contain the population, GDP, food production, and animal husbandry production; (3) household survey data, which were obtained from the household survey implemented by the research team in August and October 2012 with the stratified random sampling method performance. Three counties were selected as the sample sites to conduct the survey, which aims to identify the agents and parameterize their behaviors. The questionnaires have been modified after the pretest. Besides, an intraday examination was carried out on the result of questionnaires so as to revise the inconsistent or unreasonable answers with the retelephone investigation. This could not only save a lot of human resources and finance but also ensure the representativeness and reliability of the samples.

Identification of Agents

It is more difficult to identify the agents in the study area than in other regions where the agents such as farmers and herdsmen have inherent differences. In the study area, most people are herdsmen who rely on the grassland, and it seems that there are no absolute differences among these individuals. However, the statistical analysis and preinvestigation indicate that the herdsmen are different in the economic background and production resources, which would influence their land-use decisions. Therefore, based on the analysis of the socioeconomic background, natural environment conditions, and resources endowments, all the counties in the study area were classified into three categories, that is, the traditional development area, animal husbandry dominated area, and restricted development area. By doing this, the major agents can be extracted from each category, including the farmer agent, herdsman agent, and hominid agent.

The farmer agents are mainly located in the traditional development areas, eastern regions. They possess abundant labor force, production resources, and relatively large cultivated land area, which help them get a higher economic return from diverse income sources. The herdsman agents mainly dwell in the animal husbandry dominated areas, where the cultivated land shares a very small proportion, but the grassland occupies more than 70 % of the total area of these regions. So the herdsmen extremely rely on the grassland to develop the animal husbandry industry to get the revenue for supporting their families. Subsequently, the proportion of animal husbandry output to the primary industry output is higher than other sectors due to the intensive resource utilization and collective land-use practices. The hominid agents were identified according to the habitat properties. The hominid agents were located in the counties where the social and natural ecological systems are extremely fragile. These counties are characterized by the harsh climatic conditions, lower economic levels and human resource, and lack of production resource endowments. Therefore, the agents in these areas would heavily rely on the natural conditions rather than the land-use activities to support their life.

Agents' Behaviors

The general behavior rules of how the agents decide the land-use practices for a targeted land parcel were introduced first, and then, the individual behavior rules that varied from each agent were designed and specified with the diverse characteristics. At last, the impacts of the government agent's behaviors on other agents' decision through punishment and compensation were described.

Generally, all the agents have the same procedures for determining the land-use practices for a certain land parcel. It means that each agent will choose a land-use type for a land parcel at the beginning of each simulation step; then, the agent will calculate the maximum combined economic and social benefit at the end of each step to decide which land-use type should be chosen for this land parcel. Thereafter, the agent chooses a certain land-use type according to the land-use decision algorithm. Finally, the agent will decide whether or not to sell or buy land parcels based on the estimation of economic account (Polhill et al. 2010).

The individual behavior rules are different among agents due to the variation in economic background, natural conditions, and production resource endowments, which are vital to the design of the land-use decision algorithm. There are three categories of individual decision algorithms in terms of satisfaction, imitation, and CBR strategies (Izquierdo et al. 2004). It was assumed that agents have different probabilities to choose the decision algorithm according to the results of the field survey. Specifically speaking, if the benefits of the land-use decision have reached the agents' expectation, the agents will choose the satisfactory land-use strategy, that is, choose the same land-use type in last year. Or else, the agents will get certain probabilities to choose the imitation strategy or CBR strategy if the expectation is not realized (Izquierdo et al. 2004). As for the imitation strategy, the agents will imitate the successful land-use practices of their neighbors. But actually all the agents are more willing to determine the land use for the land parcel according to their own historical experiences. Therefore, they will get a relatively higher probability to choose the CBR strategy.

The CBR strategy is an approach to acquire the solutions to land-use decisions through analyzing the historical experience. It can simplify the knowledge, improve efficiency, and update the cases (Polhill et al. 2010). Each case records one agent's land-use experiences in a specific land parcel, which includes the year, biophysical characteristics of the land parcel, climatic conditions, land-use type, and economic and social benefits. Cases in the model can be sorted and updated through putting the latest case in the forefront of the retrieval store when running the model (Izquierdo et al. 2004). The agents will retrieve the case store to find the most appropriate case that matches the properties and conditions in the simulation year. By doing this, the agents can effectively and quickly collect detailed information such as the economic and social benefits and land-use type to decide whether or not to choose this land-use type. Actually, as for the utilization of CBR to describe the agents' behaviors, the memory capacity of the agents is limited by the memory time and size. It is assumed that the agents can always remember several latest cases but not forever. And what is more important is that the excessive cases will decelerate the retrieval speed and require advanced hardware configurations.

The government agent relies on the participation message interface to pass parameters in the model, so it is also necessary to conduct quantitative and positioning analyses. In the ABM model, it is assumed that the government agent pays more attention to the ecological and environmental problems, and the government agent's behaviors were parameterized on the basis of the ecological compensation policies and land planning. With regard to the ecological compensation policies, a specific analysis was conducted with the GFG policy as an example. The GFG policy insists that the government should offer some compensation for the conversion from cultivated land to forestry area or grassland, and there is no restriction on the total amount of compensation fee to practitioners, given that the amount of compensation fee is just related to the total land conversion area. Besides, the nature reserves and future land planning will be considered, all of which are the bases to support the parameterization of the government agent's behaviors.

Key Factors of Agents' Behaviors

Table 4.1 shows the key influencing factors of the agents' behaviors, such as socioeconomic background, bad events, and human awareness to society and

Table 4.1 The factors that impact the agents' behaviors

Categories	Variables	Impacts on behaviors
Climate	Temperature	Different climatic conditions determine the agents' land-use practices; e.g., mild climate is suitable for cropping, pasturing, etc.
	Precipitation	
Economy	Initial account	The amount of the agents' account impacts their land-use practice, such as investment, land bid, and cost control
	Nonagricultural income	
	Price	
Bad events	Natural disasters	Weight of balancing the economy and nature
Policy	GFG	The compensation can stimulate the agents to transform other land uses into grassland and forest land
	Nature reserve	Ban the agents' land-use practices in the core area
Topography	Distance to water	The conversion area of grassland to cultivated land decreases with the increment of distance to water and slope
	Slope	
Soil property	Soil nitrogen	Agents are willing to keep the fertile land parcel as grassland
	Soil phosphorous	
	Soil potassium	
Land use	Land use	The initial land use determines the results of next step The land-use type and area determine the probabilities of land-use change

natural conditions. Since these factors can trigger the agents' behaviors and destabilize the magnitude of activities, different parameters of these factors were used to specify the land-use decisions of agents.

The total economic account of agents, which comprises of initial account, agricultural income, nonagricultural income, and government compensation, can influence the agents' behaviors through the bid amount of land transformation and market prices of agricultural and animal husbandry products. The agents who own more wealth (higher purchasing power) are more willing to buy land parcels in the land transformation step (higher bid amount). The statistical data indicated that there is a difference in the initial accounts of agents, and the nonagricultural income of the farmer agents is higher than other agents. Besides, the agricultural income depends on the production and market prices; therefore, it is assumed that the agents are willing to change the cultivated land into grassland if the prices of animal husbandry products are higher than those of the agricultural products. In addition, there is also difference in the land-use properties of different agents, which will further lead to the difference in their behaviors. Generally, the land use of last simulation step determines the result of the next simulation step. For example, if one agent chooses the satisfactory strategy according to the land-use algorithm, the land use of last year will be selected for the targeted land parcels. Moreover, small land parcels are relatively easier to be converted, and therefore, their land conversion probabilities would be higher than plots within larger areas, and the agents' wealth and the area of each land-use type are not the same; different land-use types with different areas require diverse investments and cost. Therefore, the configuration of relevant parameters is different among these agents and land-use types. What is more, there are still some other factors that may influence the agents' decisions through destabilizing their balance analysis between two-dimensional benefits of economic returns and social acceptance, such as the initial human awareness and natural disasters. For example, agents who have large population and high economic revenue and lack the needed ecosystem services may pay more attention to the social benefits and ecosystem services rather than the economic return. The agents who suffered from natural disasters may also reanalyze the two-dimensional benefits.

Parameterization and Validation

The parameters of the ABM model mainly include the model parameters and agents' parameters. The model parameters include the topology, neighbor radius, and cell numbers and sizes, all of which are the same for all the kinds of agents. These model parameters were mainly set on the basis of the statistical data at the county level, while the agents' parameters were primarily set according to the results of county classification and household survey data. In addition, the time delay of land use was taken into account in this study. There is some time delay in the impacts of the social agents' adaptations and government policies on the land-use change. The agents cannot immediately change the land-use practices of targeted parcels even if the income loss appears; it will take them some time to make sure whether they will change the land-use practices or not. Therefore, the parameter of the model was adjusted on the basis of the time delay.

The land-use structure and pattern in 1995 and 2005 were first simulated on the basis of land-use data of 1988 and then compared with the actual land-use data so as to validate and calibrate the simulation model. The validation methods include the point to point comparison, Kappa index, Moran's I index, and ROC curve. The results of all the tests showed that the ABM model has a good simulation capacity, but the result of the ROC test indicates that there are also some potential uncertainties. The ROC curve showed that the simulation results of cultivated land and built-up land were not as good as those of other land-use types, which may be due to the rough simulation resolution. The ABM model was finally calibrated through changing the sensitive parameters on the basis of the trials and errors.

Results and Discussion

This study aims to explore the impacts of climate change on the land-use change and the agents' economic benefits in the TRHR. The land-use change was simulated under two scenarios, one with climate change and the other without the climate change.

Land-Use Change

Most land-use types except the cultivated land will change in a similar trend under both of the two scenarios. The simulation result suggests that the grassland and the built-up land will show an increasing trend, while the forest land and unused land will tend to shrink in the simulation period. On the whole, the area of grassland and unused land will change most greatly. More specifically, there is obvious change in the grassland area, with an increment rate of approximately 5.1 % under the scenario with climate change and 6.0 % under the scenario without climate change. On the contrary, the areas of forest land and unused land will decrease during the same period. The area of forest land will decrease by 23.4 and 33.7 % under the scenario with and without climate change, respectively, while that of the unused land will be 14.6 and 8.7 %, respectively (Fig. 4.6). Therefore, climate change will lead to the decrease of cultivated land and unused land and the increase of forest land and grassland by comparing the results of these two scenarios on the whole.

The area of the cultivated land, water body, and built-up area will fluctuate strongly under the two scenarios. The cultivated land area will decrease to some degree under the scenario with climate change and show an opposite trend under the other scenario. However, there are similar changing trends of the area of the water body and built-up land under these two scenarios but with more wide fluctuation

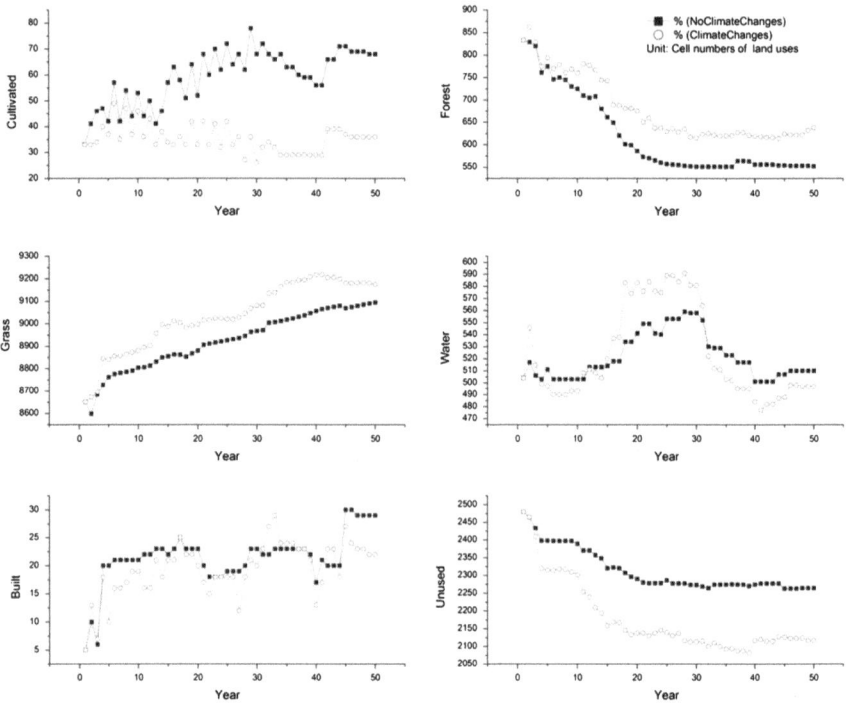

Fig. 4.6 Simulation results of the land uses under the scenario with and without climate change. The picture shows the cell numbers of each land use in every simulation step

range under the scenario with climate change. This may be due to the rough simulation resolution. The land parcel was set as the basic unit for land-use decision, which means that the agent will change the land-use practices of all the land cells in the targeted land parcel. However, there were initially only 5 built-up land cells in the 2-km resolution data, and the smallest land parcel in the study area consists of 9 grid cells; as a result, the resolution of the data leads to some errors in the simulation result. Besides, another reason may be the probability of climate change and the parameterization of the agents' behaviors, which account for the fluctuation of land-use types in the large area. The probability of climate change will influence the output of land use, which will further influence the agents' income, and therefore, the agents will change their land-use practices to adapt to the climate change. In addition, there may be amplified impacts of climate change on the land-use change on the condition that the parameters of the agents' behaviors were not set reasonably.

Economic Development

The change of the economic return was also simulated with the ABM model under different climate change scenarios. The economic return of agents mainly comes from three sources, that is, the agricultural income from the land parcel harvest, nonagricultural income, and government subsidies. The simulation results show that climate change will lead to a stable growth of the agents' wealth in the long term, and the agents' income will be higher under the scenario without climate change, showing an exponentially increasing trend.

The herdsman's annual income is more than that of the other two agents since they own more grassland, from which the major income of the animal husbandry comes (Fig. 4.7). Besides, the annual income of all agents showed a more optimistic changing trend under the scenario without climate change. The change of the annual income of the farmer agents is similar to that of the hominid agent, showing an exponentially increasing trend. The annual income of the farmer agent will show a growing trend during the initial 25 years, followed by a declining trend under the scenario with climate change. By contrast, the annual income of the hominid agent will be relatively more stable. In general, the total income of agents under the scenario without climate change is lower than that under the scenario with climate change (Fig. 4.8). However, the total income of the farmer agent and hominid agents will show an exponentially increasing trend under the scenario without climate change and a steadily growing trend under the scenario with climate change. In summary, the agents will get more wealth under the scenario without climate change in the long run.

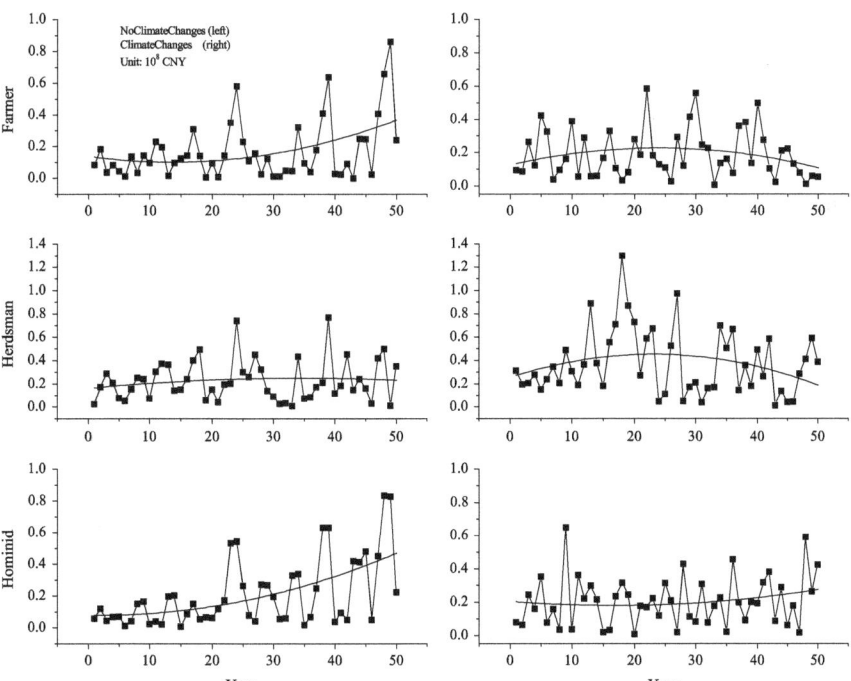

Fig. 4.7 Simulation result of annual income change of agents under the scenario with (*right panels*) and without climate change (*left panels*)

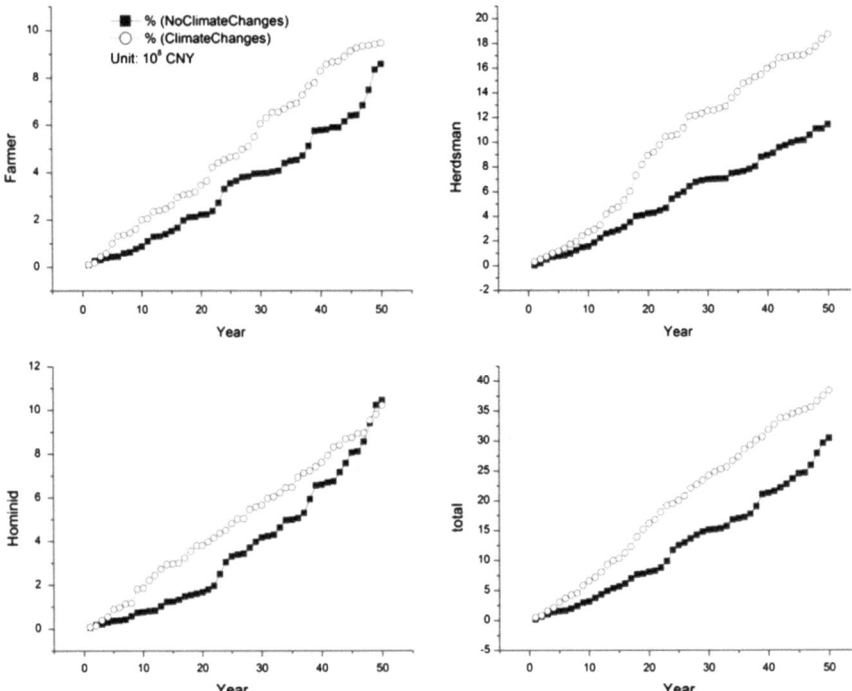

Fig. 4.8 Simulation result of the change of total income of agents under different climate change scenarios

Sensitivity Analysis

In order to analyze sensitivity of the cell numbers of cultivated land and grassland, total income of agents, and yield of land parcels to the driving factors listed in Table 4.2, the relative sensitivity was analyzed with the following formula:

Table 4.2 Results of the sensitivity analysis. Unit: %

Variables	Cell numbers of cultivated land	Cell numbers of grassland	Total income	Yield
Temperature	1.05	−0.26	−0.16	1.31
Precipitation	−0.66	1.30	1.93	2.09
Price of products of agriculture	1.19	0.16	0.21	0.95
Price of products of animal Husbandry	−0.77	3.62	8.31	1.95
Government subsidies	−1.75	2.97	2.07	0.25
Cost control	−1.97	2.02	2.00	1.20

$$S = \frac{(Y(X + \Delta X) - Y(X))/Y(X)}{\Delta X/X} \tag{1}$$

where $Y(X + \Delta X)$ is the simulated result after the change of inputs, $Y(X)$ is the background result, and ΔX is the change of input variables. The relative sensitivity represents the change of the result after the change of only one variable. The greater the change of the results is, the more sensitive to the input variables the results are. This study mainly focused on the sensitivity to the climate change, market price, government policies, and cost control.

The result of sensitivity analysis indicated that the grid numbers of the cultivated land and grassland have a higher sensitivity to the prices of products of the agriculture and animal husbandry, government subsidies, and cost control (Table 4.2). Besides, the total income is more sensitive to the prices of products of the agriculture and animal husbandry, cost control, government subsidies, and precipitation. What is more, yield has a higher sensitivity to the cost control, precipitation, the prices of products of the agriculture and animal husbandry, and temperature. In summary, the prices of products of agriculture and animal husbandry, precipitation, subsidies from government for land-use change, and cost control have significant influence on the land-use change and economic return in the study area.

Summary

Climate change can alter the land surface properties through influencing the human land-use practices, and the ABM model is a good tool to study these climatic effects. The TRHR was selected as the study area, where it is quite sensitive to climate change but has the characteristics of moderate land-use change and agents that are difficult to identify. The agents are identified and extracted and parameterized through analyzing the survey data and county-level statistical data. Then, the ABM model was established and used to simulate the dynamic land-use change under different climate change scenarios. In addition, CBR as an approach that is similar to the agents' thinking way in real life was used to parameterize the agents' behaviors in this study.

The results showed that the future trends of land-use change in the study area will be mainly characterized by the increase of grassland and decrease of unused land. A set of factors in terms of climate change, agents' decisions, and government policies will affect the dynamic process of regional land surface. Besides, the result indicated that the model is quite sensitive to the climate change, prices of agricultural and animal husbandry products, government subsidies, and cost control. Therefore, more importance should be attached to market price adjustment, stimulating policies implementation, and emerging technologies promotion, to lead reasonable land-use practices so as to promote the reasonable land-use practices, maintain the integrality and sustainability of ecosystems, and guarantee the provision of ecosystem services in the long term.

In this study, only climate change and agents' decisions on land use were taken into account. However, the household survey indicates that the agents' behaviors may be greatly affected by the governmental subsidies and the market prices that influence their economic income, which can be also illustrated with the sensitivity analysis. Therefore, more importance should be attached to the socioeconomic state when analyzing its impacts on the land-use and terrestrial surface change in the future researches.

Predicted Changes in Grassland Productivity in China

Introduction

The grassland is one of the most important land-use types in China due to its essential functions in the development of the animal husbandry and the supply of ecosystem services. Grassland is a multifunctional ecosystem, which can provide important ecological benefits as well as economic benefits. Traditionally, grasslands provide a broad range of agronomic services, including the provision of forage, milk, meat, wool, and pelts. The grassland ecosystem of China plays an important role in promoting the regional economic development, protecting the ecological environment, and conserving the biodiversity (Kang et al. 2007). The grassland of China is about 400 million hectares, accounting for about 41.7 % of the national total land area. The grassland area of China is only second to that of Australia in the world. However, the grassland productivity level of China is much lower than that in other parts of the world. Besides, the extreme climate shifts and the increasing demand for meat due to the rapid population growth have put tremendous pressure on the grassland productivity. Some recent studies show that the grassland productivity is sensitive to the climate change, especially in the Inner Mongolian Plateau, Qinghai–Tibet Plateau, and northwest of China (Baldocchi 2011). The change of precipitation regimes has profound impacts on the grassland productivity, especially in the arid and semiarid regions characterized by the limited water (Jongen et al. 2011). In the temperate grasslands, the interannual variability of the total precipitation is the primary climatic factor that causes fluctuations in the grassland productivity (Bai et al. 2004; Klumpp et al. 2011). The combined effects of the land-use and climate change further complicate the underlying change of the grassland productivity (Chen et al. 2004; Xu and Li 2010). It is necessary to explore the coupling effects of the land-use management and climate change on the grassland productivity so as to carry out the adaptive management to mitigate the climate change (Niu et al. 2011).

The grassland productivity, integrating both the fodder quality and yield level, is an important index to represent the ability of the grasslands to meet the needs of the livestock sector and the pastoral societies. It is critical to the management and planning of the grassland resources to better understand the relationship between the grassland productivity and the possible climate change scenarios, especially with regard to the livestock development. The grassland productivity can directly reflect the production capacity of the grassland in the natural environment. Besides, the grassland productivity is influenced by various factors, including the internal factors of the grassland ecosystem such as the soil, grassland area, and livestock, and the external factors such as the labor input and climate change. The Intergovernmental Panel on Climate Change (IPCC) for the fifth assessment report updated the scenario development from the SRES to a new set of integrated scenarios, the RCPs, analyzed the advantages of the new approaches of scenarios development, and depicted the characteristics of the identified four RCPs (RCP8.5, RCP6, RCP4.5, and RCP3-PD). These schedules of radiative forcing, greenhouse gas emission, concentration, and land use in every road map were proposed based on the four models including MESSAGE, AIM, MiniCAM, and IMAGE. It is of great significance to understand the changes of grassland productivity along the RCPs during 2010–2050 in China.

The grassland is widely distributed and shows great regional difference in China. The grasslands in China can be divided into seven grassland ecological–economic regions according to the ecological conditions, moisture and temperature, grassland types, grassland production, and consistence between the grassland systems and economic systems. It is necessary to manage the grassland by ecological–economic region under the condition of land-use change and climate change. The main changing trend of climate change in each grassland ecological–economic region can provide a basis for taking management measures to relieve the pressure from the climate change. But almost all the previous literatures were focusing on how and how much climate change had affected grassland. For example, Yu et al. (2012) analyzed grassland activity by evaluating remotely sensed normalized difference vegetation index (NDVI) data collected at 15-day intervals between 1982 and 2006 and found that vegetation on the Tibetan Plateau is unable to exploit additional thermal resources availed by climate change. Ni (2003) found that the numbers of C_3 species, C_4 species, grasses, and forbs had positive relationship with precipitation and aridity. On a regional basis, the combined effect of precipitation and temperature, the aridity, is more significantly correlated with the distribution of C_3 species and forbs, which are more dominant than with C_4 species, grasses and succulents in the study area. Su et al. (2007) analyzed the changes of temperature, precipitation, and climatic productivity of grasslands in Ningxia farming–pasturing ecotone based on the meteorological data from 1954 to 2004 and found that grassland productivity will increase by 10–20 % if the annual temperature increases by 1–2 °C and the annual precipitation increases by 10–20 % in the future. It is of great reference value to the reasonable development of grassland resources and the research of terrestrial ecosystem carbon cycle to accurately estimate the grassland productivity in China.

Data and Materials

This study analyzed the changes of the climate and grassland productivity from 1980 to 2010 with the correlation analysis. A series of data were collected, including the grid data of NPP, soil nitrogen and phosphorus, DEM, and the percent of plain area at the county level involving 1669 counties, which is the level of the census data such as the population, GDP, and meat (Table 4.3). Then, a panel dataset was constructed with the data of six periods, including 1985, 1990, 1995, 2000, 2005, and 2010. This study analyzed the contribution of climate change to the change of the grassland productivity on the basis of the dataset at the national and regional ecological–economic scales. The final part discusses the impacts of the future climate change on the grassland productivity from 2010 to 2050 on the basis of the GCM simulation output. These analyzed data involved grassland area, grassland productivity, and future climate data.

Grassland Data

The grassland data were extracted from the land-use database developed by the Chinese Academy of Sciences (CAS). The data were originally derived from 512 remotely sensed images with a spatial resolution of 30 m × 30 m, which were provided by the US Landsat TM/ETM satellite. The data of five periods were used in this study, that is, the data of year 1988, 1995, 2000, 2005, and 2010. A hierarchical classification system including 25 land-use types was applied to the data. Besides, the data team also spent considerable time validating the interpretation of TM images and land-use classifications against the extensive field validation. The field validation indicated that the average interpretation accuracy reached 95 % (Liu et al. 2003). The grassland was divided into three kinds, that is, the dense grass, moderate grass, and sparse grass. The grassland of china is mainly

Table 4.3 Summary of the variables used in this study

Variable	County number	Max	Mean	Min	Description
NPP	1669	1386	439	1.07	Net primary productivity (gC/m^2)
Ta	1669	25.55	12.60	−4.81	Temperature (°C)
Rain	1669	2846	926.90	19.46	Precipitation (mm)
Sun	1669	975.97	762.64	475.66	Sunshine duration (h)
Soil_n	1669	0.96	0.25	0.05	Soil n (%)
Soil_p	1669	0.25	0.13	0.02	Soil p (%)
Dem	1669	2832	732	128	Elevation (m)
Splain	1669	0.42	0.28	0.01	Plain area (%)
Pop	1669	179	43	2	Population (ten thousand people)
Gdp1	1669	1.53	0.18	0.05	GDP of the primary industry (%)
Meat	1669	14,670	3850	650	Beef and mutton (ton)

distributed along the Great Wall in Inner Mongolia, and in the region of Gansu and Xinjiang, the Tibetan Plateau, and the Loess Plateau (Fig. 4.9). The grassland area from the land-use dataset is about 346 million hectares, which is smaller than that from the comprehensive and systematic grassland surveys due to the difference of the grassland classification system.

Grassland Productivity

The grassland productivity data were derived from the regional census statistics and remotely sensed images. The regional census statistics provide the aggregated data that are collected by county. However, the statistic data also have many disadvantages. For example, they are limited by the relatively coarse spatial resolution within the administrative units, gaps in time series, and difference of the data collection methods among counties. By contrast, the data derived from the remotely

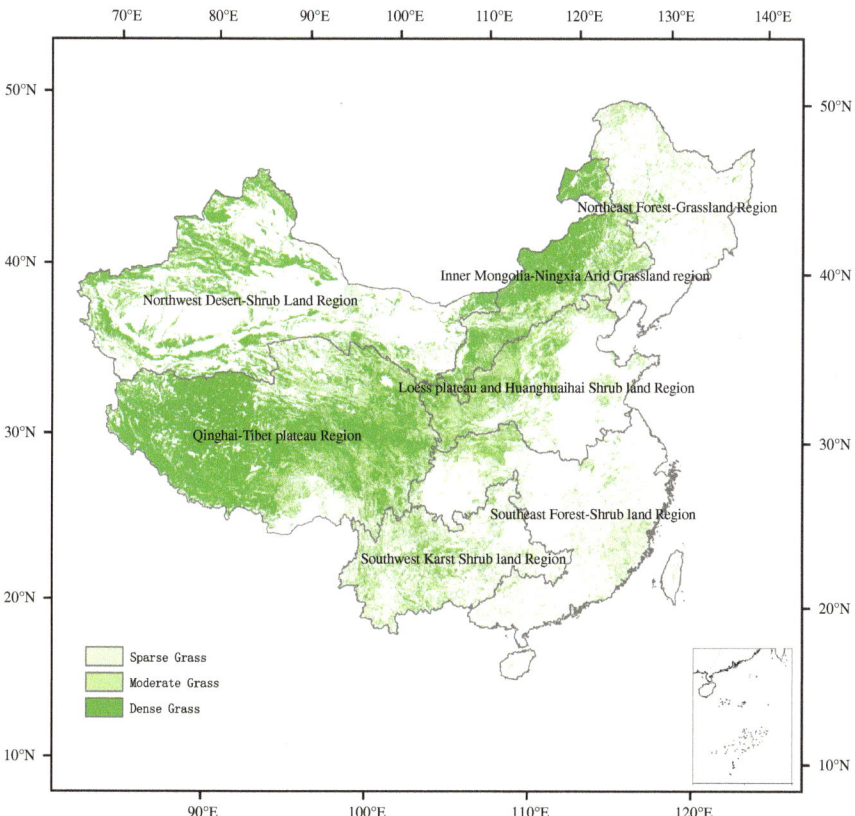

Fig. 4.9 The grassland distribution over grassland ecological–economic regions of China in year 2005

sensed images have a higher temporal continuity and higher spatial resolution. NPP is used as the indicator of the grassland productivity since it plays a key role in the vegetation growth and is closely related to the yield of the grassland. NPP was estimated with the efficiency model on the basis of the ground observation data of meteorological stations, soil quality data, and land-use data (Tucker et al. 2005). Besides, the vegetation greenness observation data from the AVHRR and the light-use efficiency model of the MODIS were used in the estimation of NPP. A simple light-use efficiency model (MOD17) is at the core of the algorithm of NPP, and it requires daily inputs of incoming photosynthetic active radiation (PAR), minimum temperature over the period of 24 h, and average vapor pressure deficit in the daytime. The NPP data during 1985–1999 came from the remote sensing data of NOAA/AVHRR and that during 2000–2010 came from the NPP product of MODIS in the study.

Meteorological Data and Future Climate Change Scenario

The projected climate data from 2006 to 2050, including the average daily temperature, precipitation, absolute minimum temperature, and daily maximum and minimum temperatures, come from the datasets of CCSM simulation. The Cancún agreements state that the future global temperature rise should be limited to below 2.0 °C (3.6 F) in comparison with the preindustrial level in 2100. The low concentration pathway of RCP3-PD has attracted the attention of international research community among RCP8.5, RCP6, RCP4.5, and RCP3-PD since the temperature rise is approaching the 2.0 °C goal under this scenario (Moss et al. 2010). These scenarios describe a world in which great emphasis is put on local solutions to economic, social, and environmental sustainability. The data of the future climate change from 2011 to 2050 were obtained by the simulation with the CCSM model. The future climate data from 2011 to 2050 can be corrected through overlying the future climate change data and historical meteorological data from 2006 to 2010.

The historical meteorological data, including the annual temperature and annual precipitation from 1980 to 2010, were acquired from China Meteorological Bureau and were interpolated into the 1 km × 1 km grid data with the Kriging algorithm and modified with the DEM data. The data of the future climate come from the scenario data of the RCPs, which were adopted by the fifth assessment report of IPCC, including four scenarios, the radiative forcing of which ranges from 2.6 to 8.5 W/m^2. The data of the four RCPs scenarios are the product of an innovative collaboration among the assessment modelers, climate modelers, terrestrial ecosystem modelers, and emission inventory experts. These four RCPs scenarios provide the global dataset of land use and greenhouse gas emission for the period extending to 2100, with the spatial resolution of 0.5° × 0.5°.

Method

The characteristics of the spatial distribution and seasonal change of the grassland NPP are as follows. The annual total and average grassland NPP in China in year 2010 were 6.8×10^9 gC/year and 490 gC/m^2/year, respectively. There was significant regional heterogeneity of the grassland productivity. The grassland NPP in the east part was higher than that in the west part, and it was higher in the south part than in the north part. On the whole, it increases from northwest China toward Southeast China except Tibetan Plateau (Fig. 4.10). On the whole, the grassland NPP increases with the longitude and decreases with the latitude although there are some fluctuations. The characteristic of longitude zonation of grassland NPP is highly correlated with the longitudinal variability of the climate. The climate becomes more favorable for the vegetation growth as the latitude decreases. The grassland productivity of Southwest karst shrubland region is the highest among these regions, reaching about 1000 gC/m^2 in year 2010.

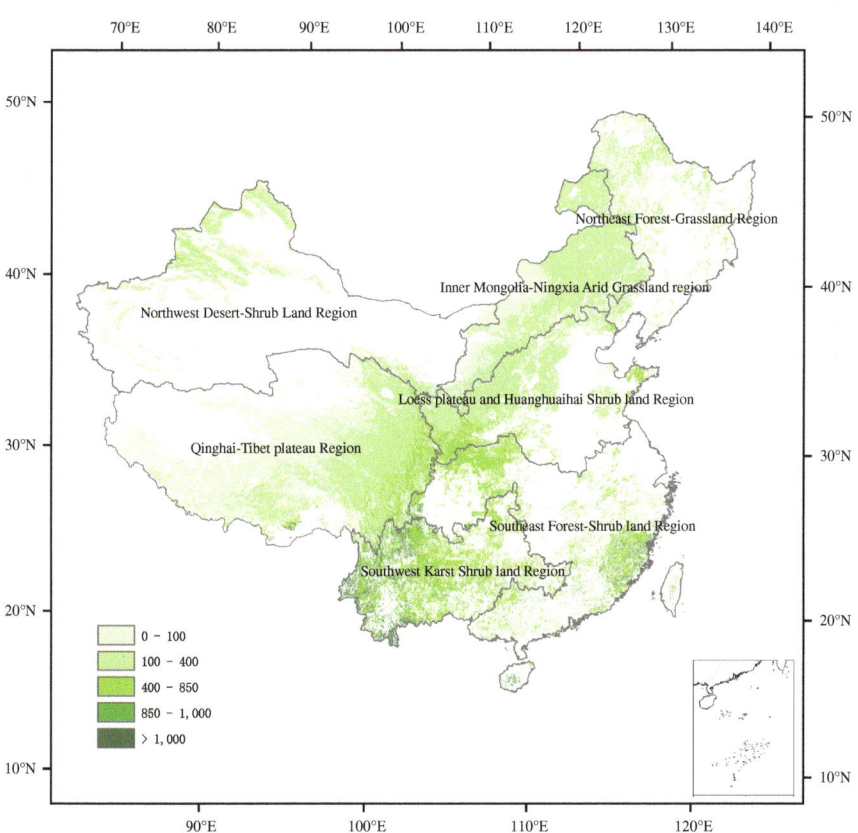

Fig. 4.10 The spatial distribution of grassland productivity over China in the year of 2010

In this study, we unified these data of different spatial scales and converted the data at the grid scale into the county scale so as to identify the reasons for the regional heterogeneity of the grassland productivity and analyze the impacts of geographical factors, climate, and human activities on the grassland productivity. It is scarcely possible to establish model in property of mechanism due to the complex relationships between grassland productivity and climate change as well as social and economic activities and geographical factors. In this condition, the econometric model becomes an alternative choice. This kind of model is increasingly applied to explore the relationships between productivity and its driving factors. In this study, we established an econometric model and analyzed the effects of variables at different levels on the grassland productivity.

Since the panel data are repeated observations of the same samples that are nested into the individual objects, we treated the repeated observations as the first level and the individual objects as the second level. Besides, the random factors over time were also introduced into the first level. A series of fixed effect models were established and then compared so as to choose a robust one as the basis model to more accurately analyze impacts of climate change on the grassland productivity:

$$\log(\text{npp})_{it} = \beta_0 + \sum_{p=1}^{P} \alpha_p X_{pit} + \sum_{q=1}^{Q} \lambda_q E_{qit} + \sum_{r=1}^{R} \gamma_r Z_{rit} + \mu_i + \varepsilon_{it}$$

where X_{pit} refers to the annual average temperature, precipitation, and sunshine duration; E_{qit} refers to available phosphorus in the soil, available nitrogen in the soil, and terrain elevation factor; Z_{rit} refers to the population, GDP, beef, and mutton; α_p, λ_q, and γ_r act as the parameters; μ_i represents the individual factor; $\varepsilon_{it} \sim N(0, \sigma_t^2)$ the composed error. The pool regression model has been chosen here since the fitting accuracy of this model is better than others (including first-difference approach and deviation from means). We analyzed the effects of variables at different levels on the grassland productivity by introducing the variables step by step (Table 4.4).

Results

Relationship Between NPP and Climate Change

As can be seen from the estimation results, the influence of the precipitation and temperature is more significant than that of the sunshine duration. Besides, the elevation has a significant negative impact on NPP, which reflects the character of vertical zones. In addition, the soil phosphorus also has significant positive effects on NPP. What is more, as for the socioeconomic factors, both the population and the demand for meat have significant positive effects on NPP, and more labor will be invested into the management and production of grassland with increase of population and demand for meat. The percentage of the primary industry in the

Table 4.4 Analysis of impacts of influencing factors on NPP based on the econometric model

Dependent variable: log (NPP)				
	Model-I	Model-II	Model-III	Model-IV
log (ta)	0.035***	0.026***	−0.022***	−0.035***
	(−11.54)	(−8.12)	(−8.03)	(−9.72)
log (rain)	0.967***	0.856***	0.928***	0.963***
	(−197.32)	(−67.74)	(−83.42)	(−58.75)
log (sun)		−0.829***	−0.016	0.169
		(−9.48)	(−0.19)	−1.72
log (dem)			−1.745***	−1.803***
			(−67.42)	(−54.68)
Soil_n			−0.119***	0.036
			(−4.09)	(−1.03)
Soil_p			7.817***	6.550***
			(−19.8)	(−14.05)
Splain			0.026***	0.025***
			(−4.79)	(−3.54)
pop_{t-1}				0.031***
				(−3.59)
$meat_{t-1}$				0.156***
				(−12.73)
$Gdp1_{t-1}$				−0.049***
				(−8.14)
Intercept	−2.127***	4.247***	12.42***	9.750***
	(−54.03)	(−6.31)	(−20.04)	(−13.06)
Adjusted R^2	0.732	0.733	0.818	0.838

t statistics in parentheses: $*p < 0.05$, $**p < 0.01$, $***p < 0.001$

total GDP has negative effects on NPP, which demonstrates that the production of primary industry would put more stress on the grassland.

The climatic factors included the temperature, precipitation, and sunshine duration in this study. The result indicated that the precipitation and temperature were closely correlated with NPP at the significant level of 0.1 %, with the coefficient of 0.96 and −0.04, respectively. According to the analysis of the changing trends of temperature and precipitation under the RCPs-PD scenarios, it would be concluded that most of the grassland regions will turn "warm," which will promote the accumulation of dry matter and consequently increase NPP.

As for the physical geographic factors, the increase of the altitude was negatively related with NPP. By contrast, the percent of plain area was positively correlated with NPP. The result showed that the temperature and the precipitation would decrease as the altitude increases, indicating the vertical zonality of vegetation. In addition, the result indicated that the higher levels of the soil nitrogen and phosphorus were more beneficial to the increase of NPP.

Among the regional socioeconomic factors, the population showed a significant correlation with NPP. On the basis of the constantly added variables, the coefficient of the population eventually stabilized at 0.03 (at the significant level of 0.1 %), suggesting that when population increases by 0.1 %, the NPP would drop by 0.03 %, which means that population growth will increase labor input and consequently lead to the increase of NPP of the grassland. In fact, the population growth will inevitably increase the input to grassland production and management. What is more, the increasing demand for food and meat may also put more attention on the existing grassland and promote the grassland productivity.

In this study, the regional economic indices, including the percentage of the primary industry in the total GDP, population, and the demand for meat, were used to analyze the effects of the regional economic development on the grassland NPP. The result suggests that the coefficient of GDP is -0.05, which shows that when GDP increases by 0.1 %, the pressure on NPP will increase by 0.05 %, indicating that the rapid growth of the primary industry can reduce the grassland productivity to a certain extent. In the regions where there is a very low level of social development, the production and living of the people mainly rely on the animal husbandry, and the socioeconomic development greatly depends on the grassland resource. The increase of the economic output will put more pressure on the grassland resource and may consequently lead to the decline of the grassland NPP.

There is an obvious difference in the characteristics of the different grassland ecological–economic regions in China, which greatly influences the grassland productivity. So the ecological–economic region is a suitable unit to analyze the relationship between the grassland productivity and climate. In this study, we analyzed the relationship between temperature, precipitation, and NPP at the regional scale with the scatter plots, fitted lines, and coefficient of determination R^2 (Fig. 4.11). The fitted lines were calculated by OLS in this study. The result indicated that NPP has good correlation with the precipitation in the Inner Mongolia–Ningxia arid grassland region. The precipitation plays a dominant role in affecting the grass quality in this region, indicating that the drought may greatly restrict the vegetation growth. Besides, in the Qinghai–Tibet Plateau region, there is a better correlation between temperature and NPP, while the correlation between rainfall and NPP is not very significant. The local climate is the plateau climate in this region, and the low average temperature is the basic factor that restricts the grass growth. In addition, the correlation between temperature, rainfall, and grassland NPP is comparatively better in the northwest desert–shrubland region. However, there is less significant correlation between temperature, precipitation, and NPP in other four grassland ecological–economic regions (Fig. 4.11), which may be due to the greater impacts of human activities on NPP in these regions.

In summary, according to the annual, monthly, and seasonal variations in grassland NPP, the suitable combination of water and thermal factors plays a key role in influencing the grassland productivity. The finding has significant implications for the vegetation succession in the grassland ecosystems with predicted changes in spatial–temporal patterns of precipitation under the influence of global climate change.

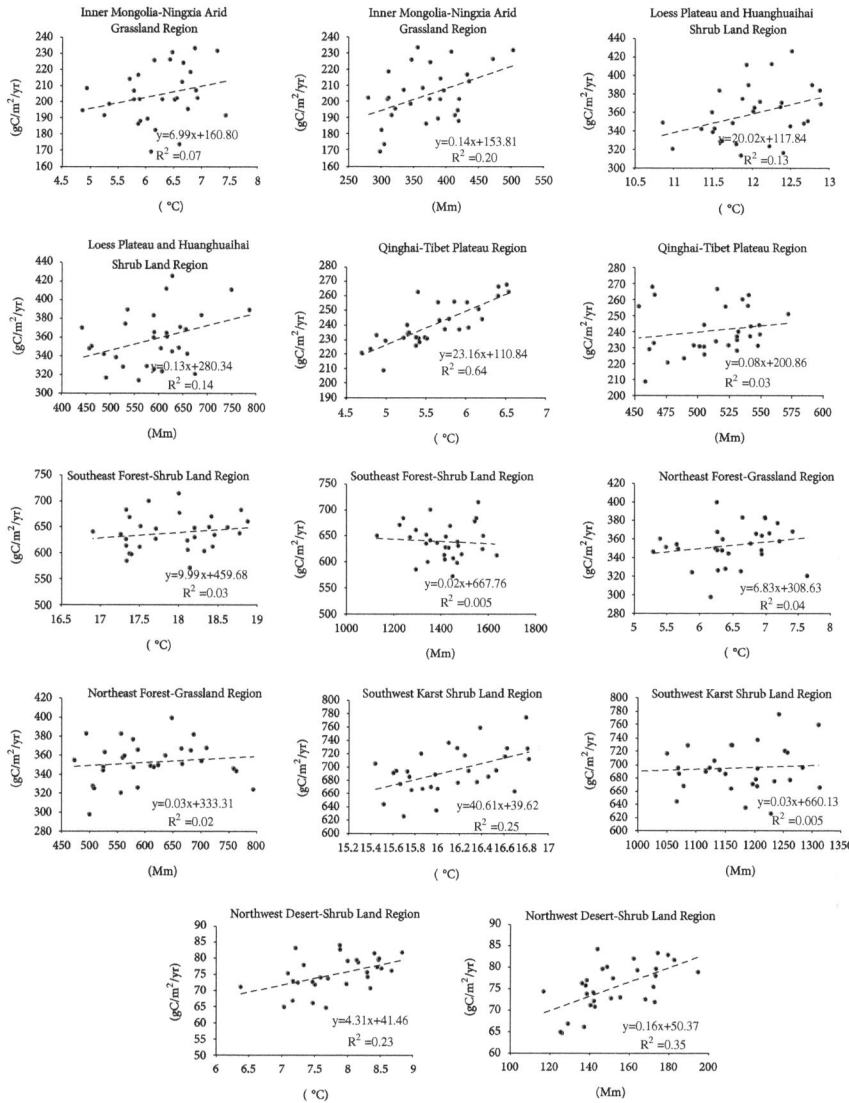

Fig. 4.11 Correlation analysis between the climate index and grassland productivity during 1985–2010 at the grassland ecological–economic zones level

Analysis of the Future Scenario

This study simulated the future land use under the four scenarios in China (Fig. 4.12). The forecasted future temperature rise in the whole world is controlled within 2 °C at present, so the result simulated with the MESSAGE model was selected as the first choice for this study. The simulation result using the MESSAGE model indicated that the fluctuation range of the grassland area is

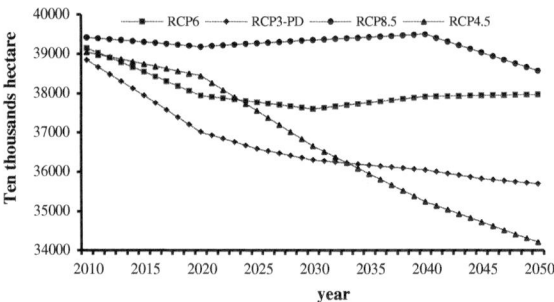

Fig. 4.12 Changing trends of the projected grassland area under the RCPs scenarios

very limited, reaching no more than 3.9 million square kilometers. The simulation results show that there would be a downward trend of the grassland area and rising of temperature at 3.2–3.8 °C in the future, but the area would be around 360 million hectares after year 2025.

With the improvement of the income level rise and more attention paid to the dietary nutrition balance, China's beef import and export trade and consumption will increase gradually. According to the trends of China's beef consumption, it can be deduced that with the improvement of income level, the consumption of beef will continue to grow and people's living standards are bound to rise to a new level. The forecast results indicated that the demand for beef will reach 9.8 million tons in 2020 and increase to 15.38 million tons in 2050. The increase of beef consumption would put more pressure on the grassland productivity. The result indicates that the annual average temperature of Qinghai–Tibet Plateau and northwest part of China will increase by 1.3 °C in the future 40 years. Besides, the annual precipitation will show a slight increasing trend, with an increment of 100 mm in the future 40 years (Fig. 4.13). In addition, the simulation result indicates that the grassland ecological–economic regions in China will become warmer and wetter in the future, which is favorable to the increase of grassland productivity. According to the simulation with the CCSM model, the change of the temperature and precipitation will make the grassland productivity in China increase by 8.3–16.7 % under the RCPs-PD scenarios. However, the grassland productivity would reduce greatly due to the decrease of moisture and the increase of the temperature and evaporation in the arid and semiarid regions of China during 2010–2050 under the RCPs scenarios.

Summary

This study analyzed the correlation between the grassland productivity and climate change on the basis of the panel data at the national levels and assessed

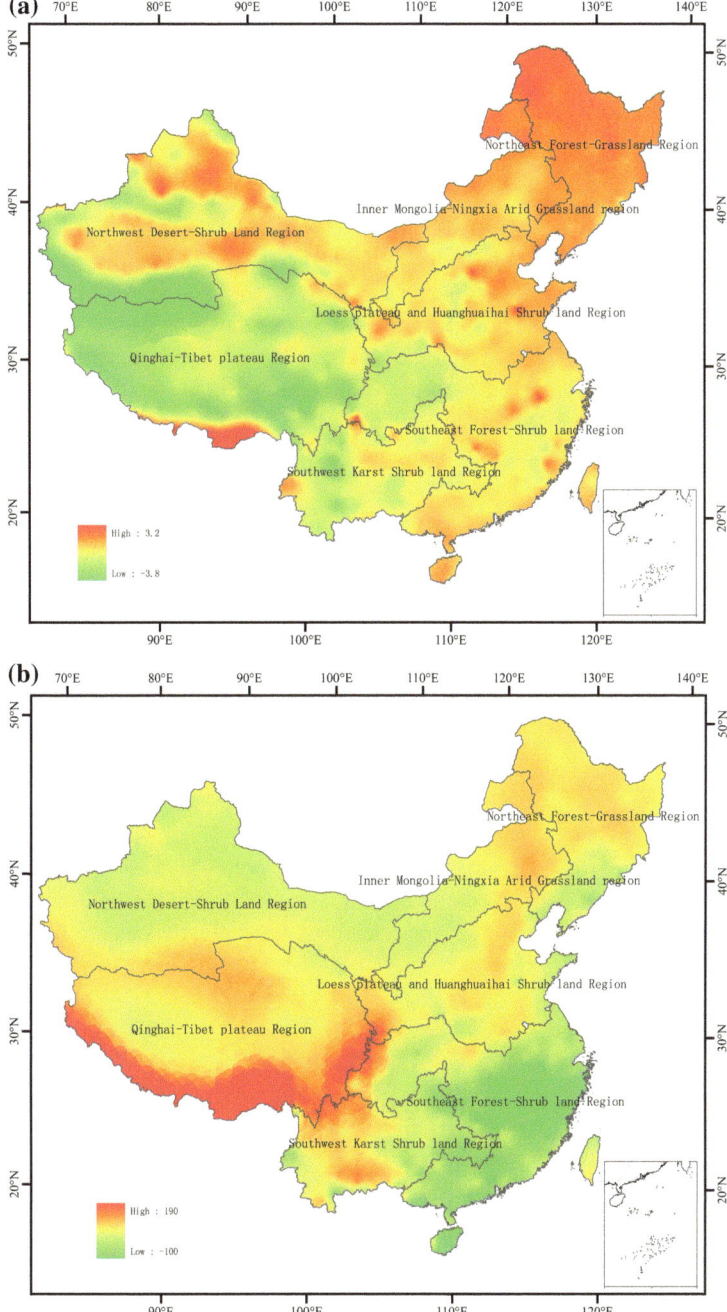

Fig. 4.13 The change of the temperature and precipitation using CMIP5 GCM from 2010 to 2050. **a** Temperature, **b** precipitation

the contribution of climate to the change of grassland productivity at the regional level. The research also employs the climate change scenarios to analyze the pressure on the grassland productivity in the future. Despite some limitations due to the quality and spatial resolution of the data, the data are still helpful in assessing the impacts of climate change on the grassland productivity. The data indicate that grassland is an important land-use type in China, accounting for approximately 40 % of the total land area of China. There is significant spatial variability of the grassland productivity in different grassland ecological–economic regions. The result of the correlation analysis shows that the precipitation is the key limiting factor of the grassland productivity in the Inner Mongolia–Ningxia arid grassland region and northwest desert–shrubland region. At the same time, we found the temperature is beneficial to the increase of grassland productivity in Qinghai–Tibet Plateau. The simulation result with the CCSM model indicates that the change of the temperature and precipitation will make the grassland productivity in China increase by 8.3–16.7 % during 2010–2050 under the RCPs-PD scenarios. The experiments in this study have been done using the GCM production, and more effects should be made in the future research to further reveal the relationship between the dynamical downscaled climate data and grassland productivity.

The sustainability of the grassland ecosystems is influenced by various factors such as the increasing demand of human for meat, increased cropland area, and increasing land-use intensity of the agricultural system. The spatial heterogeneity of the grassland productivity is well consistent with that of the stockbreeding productivity, indicating the importance of the grassland to stockbreeding in the pastoral area of China. The grassland productivity in the pastoral area is considerably lower than that in the farming–pastoral ecotone and agricultural area where there is high-intensity management and input. As the pastoral grasslands are usually located in the environmentally fragile regions where it is unsuited for intensive exploitation, the farmers, pastoralists, ecologists, policy makers, and economists all should look for a way to combine the grassland production and the grassland protection under the condition of land-use change and climate change.

References

Bai Y, Han X, Wu J, Chen Z, Li L (2004) Ecosystem stability and compensatory effects in the Inner Mongolia grassland. Nature 431(7005):181–184

Baldocchi D (2011) Global change: the grass response. Nature 476(7359):160–161

Briner S, Elkin C, Huber R, Grêt-Regamey A (2012) Assessing the impacts of economic and climate changes on land-use in mountain regions: a spatial dynamic modeling approach. Agric Ecosyst Environ 149:50–63

Cao M, Prince SD, Li K, Tao BO, Small J, Shao X (2003) Response of terrestrial carbon uptake to climate interannual variability in China. Glob Change Biol 9(4):536–546

Chen J, Ustin SL, Suchanek TH, Bond BJ, Brosofske KD, Falk M (2004) Net ecosystem exchanges of carbon, water, and energy in young and old-growth Douglas-fir forests. Ecosystems 7(5):534–544

Deng X, Zhao C, Yan H (2013) Systematic modeling of impacts of land use and land cover changes on regional climate: a review. Adv Meteorol 2013:11

Drinkwater KF, Beaugrand G, Kaeriyama M, Kim S, Ottersen G, Perry RI, Pörtner H-O, Polovina JJ, Takasuka A (2010) On the processes linking climate to ecosystem changes. J Mar Syst 79(3–4):374–388

Fleischer A, Sternberg M (2006) The economic impact of global climate change on Mediterranean rangeland ecosystems: a space-for-time approach. Ecol Econ 59(3):287–295

González MH, Cariaga ML, Skansi MDLM (2012). Some factors that influence seasonal precipitation in Argentinean Chaco. Adv Meteorol 2012:13

Izquierdo L, Gotts NM, GPJ (2004) Case-based reasoning, social dilemmas, and a new equilibrium concept. J Artif Soc Soc Simul 7(3):1

Jentsch A, Beierkuhnlein C (2008) Research frontiers in climate change: effects of extreme meteorological events on ecosystems. CR Geosci 340(9–10):621–628

Jongen M, Pereira JS, Aires LMI, Pio CA (2011) The effects of drought and timing of precipitation on the inter-annual variation in ecosystem-atmosphere exchange in a Mediterranean grassland. Agric For Meteorol 151(5):595–606

Kang L, Han X, Zhang Z, Sun OJ (2007) Grassland ecosystems in China: review of current knowledge and research advancement. Philos Trans R Soc Lond B Biol Sci 362(1482):997–1008

Klumpp K, Tallec T, Guix N, Soussana J-F (2011) Long-term impacts of agricultural practices and climatic variability on carbon storage in a permanent pasture. Glob Change Biol 17(12):3534–3545

Liu J, Chen JM, Cihlar J, Chen W (1999) Net primary productivity distribution in the BOREAS region from a process model using satellite and surface data. J Geophys Res Atmos 104(D22):27735–27754

Liu J, Liu M, Zhuang D, Zhang Z, Deng X (2003) Study on spatial pattern of land-use change in China during 1995–2000. Sci China, Ser D Earth Sci 46(4):373–384

Moss RH, Edmonds JA, Hibbard KA, Manning MR, Rose SK, van Vuuren DP, Carter TR, Emori S, Kainuma M, Kram T, Meehl GA, Mitchell JFB, Nakicenovic N, Riahi K, Smith SJ, Stouffer RJ, Thomson AM, Weyant JP, Wilbanks TJ (2010) The next generation of scenarios for climate change research and assessment. Nature 463(7282):747–756

Ni J (2003) Plant functional types and climate along a precipitation gradient in temperate grasslands, north-east China and south-east Mongolia. J Arid Environ 53(4):501–516

Niu S, Luo Y, Fei S, Montagnani L, Bohrer GIL, Janssens IA, Gielen B, Rambal S, Moors E, Matteucci G (2011) Seasonal hysteresis of net ecosystem exchange in response to temperature change: patterns and causes. Glob Change Biol 17(10):3102–3114

Piao S, Fang J, He J (2006) Variations in vegetation net primary production in the Qinghai-Xizang Plateau, China, from 1982 to 1999. Clim Change 74(1–3):253–267

Polhill JG, Sutherland L-A, Gotts NM (2010) Using qualitative evidence to enhance an agent-based modelling system for studying land use change. J Artif Soc Soc Simul 13(2):10

Rounsevell MDA, Reay DS (2009) Land use and climate change in the UK. Land Use Policy 26(Suppl 1(0)):S160–S169

Rustad LE (2008) The response of terrestrial ecosystems to global climate change: Towards an integrated approach. Sci Total Environ 404(2–3):222–235

Su Z-S, Cheng X-G, Huang F, Yang S-P (2007) Response of grassland productivity to climate change in farming-pasturing interlaced area of Ningxia. J Desert Res 27(3):430–435

Tucker CJ, Pinzon JE, Brown ME, Slayback DA, Pak EW, Mahoney R, Vermote EF, Saleous NE (2005) An extended AVHRR 8-km NDVI dataset compatible with MODIS and SPOT vegetation NDVI data. Int J Remote Sens 26(20):4485–4498

Wu F, Deng X, Yin F, Yuan Y (2013) Projected changes of grassland productivity along the representative concentration pathways during 2010–2050 in China. Adv Meteorol 2013:9

Xu X, Li K (2010) Biomass carbon sequestration by planted forests in China. Chin Geograph Sci 20(4):289–297

Xu XL, Liu JY, Shao QQ, Fan JW (2008) The dynamic changes of ecosystem spatial pattern and structure in the Three-River Headwaters region in Qinghai Province during recent 30 years. Geograph Res 27(4):829–838

Yao Y, Yang J, Xiao G, Lu D (2011) Change feature of net primary productivity of natural vegetation and its impact factor in the source region of Yellow River in recent 50 years. Plateau Meteorol 30(6):1594–1603

Yu H, Xu J, Okuto E, Luedeling E (2012) Seasonal response of grasslands to climate change on the Tibetan Plateau. PLoS ONE 7(11):e49230

Yue T-X, Fan Z-M, Chen C-F, Sun X-F, Li B-L (2011) Surface modelling of global terrestrial ecosystems under three climate change scenarios. Ecol Model 222(14):2342–2361

Zhan J, Deng X, Yue T (2004) Landscape change detection in Yulin prefecture. J Geog Sci 14(1):47–55

Zhou G, Zhang X (1996) Study on NPP of natural vegetation in China under global climate change. Acta Phytoecologica Sinice 20(1):11–19

Chapter 5
Impact Assessments on Water and Heat Fluxes of Terrestrial Ecosystem Due to Land Use Change

Xiangzheng Deng, Jiyuan Liu, Enjun Ma, Li Jiang, Rui Yu, Qun'ou Jiang and Chunhong Zhao

Abstract Water and heat fluxes of terrestrial ecosystem play an important role in the sustainable development of ecosystem services. In this chapter, we first investigated the spatial variation of heat fluxes and surface temperature in an inland irrigation area of the northern China. Irrigated agriculture has the potential to alter regional to global climate significantly. We investigate how irrigation will affect regional climate in the future in an inland irrigation area of northern China, focusing on its effects on heat fluxes and near-surface temperature. Using the Weather Research and Forecasting (WRF) model, we compare simulations among three land cover scenarios: the control scenario (CON), the irrigation

X. Deng (✉) · J. Liu
Institute of Geographic Sciences and Natural Resources Research,
Chinese Academy of Sciences, Beijing 100101, China
e-mail: dengxz.ccap@igsnrr.ac.cn

X. Deng
Center for Chinese Agricultural Policy, Chinese Academy of Sciences,
Beijing 100101, China

E. Ma
School of Mathematics and Physics, China University of Geosciences (Wuhan),
Wuhan 430074, Hubei, China

L. Jiang
School of Economics, Renmin University of China,
59 Zhongguancun Street, Beijing 100872, China

R. Yu
School of Geography, University of Leeds, Leeds LS2 9JT, UK

Q. Jiang
School of Soil and Water Conservation, Beijing Forestry University, Beijing 100038, China

C. Zhao
Department of Geography, Texas State University, 601 University Drive,
San Marcos, TX 78666, USA

© Springer-Verlag Berlin Heidelberg 2015   149
J. Zhan (ed.), *Impacts of Land-use Change on Ecosystem Services*,
Springer Geography, DOI 10.1007/978-3-662-48008-3_5

scenario (IRR), and the irrigated cropland expansion scenario (ICE). Our results show that the surface energy budgets and temperature are sensitive to changes in the extent and spatial pattern of irrigated land. Conversion to irrigated agriculture at the contemporary scale leads to an increase in annual mean latent heat fluxes of 12.10 W m^{-2}, a decrease in annual mean sensible heat fluxes of 8.85 W m^{-2}, and a decrease in annual mean temperature of 1.3 °C across the study region. Further expansion of irrigated land increases annual mean latent heat fluxes by 18.08 W m^{-2}, decreases annual mean sensible heat fluxes by 12.31 W m^{-2}, and decreases annual mean temperature by 1.7 °C. Our simulated effects of irrigation show that changes in land use management such as irrigation can be an important component of climate change and need to be considered together with greenhouse forcing in climate change assessments. Then, the spatial variation of surface temperature and precipitation due to grassland conversion to forestry area in southeast China was examined. The land use/land cover change (LUCC) is the synthetic result of natural processes and human activities; it largely depends on the surface vegetation conditions, and the mutual conversion among land cover types can accelerate or alleviate the regional and global climate change. Aiming at analyzing the regional climatic effects of the conversion from grassland to forestland, especially in the long-term perspective, we carried out the comparison simulation using the WRF model in Fujian Province, and results indicated that this conversion had a significant influence on the regional climate; the annual average temperature decreased by 0.11 °C, and the annual average precipitation increased by 46 mm after 11.2 % of the grassland was converted into the forestland in the study area from 2000 to 2008. In the future (from 2010 to 2050), the conversion from grassland to forestland is significant under two representative concentration pathways (RCPs) (RCP6 and RCP8.5); the spatial pattern of this conversion under the two scenarios is simulated by dynamics of land system (DLS); then, the regional climate effects of the conversion are simulated using WRF model. Further, the spatial variation of surface heat fluxes due to land use change across China was estimated. We estimate the heat flux changes caused by the projected land transformation over the next 40 years across China to improve the understanding of the impacts of land dynamics on regional climate. We use the WRF model to investigate these impacts in four representative land transformation zones, where reclamation, overgrazing, afforestation, and urbanization dominate the LUCC in each zone, respectively. As indicated by the significant variance of albedo due to different LUCCs, different surface properties cause great spatial variance of the surface flux. From the simulation results, latent heat flux increases by 2 and 21 W/m^2 in the reclamation and afforestation regions, respectively. On the contrary, overgrazing and urban expansion result in decrease of latent heat flux by 5 and 36 W/m^2, correspondingly. Urban expansion leads to an average increase of 40 W/m^2 of sensible heat flux in the future 40 years, while reclamation, afforestation, and overgrazing result in the decrease of sensible heat flux. Results also show that reclamation and overgrazing lead to net radiation decrease by approximately 4 and 7 W/m^2, respectively; however, afforestation and urbanization lead to net radiation increase by 6 and 3 W/m^2, respectively. The simulated impacts of projected HLCCs on surface energy fluxes will inform sustainable land management and climate change mitigation. Finally, we summarized the predicted impacts of land use change on surface temperature in the typical areas around the world. This study focuses on the potential impacts of large-scale LUCC on surface temperature from a global perspective. As important types of LUCC, urbanization, deforestation, cultivated land reclamation, and grassland degradation have effects on the climate, the potential changes of the surface temperature caused by these four types of large-scale LUCC from 2010 to 2050 are downscaled, and this issue is analyzed worldwide along with RCPs of the Intergovernmental Panel on Climate Change (IPCC). The first case study presents some evidence of the effects of future urbanization on surface temperature in the Northeast megalopolis of the USA. In order to understand the potential climatological variability caused by future forest deforestation and vulnerability, we chose Brazilian Amazon region as the second case study. The third selected region in India is a typical region of cultivated land reclamation, where the possible climatic impacts are explored. In the fourth case study, we simulate the surface temperature changes caused by future grassland degradation in Mongolia. Results show that the temperature in built-up area would increase obviously throughout the four land types. In addition, the effects of all four large-scale LUCCs on monthly average temperature change would vary from month to month with obviously spatial heterogeneity.

Keywords Heat flux · Surface temperature · Precipitation · Land use change · Scenarios · WRF model · Irrigation area · Grassland conversion

Spatial Variation of Heat Fluxes and Surface Temperature in an Inland Irrigation Area of the Northern China

Introduction

Radiation and heat fluxes on the Earth's surface play important roles in regional to global circulations and may cause significant changes to the surface climate. Human activities have been widely recognized as a key contributor to climate change. Human activities modify radiation and heat fluxes at the surface by altering the composition of the atmosphere and by changing the land cover (Biggs et al. 2008; Sahu et al. 2012). Land cover changes can affect the regional energy balance and atmospheric processes through changes in land surface attributes including albedo, roughness length, and soil moisture. There is growing evidence that these impacts of land cover changes need to be considered when evaluating the effects of anthropogenic greenhouse warming and modeling future climate change (Defries et al. 2002).

Many studies have shown that changes in land use management including irrigation can be as important as changes in land cover in terms of their influences on the climate system (Sacks et al. 2009). It is reported that globally irrigated agriculture consumes 2600 km^3 of water each year, which comprises 70 % of all human water withdrawals (Shiklomanov 2000). This huge manipulation of water for irrigation would imply the regional significance of irrigation on near-surface energy exchange and hydrologic cycle. There are a number of pathways through which irrigation may affect local and regional climates. Irrigation reduces surface albedo and increases net surface radiation (Kueppers et al. 2007). Irrigation enhanced soil moisture, which leads to greater evapotranspiration and land surface cooling through the repartitioning between sensible heat fluxes and latent heat fluxes (Puma and Cook 2010). Under given circumstances, increased atmospheric water vapor may enhance cloud cover, convection, and downstream precipitation. Irrigation creates thermal contrast between cool and wetter irrigated land and nearby hotter and drier non-irrigated land and thus can modify regional circulation patterns (Kueppers et al. 2007).

Despite the importance of irrigation for regional climate, there is limited understanding about the effects of irrigation on the heat fluxes and near-surface climate in the future, especially at the regional scale and for countries in the developing world. Previous studies have demonstrated the climatic effects and responses of irrigation using observational records (Lee et al. 2009) and climate modeling (Lobell et al. 2009). For example, based on historical near-surface temperature records, Mahmood et al. conducted pairwise comparisons of temperatures between irrigated and non-irrigated locations in Nebraska, USA. They found a decrease of 1 °C in mean maximum growing season temperature over irrigated areas (Mahmood et al. 2006). Using a regional climate model, Kueppers et al. (2007) investigated the climate effect of irrigated agriculture in California and they showed that a regional irrigation cooling effect existed and the magnitude of this effect had strong seasonal variability. Mainly due to data limitations, however, almost all of these studies have focused on the interaction between irrigation and past climate change. To our knowledge, none of these studies provide direct insights about how future climate will respond to irrigation. Understanding how irrigation amount, extent, and location will affect future climate is needed to formulate policies aimed at mitigating or adapting to climate change. In addition, to date, most studies have been devoted to irrigated areas in the USA or India where data on land use and climate are relatively ample (Lobell et al. 2009; Debele et al. 2010). There is one study that has examined the effects of irrigation on climate in Asia including China (Lee et al. 2011), but to our knowledge, no study focuses particularly on China or regions in China. More efforts are required to understand the climatic feedbacks of many regions in the developing world including China which continue to experience expansion of irrigated land.

As 80 % of the nation's food is produced on irrigated cropland, irrigated agriculture in China plays an important role in sustaining people's livelihood and the economy (Yang et al. 2003). In arid regions of northern and northwestern China where natural rainfall cannot fulfill agricultural water requirements, irrigation is used to improve soil moisture conditions and increase yields (Deng et al. 2006a). Situated in the middle reaches of the Yellow River basin in northern China, our study area has a long history of intensive farming and irrigation water use (Fu and Chen 2000). The loose texture and high erodibility of loess soil, sparse vegetation, and intensive agriculture combined have resulted in severe soil erosion and ecological deterioration over this region (Zhang et al. 2008). Given the reliance on irrigated agriculture and high ecological vulnerability, a better understanding of the land–atmosphere interactions and feedback mechanisms in this region is of both scientific and social importance.

This study examines the impacts of irrigated agriculture on the heat fluxes and near-surface temperature in the future using a next-generation mesoscale forecast model, the Weather Research and Forecasting (WRF) modeling system. One of the key components of the WRF system is a sophisticated land surface model (LSM), which allows an explicit calculation of the water and energy dynamics of the land surface. With the most recent information on land use in our study area and projected future atmospheric lateral boundary conditions, we intend to extend the existing findings about the climatic effects of irrigation to further investigate future changes in the energy budget and regional climate in response to irrigation.

Study Area

Our study area is situated in northern China, and it occupies an area of 476,100 km^2 (Fig. 5.1). The Yellow River, the river with the highest sediment concentration in the world, flows through the region. The region has a semiarid to semihumid continental climate, with extensive monsoonal influence. Winters are cold and dry, while summers are warm and in many areas hot. The average annual temperature ranges

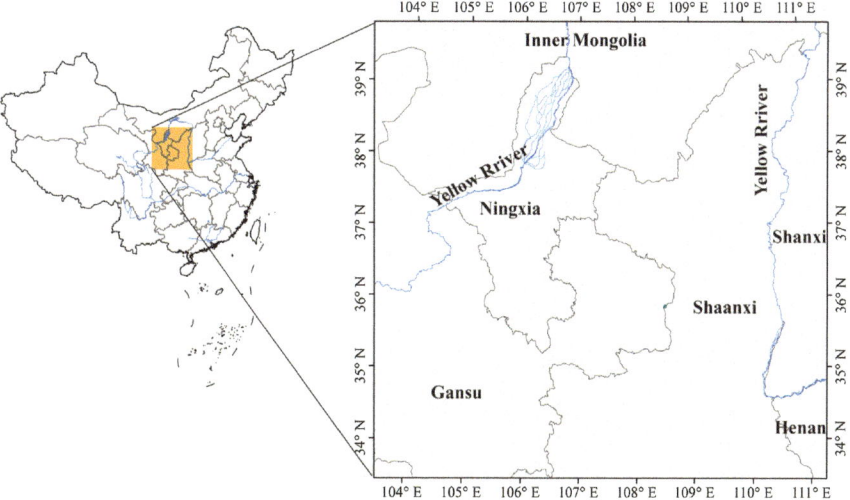

Fig. 5.1 Location map of the study area

from 4 to 12 °C. The average annual precipitation varies between 440 and 600 mm, and 70 % of the rainfall is concentrated during the summer in the form of storms (Fu and Chen 2000). Covered with highly erodible loess deposits, the region has become one of the regions with the world's highest soil erosion rates. More than 1.6 billion tons of sediment has been deposited to the Yellow River annually, posing a serious flood risk for downstream areas (Chen et al. 2001). The highly erodible soil, low vegetation cover, and centuries of unsustainable agricultural practices, combined with great population pressures, have led to severe environmental degradation. The main land use categories comprise cultivated land and sparse grassland. The typical cropping system over the area is double cropping with winter wheat rotated by maize or soybeans each year (Deng et al. 2006a). The cropping season of wheat occurs during the fall, winter, and spring (October–May), and the cropping season of maize (soybeans) occurs during the summer (June–September). The maize is sown immediately after the wheat is harvested. In order to increase grain production, farmers usually apply supplemental irrigation to their crops to overcome rainfall deficiency taking place throughout the year.

Data

Since understanding the interactions between irrigation and changes in the regional heat fluxes and temperature in the future is the main goal of the study, we require high-resolution, spatially explicit data on land use. We used a land use dataset that was derived from the National Aeronautics and Space Administration (NASA) Landsat Thematic Mapper and Enhanced Thematic Mapper (TM/ETM) satellite and analyzed by the Chinese Academy of Sciences (CAS) (Liu et al. 2003a). This national dataset, which has undergone extensive testing and development, contains the most recent information about the extent of different land uses in China for the year 2010. Original land use maps, which have been classified into 6 first-level land use categories and 25 second-level categories, exist at the scale of 1:100,000. The validation test indicates that the average interpretation accuracy for the land use maps is 95 %. The land use maps are further geocoded and stored as 1-km grid data using both the area percentage grid method and the greatest area method. For the greatest area method, if a cell contains more than one polygon and has more than one possible code, the code of the polygon with the greatest area in the cell is used. Land use data operated by the WRF system follow the US Geological Survey's (USGS) land cover classification scheme. Therefore, we implemented a spatial data mining technique proposed by Wu et al. (2013) to our study area and converted the land use data from the CAS to the USGS classification system. The USGS classification system, which consists of 24 categories, has been widely recognized and used in the simulation studies of climate change. Particularly, we produced the majority of the land cover categories including dryland and irrigated cropland in the USGS system by comparing the relevant subcategories between the CAS and USGS systems using the greatest area grid data. In this way, we generally retained the spatial distribution of land covers in the original high-resolution dataset. The new land cover dataset is a single-layer dataset with ~1 km spatial resolution, representing 15 land cover categories. Finally, to meet the input requirements of the WRF system, we aggregated the 1-km grid data of the USGS system to the 10-km grid data to facilitate the simulation of atmospheric responses to different land cover scenarios.

We constructed the atmospheric lateral boundary forcing conditions for the period of 2010–2020 using the projected output of future climate change based on the Representative Concentration Pathway 6 (RCP6) from the multimodel dataset of the Coupled Model Intercomparison Project Phase 5 (CMIP5). The RCPs provide four possible climate trajectories, all of which depend on how factors including radiative forcings, technology, economies, lifestyle, and policy will change in the future (Moss et al. 2010). The RCP6, which is developed by the Asia-Pacific Integrated Model (AIM) modeling team at the National Institute for Environmental Studies (NIES), Japan, is a stabilization scenario where total radiative forcing is stabilized after 2100 (Hijioka et al. 2008). The variables that we use to construct the atmospheric forcing conditions from 2010 to 2020 include air temperature, specific humidity, sea-level pressure, u-wind, v-wind, and geopotential height.

Methods

Land Cover Scenarios

Figure 5.2 displays the land cover map of our study area in 2010 represented by the new land cover dataset that is generated from the CAS national land use dataset. This land cover map shows that the contemporary landscape of our study area is characterized by a rich culture of agriculture and the mixture of different land covers. In 2010, the area of cultivated cropland, including dryland and irrigated cropland, takes up 37 % of the total landscape. More than 72 % of the cultivated cropland is irrigated, while the proportion of dryland is relatively small.

In order to identify the effects of irrigated agriculture on the energy fluxes and near-surface temperature, we designed three land cover datasets corresponding to three land cover scenarios (Fig. 5.3). The control scenario (CON) assumes that there is no irrigation at all. The irrigation scenario (IRR) assumes that irrigated agriculture maintains at the contemporary scale. The irrigated cropland expansion scenario (ICE) assumes an expansion of the irrigation to all cultivated cropland. For the IRR, we used the new land cover dataset derived from the CAS dataset. For the CON, we selected any grid cell in which the land cover was irrigated cropland in the new land cover dataset and replaced it with the cell of dryland. For the ICE, we selected any grid cell specified as dryland in the new land cover dataset and replaced it with the cell of irrigated cropland. In reality, we expect the potential expansion of irrigated cropland to be limited because we consider that further

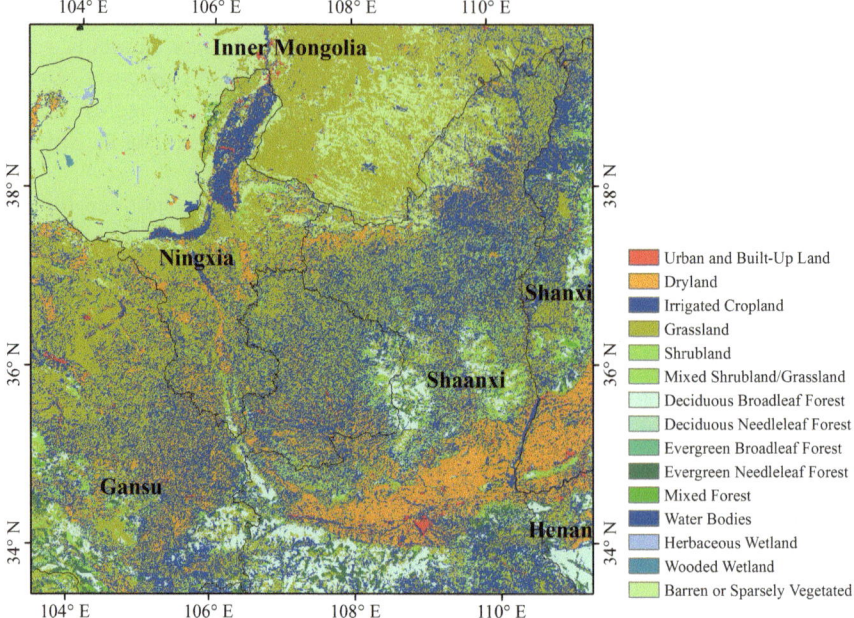

Fig. 5.2 Land cover map of our study area in 2010 represented by the new land cover dataset, with a spatial resolution of 1 km

expansion of irrigation would be largely constrained by the shortage of water resources in northern China. The land cover scenarios defined here are relatively extreme and are intended to help understand the nature and potential magnitude of climate effects of irrigated agriculture. However, based on the modeling results of these scenarios, we can derive useful insights about the actual climate effects of past and future expansion of irrigation over existing cropland.

Model Description

We conducted numerical simulations using the WRF modeling system version 2. The WRF model is a fully functioning, next-generation mesoscale forecast model and data assimilation that will advance both the understanding and prediction of mesoscale climate systems and promote the interactions between the atmospheric research and operational communities. It is a three-dimensional, non-hydrostatic, and primitive equation atmospheric modeling system. The WRF model version 2 uses the Eulerian mass coordinate and is referred to as the Advanced Research WRF (ARW) system (Kumar et al. 2008). Compared to the previous versions, changes and updates to the ARW include mixed-phase physics schemes suitable for cloud-resolving modeling, multiple-nesting capabilities that enhance resolution over the areas of interest, and enhanced model coupling Application Program

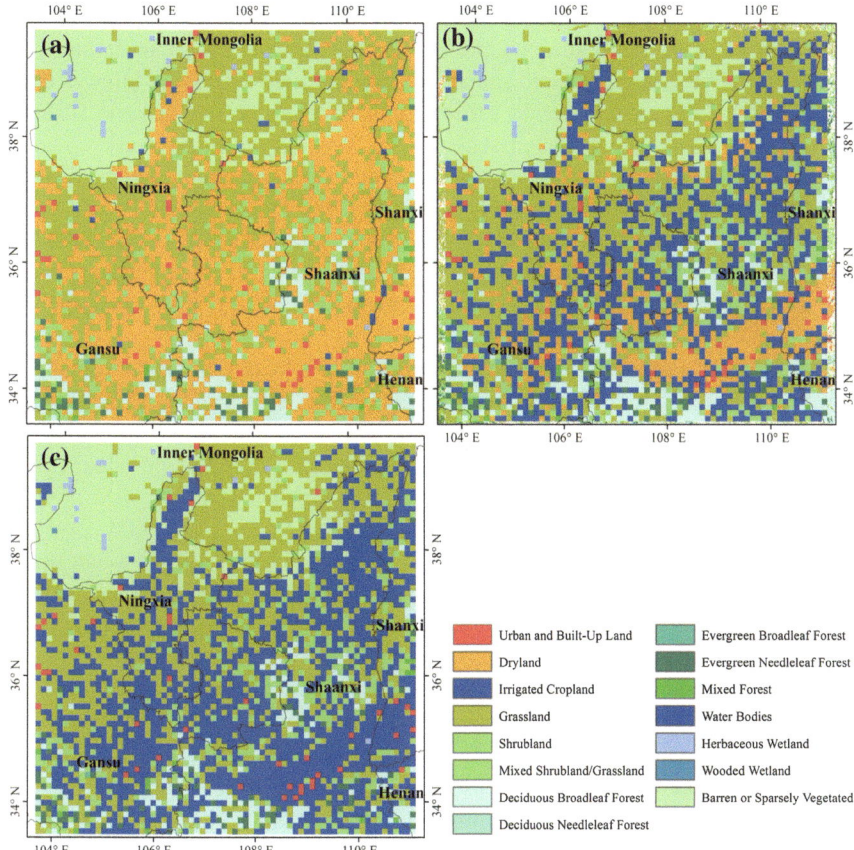

Fig. 5.3 Three land cover scenarios of our study area, with a spatial resolution of ~10 km: **a** control scenario (CON), **b** irrigation scenario (IRR), and **c** irrigated cropland expansion scenario (ICE)

Interface (API) enabling WRF to be coupled with other models such as ocean and land models. The WRF model has been utilized extensively for regional climate, operational numerical weather prediction, and hurricane and storm prediction (Kumar et al. 2008; Mölders 2008). Validation against observational records indicates its high performance in simulating spatial and temporal climatic features.

For all simulation runs, we used a domain centered at 36.7°N, 107°W, with a horizontal resolution of 10 km and with 31 vertical layers. The soil column has four soil layers, with the bottom layer extending to a maximum depth of 3.5 m. In the WRF model, we simulated water vapor and cloud processes using the Single-Moment 3-class (WSM3) scheme (Hong et al. 2004). We used the Grell–Devenyi ensemble cumulus parameterization scheme to account for the subgrid-scale effects of convective and shallow clouds. The surface layer physics followed similarity theory. We parameterized atmospheric boundary layer processes using the

Yonsei University (YSU) planetary boundary layer, a non-local K scheme with an explicit treatment of the entrainment layer and parabolic K profile in the unstable mixed layer. We chose the rapid radiative transfer model (RRTM) (Mlawer et al. 1997) as the longwave radiation scheme and MM5 shortwave scheme (Dudhia 1989) as the shortwave radiation scheme. We selected the Smagorinsky scheme for horizontal diffusion and Mellor and Yamada (1982) scheme for vertical diffusion. We used the Noah LSM (Chen and Dudhia 2001) to model land surface processes. The Noah LSM is a four-layer soil temperature and moisture model which calculates thermal and moisture stocks and fluxes at the land–atmosphere interface and may handle vegetation, root, and canopy effects and surface snow-cover prediction.

For irrigated grid cells representing irrigated cropland, our model forces soil moisture at root zone to field capacity at every time step (time step = 30 min), all year round. Field capacity, defined as the maximum amount of water that the unsaturated zone of a soil can hold, is a parameter prescribed based on soil type. Irrigation practices and technologies vary in real cases. However, since most of our study area supports a whole year growing season (Deng et al. 2006a), we use this as a reasonable approximation to actual irrigation. The irrigation water either leaves the surface through evapotranspiration or gets lost as runoff or subsurface drainage. Our model separately simulates three processes of evapotranspiration—ground evaporation, vegetation transpiration, and vegetation evaporation, i.e., evaporation of water intercepted by vegetation canopy. The amount of evapotranspiration is calculated as the sum of the three. For non-irrigated grid cells representing dryland and the other land covers, the LSM determines soil moisture as a function of precipitation, evapotranspiration, infiltration, and soil properties. Initial soil moisture conditions and vegetation parameters specific to different land covers were taken from the parameter files in the Noah LSM. All vegetation interacts with soil moisture from the single vegetated soil column. Like Lobell et al. (2006) and Kueppers et al. (2007), we set atmospheric CO_2 concentrations constant at 355 ppm. To study the impacts of land surface changes related to irrigation, we simulated the regional atmospheric dynamics for a period of 11 years (January 1, 2010–December 31, 2020), based on the three land cover scenarios described earlier. We discarded the first year of each simulation as spin-up and compared results from the final 10 years among different scenarios.

Results

Effects of Irrigation on Heat Fluxes

Over the 10-year period of simulations, irrigation displays significant impacts on surface energy budgets, evidenced by the differences in the partitioning of this energy between latent heat and sensible heat for the three land cover scenarios (Fig. 5.4). The conversion from CON to IRR shows an increase in annual mean

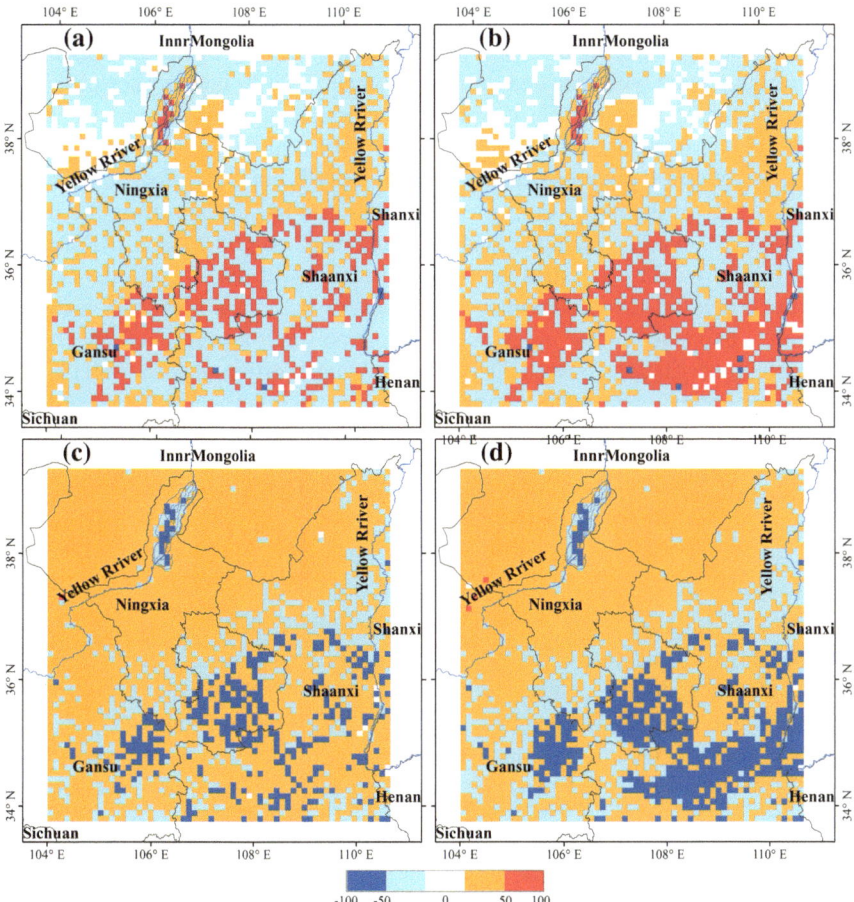

Fig. 5.4 Annually averaged differences in the latent heat fluxes (W m^{-2}) **a** from CON to IRR and **b** from CON to ICE. Annually averaged differences in the sensible heat fluxes (Wm^{-2}) **c** from CON to IRR and **d** from CON to ICE

latent heat fluxes of 12.10 W m^{-2} and a decrease in annual mean sensible heat fluxes of 8.85 W m^{-2}. The largest increases in annual mean latent heat fluxes occur in the central and eastern parts of Gansu Province. For large areas in the middle and south section of the model domain, where irrigated cropland is densely distributed in IRR, latent heat fluxes increase by 40–80 W m^{-2}. Surface energy budgets appear to change little for areas in the north section of the model domain, where much of the landscape is made up of barren land and grassland. More differences can be identified when irrigation extent expands to all cultivated cropland. The conversion from CON to ICE results in an increase in annual mean latent heat fluxes of 18.08 W m^{-2} and a decrease in annual mean sensible heat fluxes of 12.31 W m^{-2}. Particularly, the latent heat fluxes increase by 80 W m^{-2} or more for a broad region in Middle Shaanxi and Shanxi provinces, where large areas of dryland are converted to irrigated cropland. The shift of energy balance away from sensible heat and toward latent heat is primarily due to increased soil moisture associated with irrigated grid cells, which leads to higher rates of evapotranspiration. The annual mean increase in evapotranspiration between the CON and IRR is 0.41 mm/day and is 0.61 mm/day between the CON and ICE.

Interestingly, further expansion of irrigation in ICE not only affects the surface energy budgets of the newly converted land, but also has an impact on the larger surrounding region. For example, many existing irrigated grid cells in IRR experience a further increase in latent heat fluxes in ICE. Our results indicate that the impact of irrigation can move beyond the scope of individual grid cells. This can be partially explained by the processes of irrigation water loss and a likely change in regional precipitation patterns. Despite that a large portion of extra water added from irrigation leaves the surface through evapotranspiration, the rest is lost in the form of runoff and subsurface drainage, both of which may affect hydrology of the areas nearby. Compared to CON, irrigation increases runoff by 37 and 38 mm/year in IRR and ICE, respectively. In addition, previous studies have demonstrated that precipitation feedbacks are not confined at local level (Sacks et al. 2009).

Substantial seasonal variation exists in terms of the magnitude of change in surface energy budgets (Tables 5.1 and 5.2; Fig. 5.5). The increases in latent heat fluxes due to irrigation peak during the summer in both IRR and ICE. From CON to IRR, irrigated land conversion increases latent heat fluxes by 27.79 W m^{-2} for the summer months (June, July, and August) and by merely 0.50 W m^{-2} for the winter months (December, January, and February). From CON to ICE, further expansion of irrigated land leads to an increase in latent heat fluxes of 39.41 W m^{-2} in the summer and only 1.47 W m^{-2} in the winter. In typical

Table 5.1 Domain-averaged differences for climate variables between CON and IRR

	T_{mean} (°C)	T_{max} (°C)	T_{min} (°C)	ET (mm/day)	LH (W m^{-2})	SH (W m^{-2})	DR (W m^{-2})
Jan	−0.58	−1.10	−0.01	0.01	0.20	−0.12	−0.23
Feb	−0.75	−1.27	−0.09	0.03	0.85	−0.57	−0.78
Mar	−1.04	−1.53	−0.17	0.10	2.95	−2.21	−1.83
Apr	−1.43	−1.89	−0.46	0.29	8.47	−6.45	−3.30
May	−1.53	−1.96	−0.71	0.63	18.49	−10.62	−5.70
Jun	−2.02	−2.56	−0.98	0.89	26.06	−19.07	−7.46
Jul	−2.22	−2.80	−1.16	1.02	29.88	−23.08	−8.88
Aug	−2.17	−2.75	−1.15	0.93	27.42	−21.82	−8.36
Sep	−1.92	−2.47	−0.85	0.65	19.16	−14.28	−4.99
Oct	−1.06	−1.55	−0.39	0.28	8.42	−5.66	−3.16
Nov	−0.77	−1.29	−0.15	0.09	2.87	−2.05	−1.55
Dec	−0.59	−1.12	−0.03	0.01	0.45	−0.23	−0.36
Year	−1.34	−1.86	−0.51	0.41	12.10	−8.85	−3.97

Note T_{mean} mean temperature; T_{max} maximum temperature; T_{min} minimum temperature; ET evapotranspiration; LH latent heat fluxes; SH sensible heat fluxes; DR downwelling radiation

Table 5.2 Domain-averaged differences for climate variables between CON and ICE

	T_{mean} (°C)	T_{max} (°C)	T_{min} (°C)	ET (mm/day)	LH (W m^{-2})	SH (W m^{-2})	DR (W m^{-2})
Jan.	−0.57	−1.24	−0.09	0.03	0.91	−0.45	−0.40
Feb.	−0.79	−1.48	−0.13	0.06	1.98	−1.03	−1.12
Mar.	−1.23	−1.88	−0.23	0.19	5.77	−3.32	−2.80
Apr.	−1.82	−2.42	−0.55	0.51	14.90	−10.63	−4.44
May	−2.54	−3.30	−1.06	0.97	28.49	−19.71	−7.67
Jun.	−2.67	−3.42	−1.23	1.27	37.09	−24.46	−10.03
Jul.	−2.96	−3.77	−1.45	1.44	42.23	−33.61	−11.94
Aug.	−2.88	−3.68	−1.44	1.32	38.91	−29.02	−11.25
Sep.	−1.97	−2.58	−0.88	0.93	27.49	−16.43	−6.71
Oct.	−1.32	−1.97	−0.45	0.42	12.59	−6.31	−4.25
Nov.	−0.88	−1.54	−0.16	0.16	5.03	−2.12	−2.14
Dec.	−0.61	−1.26	−0.10	0.05	1.52	−0.59	−0.50
Year	−1.69	−2.38	−0.65	0.61	18.08	−12.31	−5.27

Note T_{mean} mean temperature; T_{max} maximum temperature; T_{min} minimum temperature; ET evapotranspiration; LH latent heat fluxes; SH sensible heat fluxes; DR downwelling radiation

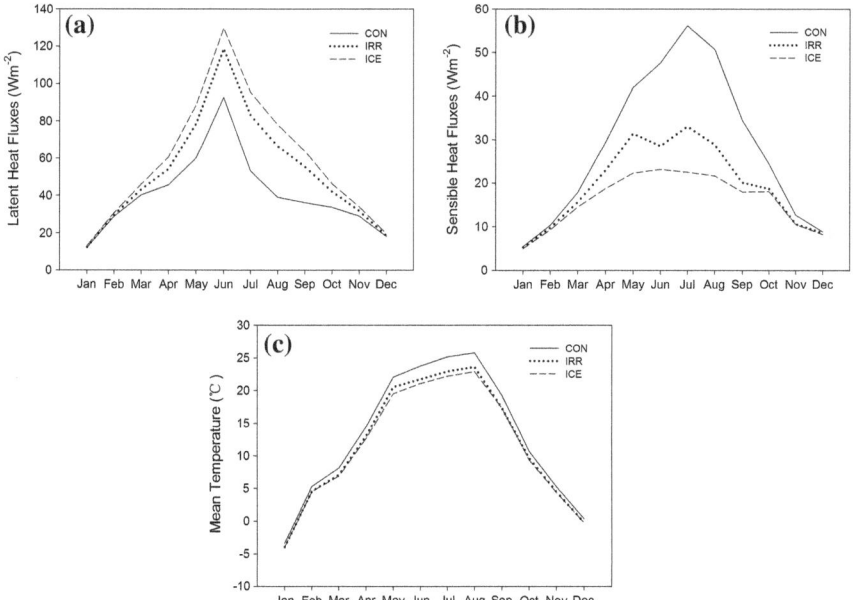

Fig. 5.5 Annual cycles of **a** latent heat fluxes; **b** sensible heat fluxes; and **c** mean temperature for three land cover scenarios

semiarid regions of northern China, as water vapor released into the atmosphere from the water-stressed landscape is limited, surface net radiation is mainly partitioned into sensible heat fluxes. However, irrigation and increased soil moisture can introduce significant latent heat fluxes through surface evaporation and transpiration. Both of these two processes usually peak during the summer, when temperature is relatively high and crop growth peaks. That basically explains why the effect of irrigation on surface energy budgets is much stronger in the summer than in the winter. As expected, there is also significant seasonal difference associated with the change in evapotranspiration and the pattern of this difference agrees with the seasonality of the change in energy fluxes. Similarly, the largest increases of evapotranspiration occur during the summer, with an average increase of 0.95 mm/day for IRR and 1.35 mm/day for ICE, while evapotranspiration increases are small during the winter.

Effects of Irrigation on Near-Surface Temperature

There are significant changes in near-surface temperature due to irrigation (Fig. 5.6). Converting from CON to IRR, the annual mean temperature of our study area is 1.3 °C lower on average, when a considerable proportion of dryland is converted to irrigated cropland. The irrigation cooling effect is the most significant for areas where irrigated cropland heavily concentrates. Particularly, parts of the middle section of the model domain, including portions of the central and eastern parts of Gansu Province, portions of the southern Ningxia Province, and portions of the central Shaanxi Province, are cooled by 5 °C averaged over the year. Most of the irrigated areas in the north and northeast sections of the model domain

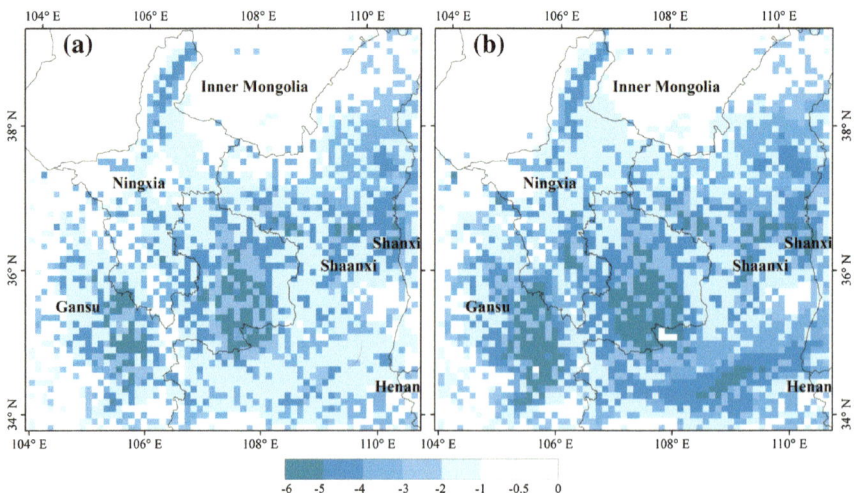

Fig. 5.6 Annually averaged differences in near-surface temperature (°C) **a** from CON to IRR and **b** from CON to ICE

are cooled by 3 °C. On the other hand, most desert and grassland areas in the north section of the model domain experience little temperature change. On average, the decrease in annual mean temperature is eight times greater over irrigated areas than over non-irrigated areas. The cooling effect becomes more significant with the further expansion of irrigated land to all cultivated cropland. The conversion from CON to ICE decreases the annual mean temperature of our study area by 1.7 °C. Expansion of irrigated land in the central parts of Shaanxi and Shanxi provinces suppresses the local annual mean temperature by 4–5 °C. Many existing irrigated areas in IRR show a further reduction of annual mean temperature in ICE. Our modeled decreases in near-surface temperature are accompanied by large increases in evapotranspiration and a shift of energy partitioning toward latent heat fluxes. Comparisons between CON and IRR, and between CON and ICE demonstrate that generally areas experiencing significant cooling are consistent with areas that have large increases in latent heat fluxes.

There are two mechanisms that could explain the surface cooling effect associated with irrigation. Irrigation increases the latent heat fluxes and decreases the sensible heat fluxes, thereby leading to a direct evaporative cooling of the surface. This is the dominant mechanism that contributes to the cooling effect of irrigation, as suggested by our modeled results about the changes in heat fluxes and near-surface temperature across different land cover scenarios. Furthermore, other factors can strengthen this cooling effect. Increased water vapor in the atmosphere causes an increase in cloud cover, which in turn can result in a greater fraction of reflected solar radiation and a decrease in total downwelling radiation. This non-local mechanism of irrigation is usually called the indirect cooling effect of irrigation (Sacks et al. 2009). Compared to CON, annual mean downwelling radiation of our study area decreases by 3.97 and 5.27 W m^{-2} in IRR and ICE, respectively. Very likely, this is due to enhanced cloud cover resulting from additional water vapor introduced from irrigation. Clouds can affect the energy distribution in the atmosphere through two mechanisms. Lower-level clouds cool the surface by reflecting solar radiation, while upper-level clouds warm the atmosphere by partially capturing emitted thermal radiation (Stubenrauch et al. 2002). Based on this, it is likely that the increased cloud cover due to irrigation would be lower-level cloud cover. In addition, our modeling results show that the irrigation cooling effect is not restricted to the near-surface atmosphere associated with irrigated areas, but spread to adjacent non-irrigated areas. For example, from CON to ICE, in the central and eastern parts of Gansu Province, and the central Shaanxi Province, even non-irrigated grid cells undergo various degrees of cooling, although the magnitude of this cooling is smaller compared to that of the nearby irrigated grid cells. This indicates that non-local processes such as enhanced cloud cover are probably also important in determining regional surface temperature.

Differences in monthly mean surface temperature for the three land cover scenarios reveal that the impact of irrigation on regional temperature varies substantially across seasons. The cooling effect of irrigation is much stronger in the summer than in the winter. For instance, from CON to ICE, the largest cooling effect occurs in July, with a decrease in mean surface temperature of 3.0 °C, while the smallest cooling effect occurs in January, with a decrease in mean surface temperature of 0.6 °C. This seasonal pattern of the cooling effect is consistent with the seasonality of the changes in evapotranspiration and heat fluxes due to irrigation. This is expected because our results have shown that the dominant cooling effect, the evaporative cooling, reaches its maximum during the summer growing season. In addition, irrigation exerts a stronger cooling effect during the day than night. With the expansion of irrigated cropland in both IRR and ICE, the annual average change in daily maximum temperature is much greater than that in daily minimum temperature, leading to a decrease in the diurnal temperature range. For example, from CON to IRR, the annually averaged maximum temperature decreases by 1.9 °C, while the annually averaged minimum temperature decreases by 0.5 °C. This regional result confirms to the previous finding of Sack et al. (2009) on effects of global irrigation that irrigation mostly affects daytime climate. Since both of the two cooling mechanisms—the evaporative cooling and increased reflected solar radiation due to more cloud cover—tend to peak during the day, the irrigation cooling effect is mainly reflected by a substantial decrease in daily maximum temperature.

Theoretically, lower albedo of irrigated soil may also affect the surface energy balance and regional climate. However, comparing among different scenarios, we find that albedo changes only slightly due to irrigation. For example, the average albedo for areas of cultivated cropland decreases by 0.0073 from CON to IRR, and by 0.001 from IRR to ICE. This magnitude of change in albedo due to irrigation is much smaller relative to that associated with the land conversion between natural vegetation (such as forest and grassland) and cropland. Field measurements have also verified that the differences between irrigated and non-irrigated albedos in our study region are small (Guan et al. 2009). This indicates that in our case, the change in net radiation due to lower albedo of irrigated land accounts for a minimal portion of the change in energy balance and therefore plays a less important role in regional climate. Our finding for this particular region is consistent with the global-scale experimental study of Lobell et al. (2006), which shows that the global temperature changes were highly correlated with changes in net surface radiation, but not with changes in surface albedo.

Our simulated effect of irrigation on the latent heat fluxes in the inland irrigation area of northern China is smaller than estimates in regional studies such as Kuepper et al. (2007), in which irrigation increased the latent heat fluxes in California year-round by 70.7 W m^{-2}, but close in the magnitude to estimates about India in Douglas et al. (2009), southern India in Biggs et al. (2008), and US High Plains in Adegoke et al. (2003). Similarly, although many regional studies have shown significant cooling effects of irrigation, they have disagreed about the magnitude of these effects. Our simulated cooling effect is close in the magnitude to estimates about Nebraska in Mahmood et al. (2006), California in Kueppers et al. (2007), and India in Douglas et al. (2009). Our simulated irrigation cooling in our study region is also consistent with Lee et al. (2011), which found qualitatively similar but smaller temperature decreases over Asia due to irrigation. Differences in the magnitude of the effects of irrigation between these studies can be explained by differences in how irrigation is modeled, the selection and setup of the atmospheric model, the scale of the study region, and other regional heterogeneities.

The design of our WRF model has some limitations. As different crops have distinct irrigation water requirements and timing of irrigation, ideally there could be a crop model that deals with the irrigation status of different crop types. This requires spatially explicit information about crop-type distributions and crop-specific information about agricultural management. Given the data-sparse environment, those data are not available for our study region. Using an average land cover in an irrigation simulation scheme could cause over- or underestimated irrigation water input (Ozdogan et al. 2010). Development of high-resolution land use data about crop types and incorporation of a crop model in the irrigation scheme could be a direction of future work for researchers who intend to explore climatic effects of irrigation.

Summary

Population growth and Westernization of dietary patterns have increased the demand for food, resulting in intensification of agricultural practices in many regions around the world. Expansion of irrigated agriculture has the potential to alter regional to global climate significantly. In this section, we investigate the impacts of irrigation on the heat fluxes and near-surface temperature in the future for a specific region. We focus our study on an inland irrigation area of northern China, an area in the developing world with a long history of irrigated agriculture and high ecological vulnerability. Based on the modeling results of the WRF system, we examine the changes in heat fluxes and near-surface temperature among three land cover scenarios: the CON, the IRR, and the ICE. Our results indicate that the surface energy budgets and temperature are sensitive to changes in the extent and spatial pattern of irrigated land. Conversion to irrigated agriculture at the contemporary scale (as in IRR) leads to an increase in annual mean latent heat fluxes of 12.10 W m^{-2}, a decrease in annual mean sensible heat fluxes of 8.85 W m^{-2}, and a decrease in annual mean temperature of 1.3 °C across the study region. Further expansion of irrigated land to all cultivated cropland (as in ICE) increases annual mean latent heat fluxes by 18.08 W m^{-2}, decreases annual mean sensible heat fluxes by 12.31 W m^{-2}, and decreases annual mean temperature by 1.7 °C. The cooling effects of irrigation tend to be greatest for areas where irrigated cropland heavily concentrates and are much stronger in the summer growing season than in the winter. The seasonality of the cooling effect is consistent with the seasonality of the changes in evapotranspiration and heat fluxes due to irrigation. In addition, we identify that irrigation exhibits a greater cooling effect during the day than night, leading to a decrease in the diurnal temperature range. While direct evaporative cooling plays a dominant role, indirect non-local processes such as enhanced cloud cover are probably also important in causing the cooling of the surface.

Our modeled effects of irrigation show that changes in land use management such as irrigation can be an important component of climate change. Several studies have shown that the regional climate of northern China is becoming warmer and drier and this trend is likely to continue into the future. However, those projections do not account for the climate impacts from land cover and land management change. Regional climate forcings including irrigation expansion can introduce a countervailing cooling effect, weakening the projections of future regional warming from greenhouse gases. Our study has simulated the scenario of expansion of irrigated area, which is likely to take place due to population growth and economic development. In other circumstances, irrigated area may also decline due to lack of sufficient water supply or conversion of irrigated agriculture to urban land. It is expected that these irrigation changes can modify future regional climate and need to be considered together with greenhouse forcing in climate change assessments.

Spatial Variation of Surface Temperature and Precipitation Due to Grassland Conversion to Forestry Area in Southeast China

Introduction

Under the coupled impact of natural processes and human activities over the past century, the global vegetation system has changed correspondingly, which in turn influenced the regional and global climate. The nature of the vegetation affects the surface fluxes of radiation, heat, moisture, and momentum, so modifying the vegetation cover can change the lower boundary conditions of the atmosphere and hence impact the climate (Pielke et al. 1998). Besides, changes in vegetative cover, especially the change of vegetation types, are associated with changes in plants' morphology and physiology, which could, in the absence of other force, alter the surface fluxes and consequently the climate both at regional and at global scales (Bounoua et al. 2002). In addition, forests and grasses are the most two typical vegetation types on the Earth; however, mankind has significantly changed the Earth surface by transforming natural ecosystems (forests and grasslands) (Brovkin et al. 1999). Furthermore, the conversion of forestland and grassland plays an important role in the climate system, and redistributions of forest and other vegetation (due, e.g., to extensive logging) could initiate important climate feedbacks (Bonan et al. 1992). And a significant difference in surface parameter (such as albedo) between forest and grassland results in reduced absorbed radiation for the grasslands, especially during the snow-thawing season (spring). Also, a forested landscape generally has a lower surface albedo than grassland, particularly in conditions of lying snow when shortwave radiation is trapped by multiple reflections within the forest canopy (Betts and Ball 1997). In fact, the assessment reports of Intergovernmental Panel on Climate Change (IPCC) specifically took agriculture, forestry, and other land use (AFOLU) as a major path to slow the process of global climate change (Landman 2010), and the research on the impact of forestland and grassland transition on climate is of great significance to find the optimal way to mitigate the climate change.

Numerous studies have reported the climate impact of vegetation changes such as deforestation and grassland degradation at different scales. However, little is known about the impact of large-scale conversion from grassland to forestland on climate, especially the scale and degree it may bring about in the future. The deforestation experiments with atmospheric general circulation models (AGCMs) generally agreed in significant cooling in the spring season within the models and argued that different vegetation cover plays the most significant role in determining the climate–vegetation state of the region (Chalita and Le Treut 1994). Further studies carried out in the experiments with the modified Community Climate Model (CCM2) and prescribed sea surface temperature (SST) showed a significant summer cooling effect of grass and forest cover changes on the climate in the USA (Bonan 1997). And the cooling was also observed in a simulation using a nested mesoscale model in northeast Colorado (Chase et al. 2000) and in the central plains. In addition, experiments with the Hadley Centre Climate Model (HadCM3) with prescribed SST showed that historical vegetation cover change results in a reduction in midlatitude annual mean temperatures within the model. And extreme scenarios over Amazonia replacing tropical forests with grass lead to warmer, drier climates in several studies (Sud et al. 1996; Bonan 1997). Since 1990s, numerical simulations of impact of vegetation changes on climate in China or East Asia have been conducted by Chinese scientists; results showed that the regional temperature, precipitation, and other indicators have been obviously influenced by the change, which could even affect the summer monsoon intensity of Asia in the heavily changed regions (Jin et al. 2010). However, due to the various scales and degrees of forestland and grassland change and different models that had been used, the simulation results were in poor comparability. Meanwhile, the limited calculation condition has shorter integration time of simulation, and the result cannot reflect the long-term climate effect triggered by the change, especially the medium- and long-term climate effects in the future.

How to find a systematic quantitative way to explore the future climate effects caused by the conversion from grassland to forestland? The comparative model experiments need to be carried out. In this section, we used the WRF model to quantify the regional climate effect triggered by the conversion from grassland to forestland in Fujian Province of southeast China. Based on the RCPs scenario data, future land transition of forestland from grassland in the study area was simulated by using dynamics of land system (DLS) model, and then, the space and timescales of regional climate effects were analyzed.

Data and Methodology

Study Area

The southeastern coast of China is one of the top-level regions with abundant water, heat, and sunlight resources. It plays an important role in the globalization and the Pacific Rim economic circle in China, which occupies about 5.33 % of the total land area, accounts for about 21 % of the national population, and creates 35.7 % of national GDP and 70 % of the total import and export of China. With the development of economy and the policy implementation such as "returning farmland to forest" ecological engineering project, the area of forestland expanded sharply, especially in Fujian Province, which is the most typical afforestation region in China. The forest coverage rate of Fujian is the highest in China, reaching 62.96 % in 2005. In addition, Fujian is the main region of conversion of forestland from grassland in China (Liu et al. 2010), and the conversion from grassland to forestland is obvious (Fig. 5.7). Statistical analysis indicated that during 2000–2008, 11.2 % of the grassland had been converted into forestland, accounting for 73 % of the total area of forest expansion. Besides, the dominated humid middle and south subtropical maritime monsoon climate, with various heat resources and plenty of water resources, played an auxiliary role in the conversion process. In

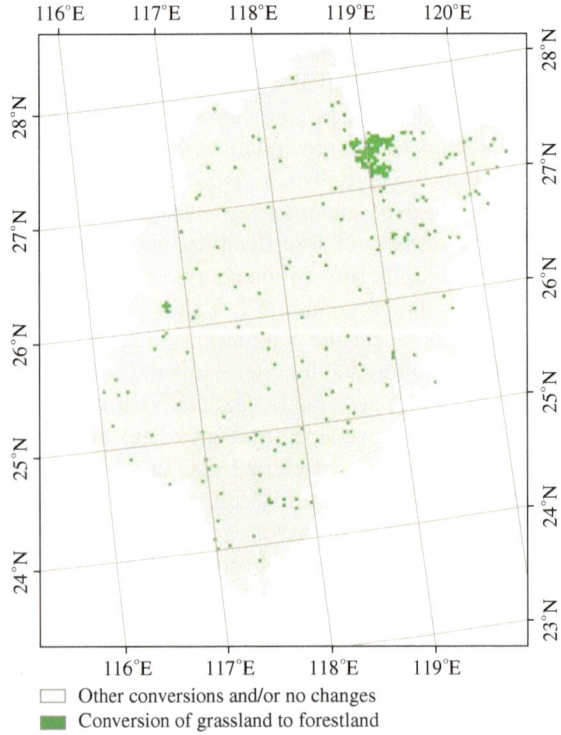

Fig. 5.7 Pattern of the conversion (measured in 3 km) from grassland to forestland in Fujian Province, 2000–2008

2005, the forest coverage rate in Fujian had reached 62.96 %, which was one of the highest forest coverage rates in the nation. In addition, the most part of grassland in Fujian Province is the secondary degenerative environment, which evolves from wasteland after the forest system damaged, and thus, it is beneficial for the conversion from grassland to forestland.

Data Sources

The lateral boundary force data were mainly derived from NCEP FNL and NCEP/NCAR reanalysis datasets. Forecasting data were derived from the multi-modal data of WCRP's Coupled Model Intercomparison Project Phase 5 (CMIP5). Chinese National Climate Center had finished the downscaling work of multi-modal simulation datasets, which was unified to the same resolution and had been used to undertake the numerical simulation in the eastern Asia. The weighted average reliability value was used to undertake the multimodel ensemble (MME); then, the annual and monthly average data were made for 2000–2050, namely climate change forecast dataset in China. Data resolution was $1° \times 1°$, which only included the continental land, with a spatial scale of 15.5°–55.5°N and 72°–135°E.

Land cover data were extracted from the Chinese subset of the global land cover characteristics database which was developed based on Landsat TM scenes for 1999/2000 and the Chinese subset of the MODIS land cover data product in 2008. The two datasets have the spatial resolution of $1 \times 1 \text{ km}^2$ and an overall accuracy of 81 % (Liu et al. 2003b). DLS simulation requires spatial data of natural environment conditions such as climate, elevation, slope, plain area percentage, soil properties, and the geographic data, and thereinto, RCPs data were derived from land use harmonized data products (http://luh.umd.edu/data.php), and the climate observation data were acquired from records of 23 weather stations of the National Meteorological Information Center of China, including the moderate average annual rainfall during 2000–2008. Data about slope, aspect, and plain area ratio were extracted from elevation data based on 1 : 250,000 digital elevation model (DEM) for each $1 \times 1 \text{ km}^2$ grid. Soil property data came from the Second National Soil Census, and by using Kriging interpolation method, the spatial distribution of soil property was acquired. Regional condition data included the distance of each grid center to highway, provincial highway, main roads, and the provincial capital, which were calculated by distance measuring tools based on the national 1 : 250,000 fundamental geographic maps.

Estimation Model of Climate Effects

The regional influence of the conversion from grassland to forestland could be described by effect index (EI), which is defined as follows:

$$\text{EI}(x) = \frac{A_{\text{change}}(x)}{A_{\text{forest}}}$$

where x is any meteorological parameter such as annual average temperature, precipitation, water vapor mixing ratio, or wind speed. $A_{change}(x)$ is the grid number within the region where x is changed, A_{forest} is the grid number of the area converted from grassland to forestland, both of them are calculated at the regional scale, and a bigger EI reflects a higher efficiency of its climate effect.

We used two nested grids with the spatial resolutions (grid numbers) of 3 km (162×177). The vertical grid consisted of 35 full sigma levels from the surface to 100 hPa, of which the lowest 13 levels were below 1 km to show a finer resolution in the planetary boundary layer. Both the initial and the boundary conditions with a resolution of $1.0° \times 1.0°$ were from the National Centers for Environmental Predictions. The simulations were divided into two stages. Firstly, simulation was integrated for 8 years from January 1, 2000, to December 31, 2008. The initial 15-day period was considered as a spin-up period to minimize the effect of the initial conditions. In the control experiment, the land cover data were updated from the 2008 MODIS satellite observations to represent current land cover conditions (TFG case). For the sensitivity experiment, all newly expanded forest cover from grassland during 2000–2008 in domain 2 was replaced by grassland, which is the most common land cover type surrounding the forest area in the simulated domain (NOTFG case). These two cases were compared to analyze the influence of historical conversion from grassland to forestland (2000–2008) in Fujian. Secondly, we were moving on to the future stage, which was simulated from January 1, 2010, to December 31, 2050, and based on the hypothesis that the future climate effect simulation of the land use and land cover change would be reliable as that of the historical one, the newly expanded forestland from grassland was updated from the results of DLS simulation to represent the future forestland transition from grassland in the NOTFG case. The parameterization schemes used for the two stages are as follows: Noah land surface processes, CAM3 radiative scheme, WSM3 simple ice microphysics scheme, Grell 3D cumulus convection schemes, and YSU boundary layer scheme.

Simulation Model of Land Transition

Based on the conversion data under the RCP6 and RCP8.5, which were produced by AIM and MESSAGE model from 2009 to 2050, respectively, the total area of land transitions from grassland to forestland was calculated from 2010 to 2050. They were used by DLS to simulate the future spatial pattern of the conversion of forestland from grassland in Fujian. The DLS model is theoretically based on restrictions of the distribution of land types. It dynamically simulates the macroscopic pattern changes in land cover by classifying the driving factors that influence this pattern. Considering the links among related models of nature, ecology, and economy, we used DLS to spatially allocate the area change in the forestland transition from grassland. Based on the spatial statistics data, the probabilities of the transition of grassland to forestland and the probabilities of the distribution of this transition were predicted at the pixel level. In the simulation process, we also considered the influence of macroscopic factors such as topography, environment, trade, and institutional arrangement and land management policies to more accurately simulate possible scenarios of pattern changes in land system (Deng et al. 2006b).

Scenario Building and Data Analysis

We took RCP6 as the moderate development scenario of Fujian; it is a stabilization scenario where total radiative forcing is stabilized after 2100 without overshoot by employment of a range of technologies and strategies for reducing greenhouse gas emissions. Grassland area declined, while total forested area extent remained constant throughout the century at the global scale.

While RCP8.5 is an emission pathway which reaches a radiative forcing of 8.5 W/m² and rising in 2100, we made it a fast development scenario of Fujian, with a high magnitude of climate change and factors related to higher vulnerability (e.g., higher population growth and lower levels of economic development). It is characterized by increasing greenhouse gas emissions over time representative for scenarios in the literature leading to high greenhouse gas concentration levels. The underlying scenario drivers and resulting development path are based on the A2r scenario. An important feature of the RCP8.5 was the increase in cultivated land by about 185 million hectares (Mha) from 2000 to 2050; forest cover declined over the century by 300 Mha from 2000 to 2050.

The transition from grassland to forestland in Fujian is significant under the RCP6 and RCP8.5 scenarios with a spatial resolution of $0.5° \times 0.5°$ (Fig. 5.8). Firstly, the downloaded data were downscaled to $1 \times 1 \text{ km}^2$ grid, and then, the overlay analysis was taken with the administrative boundary data to calculate the total amount of the transition from grassland to forestland. During 2010–2050,

Fig. 5.8 Conversion from grassland to forestland under the RCPs scenarios in Fujian Province, 2010–2050

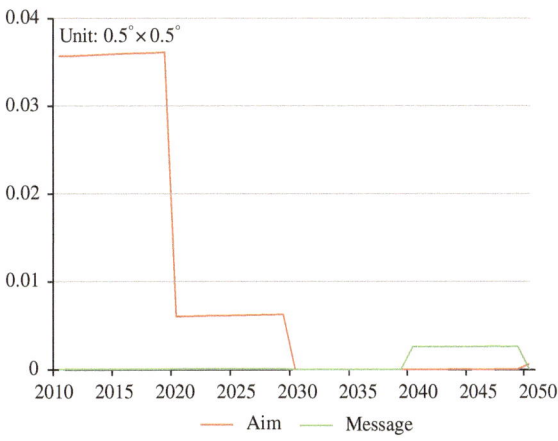

the transition of forestland from grassland presents a soaring trend in 2010–2020, followed by a sharp decline in 2020, when a stable decrease occurs in the next 30 years under the RCP6 scenario, but it is pervasively higher than that of RCP8.5.

Results

Impact on Temperature

Temperature differences of near-surface temperature in 2 meters height (T_{2m}) of forestland conversion from grassland and non-conversion experiments show the climate effect of the conversion (Fig. 5.9). In summer, the cooling effect has emerged in the conversion area of grassland to forestland. Compared with the other area, the maximum difference of the seasonal average T_{2m} is approximately 0.60 °C in the conversion area of grassland to forestland, and the regional average T_{2m} decreases by 0.16 °C with $EI(T_{2m}) = 2.23$. While in winter, there is a heating effect in the conversion area of grassland to forestland, with the maximum difference of the seasonal average T_{2m} increasing by 0.08 °C, and the average T_{2m} is rising by 0.03 °C with $EI(T_{2m}) = 1.93$. On the whole, the annual average T_{2m} decreases by 0.11 °C in Fujian.

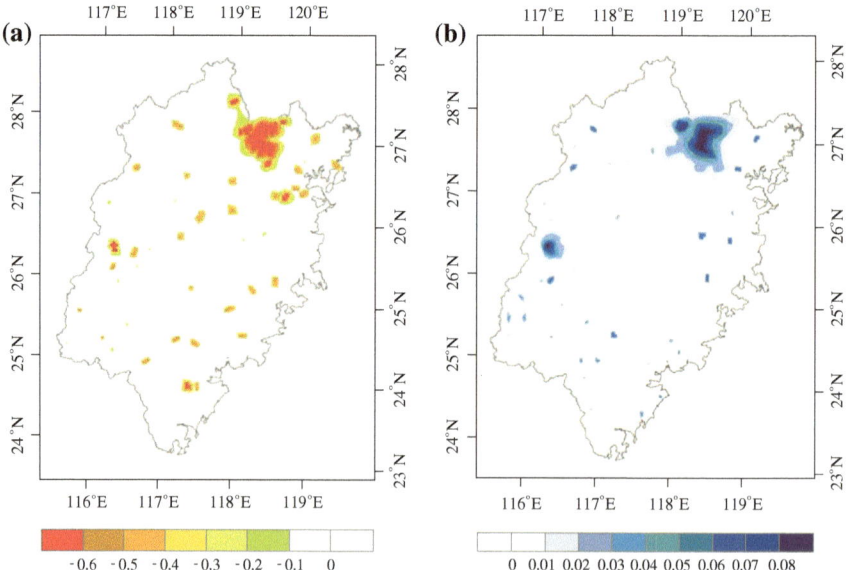

Fig. 5.9 Seasonal difference of near-surface temperature at 2 m height (*unit* °C) in the experiment of conversion from grassland to forestland in Fujian Province, 2000–2008. Panel **a** and Panel **b** identify the temperature difference in summer and winter, respectively

Fig. 5.10 Annual average precipitation difference (*unit* mm) in the experiment of conversion from grassland to forestland in Fujian Province, 2000–2008

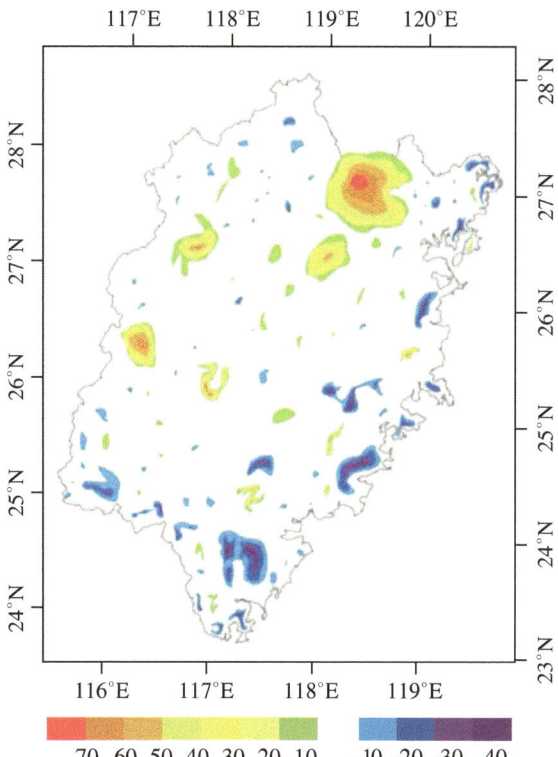

Impact on Precipitation

Precipitation varies from one region to another (Fig. 5.10), the northwest Fujian was the main area where the amount of precipitation increased, while the southeast reverse the case. According to the spatial statistical analysis, the range of the difference of precipitation between TFG case and NOTFG case is −40 to 70 mm, and the majority of the change ranged from −30 to 50 mm, while the annual average precipitation increased by 46 mm with EI ($P = 4.68$).

Validation of the Control Case

We selected 23 meteorological stations' continuous observation data from 2000 to 2008; all of them have passed the homogenization test. The monthly mean temperature data are provided by the National Meteorological Information Center of China. When assessing its simulation capability, the simulation results were bilinearly interpolated to 23 weather stations, to calculate the average and seasonal temperature of each site.

Table 5.3 Comparison of the meteorological station observed (OBS) and model-simulated annual near-surface temperature at 2 m height (*unit* °C) and precipitation error (*unit* mm) over 23 weather stations in Fujian Province in 2008

No.	Station	Latitude	Longitude	T_{err}	P_{err}
58726	Qixianshan	27.95	117.83	1.23	129.55
58730	Wuyishan	27.77	118.03	0.12	8.43
58731	Pucheng	27.92	118.53	0.04	−2.87
58734	Jianyang	27.33	118.12	1.56	51.79
58737	Jianou	27.05	118.32	−0.50	48.72
58754	Fuding	27.33	120.20	0.76	−77.27
58820	Taining	26.90	117.17	0.84	84.33
58834	Nanping	26.65	118.17	−0.32	35.78
58846	Ningde	26.67	119.52	1.34	67.00
58847	Fuzhou	26.08	119.28	0.98	12.47
58853	Taishan	27.00	120.70	0.25	−28.38
58911	Changting	25.85	116.37	0.56	55.00
58918	Shanghang	25.05	116.42	−2.01	34.00
58921	Yongan	25.97	117.35	0.73	74.74
58926	Tanping	25.30	117.42	0.34	33.11
58927	Longyan	25.10	117.03	1.62	−62.94
58931	Jiuxianshan	25.72	118.10	0.28	32.61
58933	Pingnan	26.92	118.98	−0.15	145.00
58944	Pingtan	25.52	119.78	0.37	34.49
59126	Zhangzhou	24.50	117.65	0.99	10.38
59133	Chongwu	24.90	118.92	0.33	−27.74
59134	Xiamen	24.48	118.07	−0.56	54.83
59321	Dongshan	23.78	117.50	0.76	74.70

The errors between the simulations and observations of surface temperature and precipitation are listed in Table 5.3, and the *P* test values are 0.81 and 0.76, respectively. The errors between the simulations and observations are less than 1.00 °C except for the Qixianshan, Jianyang, Ningde, Shanghang, and Longyan stations. The relative errors of the annual precipitation are all less than 12 %; therefore, we think that the simulation result is reliable and presume that it will work accordantly well in the future.

Predicted Effects on Surface Temperature and Precipitation in the Future

Under the two RCPs scenarios (RCP6 and RCP8.5), the spatial pattern of transition of forestland from grassland was simulated with DLS model (Fig. 5.11). It can be seen from Fig. 5.11a that the northeast and southwest of Fujian would

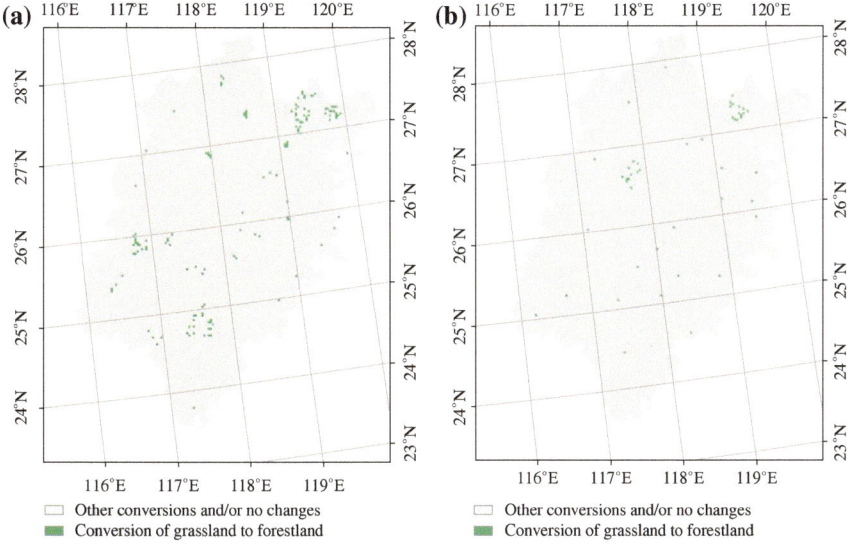

Fig. 5.11 Spatial pattern of the conversion (measured in 3 km) from grassland to forestland in 2010–2050 under RCP6 (**a**) and RCP8.5 (**b**) in Fujian Province

experience more intensive transition of grassland to forestland than other regions under the RCP6 scenario, and this conversion occurs broadly on the whole regional perspective. While under the RCP8.5 scenario appeared in Fig. 5.11b, the transition of grassland to forestland is not as intensive as that of RCP6 scenario; it is mainly concentrated in the northern Fujian.

Then, the two projections of land transition of forestland from grassland data under RCP6 and RCP8.5 were used to estimate the climate effects of that transition in the future. And the seasonal T_{2m} change in winter and summer of the two scenarios is calculated between 2010 and 2050 to show the spatial scale of the transition of grassland to forestland (Fig. 5.12). In summer, the conversion of forestland from grassland has a cooling effect under the two climate scenarios. Under the RCP6 scenario, the change of mean T_{2m} is mainly distributed in the north and southwest Fujian and that of the RCP8.5 scenario is primarily concentrated in the central and north regions. The change of average T_{2m} is -0.69 °C with $EI(T_{2m}) = 3.21$ under RCP6 scenario and -0.43 °C with $EI(T_{2m}) = 1.36$ under RCP8.5 scenario in the conversion area of grassland to forestland. While in winter, the average T_{2m} increases by 0.06 °C with $EI(T_{2m}) = 2.86$ under the RCP6 scenario scenario in the converted area, whereas under the RCP8.5 scenario, the average T_{2m} increases by 0.01 °C with $EI(T_{2m}) = 1.14$, though winter cooling occurs across the region, there is no obvious concentration area.

In order to explore the timescale of the impact of the conversion from grassland to forestland, the every 10-year difference of the T_{2m} and precipitation was calculated from 2010 to 2050 (Fig. 5.13). Both in RCP6 and RCP8.5, the annual

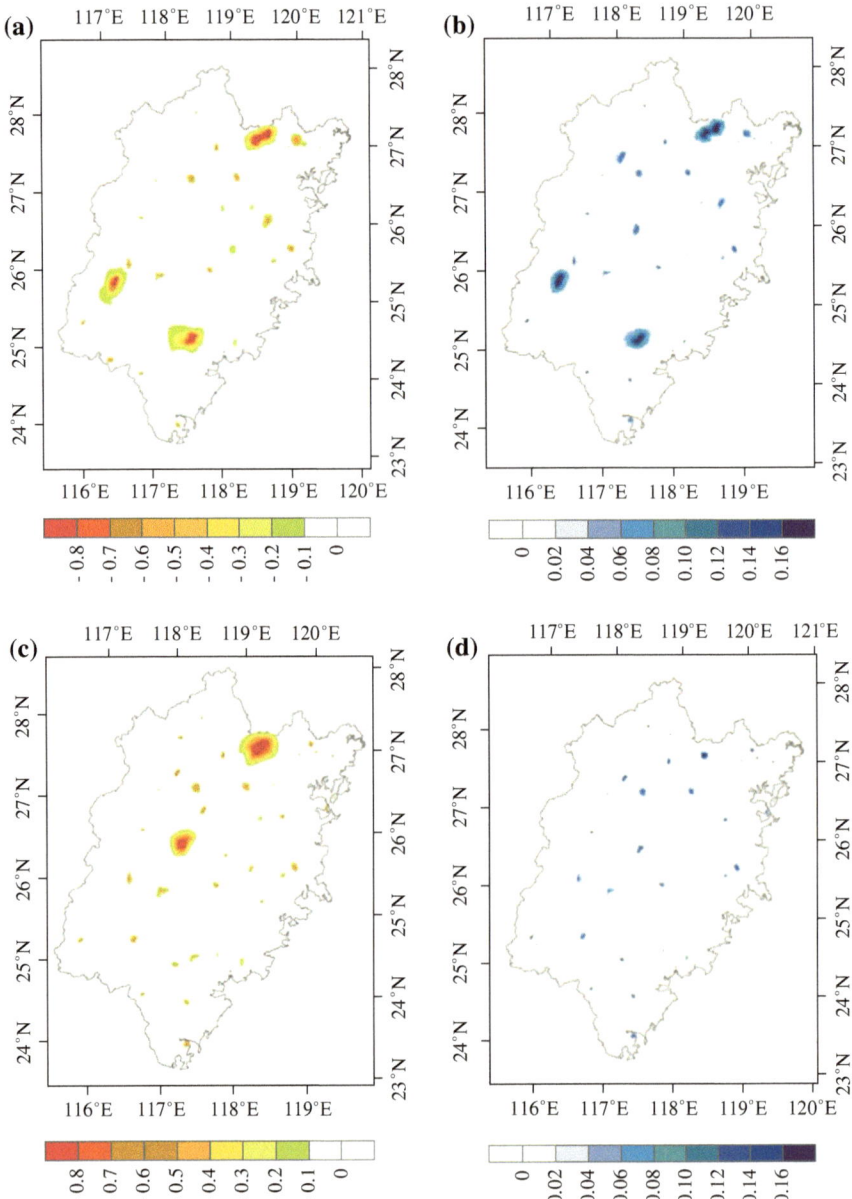

Fig. 5.12 Seasonal average near-surface temperature change at 2 m height (*unit* °C) in 2010–2050. **a** Summer temperature change under RCP6 scenario; **b** winter temperature change under RCP6 scenario; **c** summer temperature change under RCP8.5 scenario; **d** winter temperature change under RCP8.5 scenario

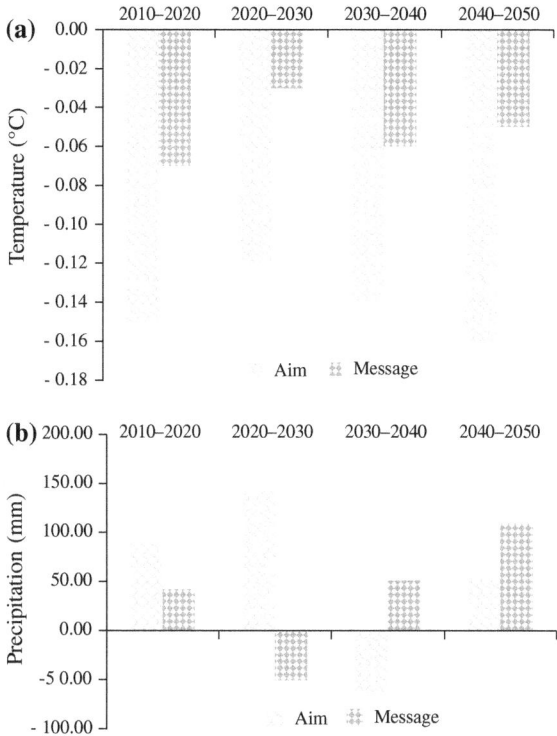

Fig. 5.13 Climate impacts of the conversion from grassland to forestland under the RCP6 and RCP8.5 scenarios from 2010 to 2050 in Fujian Province. **a** Near-surface temperature change at 2 m height (*unit* °C); **b** precipitation change (*unit* mm)

average T_{2m} decreases as the conversion from grassland to forestland occurs. On the whole, the annual average T_{2m} decreases by 0.57 °C and 0.21 °C under the scenarios of RCP6 and RCP8.5, respectively, from 2010 to 2050. The maximum decrease of T_{2m} would reach 0.16 °C in ten-year differences, and the minimum value is 0.03 °C in the two scenarios. Although the overall decline of T_{2m} under the RCP8.5 scenario is lower than that of RCP6, the minimum decrease of T_{2m} under RCP8.5 is higher than 0.03 °C. In terms of the precipitation, there are small fluctuations under the two scenarios, ranged between −63 to 142 mm and −51 to 109 mm. Meanwhile, the two scenarios would experience an overall increase of precipitation during 2010–2050.

Summary

In this section, climate impacts of historical (2000–2008) and future (2010–2050) land conversion from grassland to forestland were simulated using WRF model, whose result showed that this conversion has a significant impact on regional climate. The result shows that during 2010–2050, the conversion from grassland to forestland leads to the decrease of annual average T_{2m} by 0.57 and 0.21 °C under the RCP6 and RCP8.5 scenarios, respectively, and the value of EI (T_{2m}) in summer (represents cooling) is higher than that of winter (represents heating) in the experiments, indicating that the summer cooling efficiency is higher than winter heating efficiency in our study. The mechanism is that grassland reflects more solar radiation than forest in Fujian does, which is a very efficient absorber and scatterer of short-wavelength radiation (Lean and Warrilow 1989). Consequently, the net energy absorbed by the surface was increased as a result of the conversion from grassland to forestland, which would result in a temperature increase, but this is offset by an increase in evaporative cooling for the fact that the area is abundant with sunlight and heating resources. So, there is an overall decrease in temperature.

In spite of the change of precipitation fluctuates from one region to another as the conversion from grassland to forestland occurs in both historical and future land conversion experiments in Fujian, it is increased on the whole. It is for the fact that the precipitation within Fujian originates from two sources, one is the recycling of water vapor released during evapotranspiration and the other is moisture flux convergence into this area. The water vapor released is increased as the conversion from grassland to forestland occurs because of the enhanced evapotranspiration in Fujian, while the increase in moisture flux convergence is due mainly to decreased albedo caused by the transition of forestland from grassland, thus affecting the reflectivity of a surface by absorbing more heat, and decreased albedo reduces the available energy for upward turbulent transfer of latent energy (Lean and Warrilow 1989), which in turn carries moisture from forest trees into atmosphere, where it condenses as rain and the convergence zone is induced. So, the precipitation increases on the whole.

We have only focused on the physical impact of forestland transition from grassland change in this section, but climate change is a very complicated process and many factors may influence it. Summarizing the effects of land cover change on climate has been difficult because different biogeophysical effects offset each other in terms of climate impacts, and on global and annual scales, regional impacts are often of opposite sign and are therefore not well represented in annual global average statistics (Feddema et al. 2005). Besides, the uncertainty will remain in projections of future climates for the foreseeable future. The climate modeling community is aware of these uncertainties, and some innovative approaches to assess their magnitude have recently been explored (Stainforth et al. 2005). Sensitive study in different land surface schemes has been conducted to better estimate precipitation in WRF, but more extensive modeling experiments are needed in the WRF simulation process. Besides, the precise contribution of the land cover change to the global climate change remains a controversial but growing concerned issue (Wu et al. 2013). In addition, pollutant release is another important factor that affects the radiation process and thus affects precipitation over afforestation areas. Future studies utilizing a wide range of scenarios for climate, land use, and realistic vegetation parameters to quantify the effects of different factors on regional climate are essential. Additionally, with the increasing available computational resources, more land transition runs would be the optimal way to assess the large-scale climatic effects of LUCC.

Spatial Variation of Surface Heat Fluxes Due to Land Use Change Across China

Introduction

Land use/land cover change (LUCC) affects climate system at local, regional, and global scales through various biogeochemical and biogeophysical processes (Foley et al. 2005). LUCC would alter the physical properties of land surface, such as albedo, fractional vegetation coverage, moisture content, and surface roughness. These physical properties determine the absorption, emission, and exchange of energy at the Earth's surface (Dümenil Gates and Ließ 2001; Foley et al. 2005), represented by the fluxes of energy, radiation, and moisture into atmosphere, etc., and consequently affect climate system. Furthermore, the interaction between land surface and atmosphere mainly works through the exchanges of momentum, energy, and water, among which the exchange of energy plays a key role in affecting climate system. In this context, it is of great importance to investigate the effects of LUCC on surface energy fluxes.

The effects of LUCC on climatic change have raised growing concern and have been widely studied (Foley et al. 2005; Deng et al. 2013). There have been many studies that attempted to reveal the impacts of human-induced land cover changes (HLCCs) on the regional climate in China (Wessels et al. 2007; Deng et al. 2013). However, most of these studies only focused on the effects of HLCCs on temperature and precipitation; little is known about the effects on surface energy fluxes so far, such as latent heat flux and sensible heat flux. Moreover, previous studies on China just concentrated on one typical case study areas, for example, northeast China, northwest China, Jiangxi Province, and the Pearl River Delta, which are characterized by different types of LUCC. To date, an integrated analysis at the national level in China remained underexplored, which is needed to make comprehensive policies regarding land planning and management.

In order to address this knowledge gap, we estimate the heat flux changes caused by land transformation over the next 40 years across China to improve the understanding of the impacts of land dynamics on regional climate. In this study, we use the WRF model to investigate these impacts in four representative land transformation zones where reclamation, overgrazing, afforestation, and urbanization dominate the LUCC separately. The remainder of the study is structured as follows. First, we describe the study area and introduce the main datasets used in this chapter (Section "Study Area and Data"). Next, we present the methodology in Section "Methodology". We then analyze and discuss the simulation results in detail (Section "Results and Discussion"). Finally, we highlight the major findings in Section "Summary".

Study Area and Data

Selection of the Study Area

We measured the regional variance of LUCC by land use dynamic degree (Liu et al. 2010). Specifically, we overlap the vector data of land use dynamics from 2000 to 2008 on the 10 km × 10 km grid data, and then, we calculate and summarize the percentages of each changing type of land use within each grid. To present the dominant changing types of land use in the study areas, four zones are selected as the representatives for four types of land transformation: cultivated land reclamation, overgrazing, afforestation, and urbanization. Land use change in China has obvious regional differentiation characteristics during 2000–2008 (Fig. 5.14). Specifically, the grassland was reclaimed to cultivated land in northeast plain, eastern Inner Mongolia, northwest arid areas as well as part of traditional oasis agricultural region (Zone I). In western Inner Mongolia, loess plateau area was mainly characterized by grassland degradation (Zone II). Land use change in mountainous southern China, especially Jiangxi Province, was dominated by reforestation or expansion of grasslands due to farmland abandonment (Zone III). In the Huang-Huai-Hai Plain and the southeastern coastal areas (especially the Bohai Economic Rim and the Yangtze River Delta), there is an apparent expansion tendency of residential, industrial, and mining land uses (Zone IV). The four zones are of heterogeneous structure, and several types of land use change occurred in each zone simultaneously. However, we only presented the impacts of dominant type of land use change on surface heat flux in each zone. In other words, we merely take account of reclamation, overgrazing, afforestation, and urbanization in Zones I–IV correspondingly.

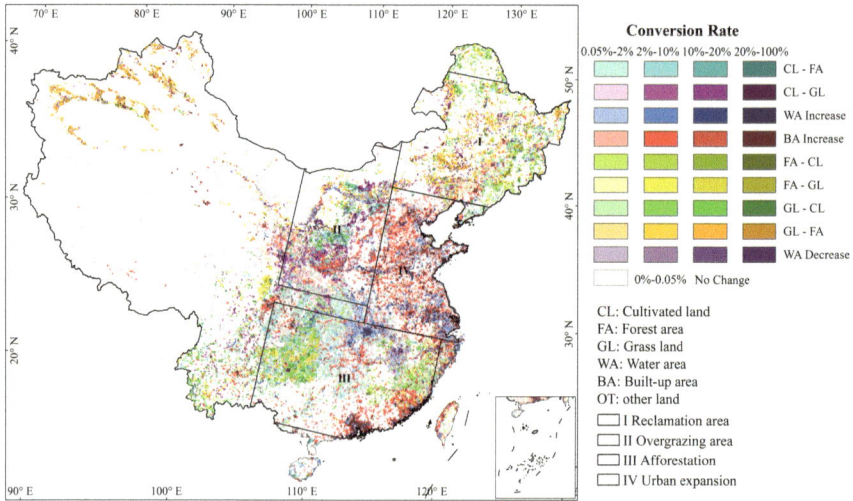

Fig. 5.14 Main land use/land cover change in China, 2000–2008

The reason why we selected these four representative types of land use change is twofold. On the one hand, these four typical types of land use change happened widely in China. On the other hand, these land transformations have significant climatic impacts according to the previous studies. In detail, the large-scale cultivated land reclamation in China started since the last century due to the continual population growth and economic development. Such rapid expansion of cultivated land suggests an intense clearance of natural vegetation, which exerts significant impacts on the climate. Grassland accounts for 40 % of the total land area of China, and the grassland degradation and desertification in China have been greatly intensified due to overgrazing, overexploitation of the natural resources, rapid population growth, and expansion of road network in the past decades (Xu et al. 2012). The serious overgrazing led to grassland degradation and caused severe grassland productivity degradation and increased the frequency of extreme climate events, such as droughts and fierce freeze-up, and consequently led to acute contradiction between the human and nature in this region (Deng et al. 2011). A number of afforestation projects have been carried out in China, such as "Grain for Green Project" and "One Big Four Small Project," which have made great contribution to the forestry area. Afforestation is one of the most noticeable human activities and has affected the climate not only as a carbon sink but also by changing the land surface thermal properties. Rapid urbanization dramatically alters the land surface properties and thus affects the surface energy fluxes and consequently exerts great influence on the regional climate (Seto and Shepherd 2009). To sum up, impacts of typical HLCCs on regional climate and environment should be explored in detail, which can shed light on sustainable land management and climate change mitigation (Bonan 2008).

Data Source

The underlying surface data of the four case study areas are presented in this section. The dataset of future land use and land cover from 2010 to 2100 in China was simulated with the dynamics of land system (DLS) model (Deng et al. 2010) and derived from the study by Yuan et al. (2013). The land use and land cover dataset of 2010 and 2050 was classified according to the land use and land cover classification system of the US Geological Survey (USGS). In this study, the 1 km × 1 km grid data were integrated into the 10 km × 10 km grid data by resampling.

In Fig. 5.15, four bars show the dynamic area of the dominant LUCC in these four study areas from 2010 to 2050, and the percentages attached to the bars indicate the proportions of the representative land use types to the total area of this region in 2010 and 2050 correspondingly.

For Zone I, cultivated land in 2010 is approximately 450,000 km^2, which accounts for 36.7 % of the total land area of this zone, with an increasing rate of 0.3 % from 2010 to 2050. In Zone II, the percentage of grassland decreases from 37 to 36 %. While the forest increases by about 0.22 % in Zone III, the proportion of urban area soars from 8 to 9.5 % in Zone IV.

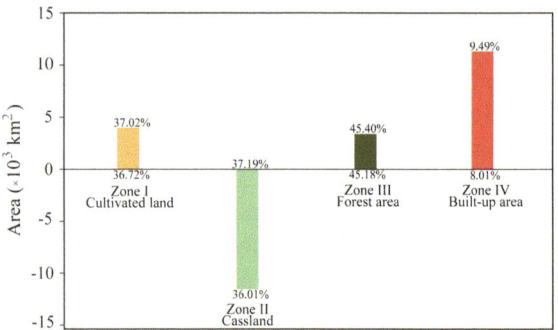

Fig. 5.15 Area and percentage of dominant land use and land cover in each zone

The dataset of RCP6.0 simulated by the fifth phase of the CMIP5 was used as the atmospheric forcing data for the WRF-ARW model. This dataset includes the data such as air temperature, specific humidity, sea-level pressure, eastward wind, northward wind, and geopotential height of years 2010–2050.

Methodology

General Procedure

The WRF Model is a state-of-the-art atmospheric simulation system based on the Fifth-Generation Penn State/NCAR Mesoscale Model (MM5). The Advanced Research WRF (WRF-ARW) model is applied to simulate the impacts of LUCC on the surface energy balance in this study (Fig. 5.16). The future spatial variation of surface energy fluxes due to land use change can be simulated based on different underlying surface data, and the impacts of land use change on the climate can be quantified through the difference of the two simulation results between the baseline test and sensibility test as follows:

$$E_i = P_i - B_i \tag{5.1}$$

where E refers to the effects of future land use change on the surface energy fluxes; i indexes the annual or monthly average surface energy fluxes; and P and B denotes the simulation results obtained with the WRF-ARW model based on underlying surface in 2010 and 2050, respectively, in this study.

Solar Radiation Algorithm

According to the surface energy budget equation, there is a close relationship between the surface net radiation, land surface albedo, downward shortwave radiation, downward longwave radiation, and land surface emissivity (Tao et al. 2013). Energy comes into the system when sunlight penetrates the top of the atmosphere,

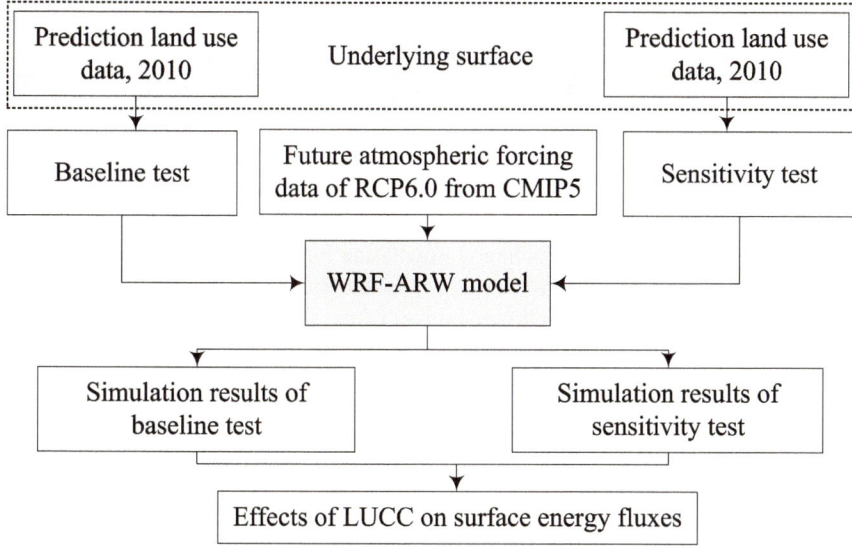

Fig. 5.16 General procedure followed in the study

and it goes out in two ways: reflection by clouds, aerosols, or the surface; and thermal radiation—heat emitted by the surface and the atmosphere, including clouds (Fig. 5.17). The net radiation can be expressed as follows:

$$R_n = (1 - \text{albedo})\text{SWDOWN} + \text{GLW} + \text{LW}_{\text{out}} \tag{5.2}$$

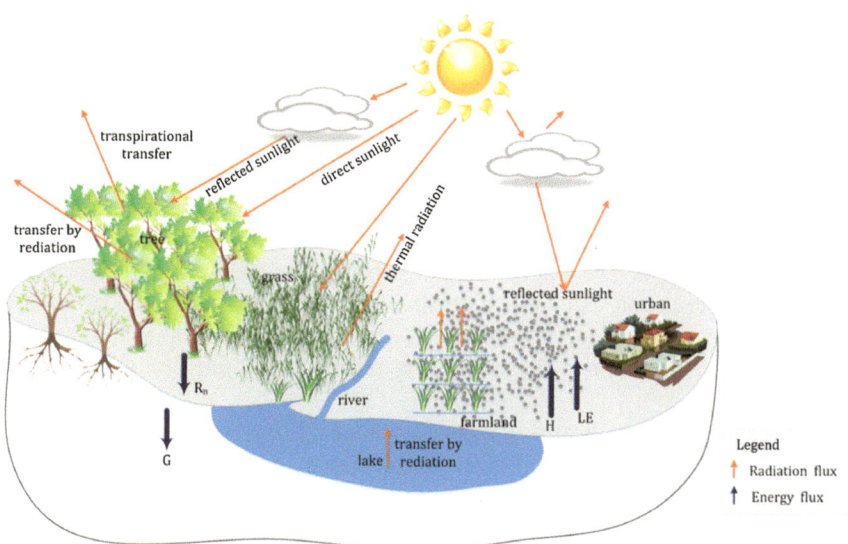

Fig. 5.17 Diagram of the energy balance

$$LW_{out} = \text{emissivity} \times \sigma \times T^4 \tag{5.3}$$

where R_n is the net surface radiation as a function of the surface albedo, incoming shortwave flux (SWDOWN) and longwave flux (GLW) radiation, and outgoing longwave radiation (LW$_{out}$), σ is the Stefan–Boltzmann constant ($\sigma = 5.67 \times 10^{-8}$ J s^{-1} m^{-2} K^{-4}), and T is the land surface temperature. The net radiation is the balance between incoming and outgoing energy at the top of the atmosphere. It is the total energy that is available to influence the climate. This study focuses on how the underlying surface change affects the land surface albedo, downward shortwave radiation, downward longwave radiation, and land surface emissivity to clarify the key influencing mechanism of the future land use change on the regional surface radiation across China.

Surface radiation and energy balance are coupled through R_n. The principal use of this energy is in the phase of changing of water (latent heat flux), changing the temperature of the air (sensible heat flux), and subsurface (ground heat flux). The surface energy budget can be expressed as follows:

$$R_n = \text{HFX} + \text{LE} + \text{GFX} \tag{5.4}$$

where R_n is the net radiation (downwelling minus upwelling solar and infrared radiation), HFX is the sensible heat flux, LE is the latent heat flux, and GFX is the heat flux into the soil.

Results and Discussion

Spatial Variation of Flux Changes

During 2010–2050, there are various changes for heat flux among these four representative zones. Figure 5.18 illustrates the spatial changes of latent heat flux, sensible heat flux, and surface net radiation (red represents increase, and blue represents decrease). Over all three maps, in the overgrazing area (Zone II), each type of heat flux shows a downward trend, while in other three zones, they present upward trends. As expected, changes in surface albedo caused by urban expansion may actually be greater than that caused by cultivated land reclamation, overgrazing, and afforestation. The change of energy balance is mainly about decrease of net shortwave radiation and surface sensible heat flux due to increase of surface albedo. As a consequence of widespread afforestation, plant transpiration and surface evapotranspiration will increase and further latent heat flux will increase correspondingly.

It is found that latent heat flux increases about 2 W/m^2 in the northeast China (Zone I) due to conversion from other land to cultivated land, while in the overgrazing area, it mostly decreases by about 5 W/m^2 (Zone II) (Fig. 5.18). In the afforestation area (Zone III), the southeast China, the increase in latent heat flux will be mostly less than about 21 W/m^2. Meanwhile, latent heat flux decreases by mostly less than about 36 W/m^2 in the surrounding urban expansion area

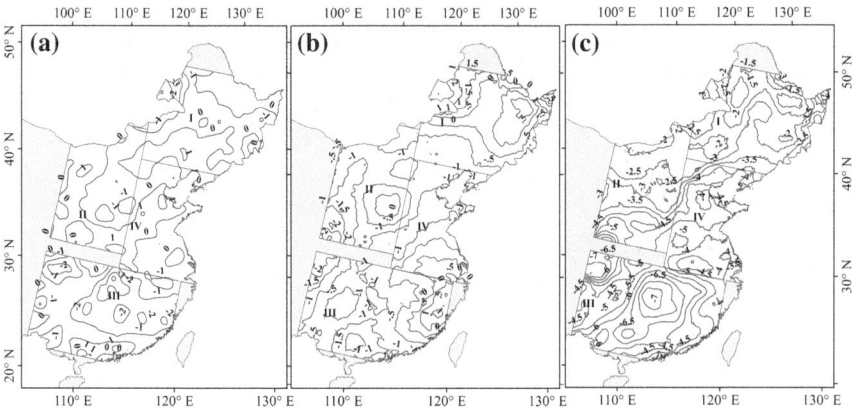

Fig. 5.18 Changes in latent heat flux (**a**), sensible heat flux (**b**), and surface net radiation (**c**) in the four case zones during 2010–2050 (*unit* W/m²)

(Zone IV) due to conversion from other land to built-up area. During 2010–2050, sensible heat flux will decrease by approximately 4, 2, and 13 W/m² in the reclamation area (Zone I), the overgrazing area (Zone II), and the afforestation area (Zone III), respectively. However, in urban expansion area (Zone IV), sensible heat flux will increase by about 40 W/m² and the increase of albedo is associated with sensible heat flux increasing. For Zones I and II, the net radiation will decrease by 4 and 7 W/m², respectively. On the contrary, in Zones III and IV, it will increase by 6 and 3 W/m², respectively, during 2010–2050.

Annual Average Heat Flux

Since each case zone has different types of LUCC, we simulate the changes of heat flux under different types of LUCC in the four case zones. We compare and analyze the differences of heat flux simulations in specified grids with and without changes of land use types. Specifically, in the cultivated land reclamation area, we figure out the grids with the conversions from grassland and forest area to cultivated land, and in the overgrazing area, the specified grids are mainly with the changes to cultivated land, forest area, and built-up area at the expense of grassland. In afforestation area, we focus on the land transformation to forest area (such as from cultivated land, grassland, and built-up area), while in the urbanization area, the statistics are mainly about the area with the cultivated land, grassland, and forest converted to built-up area. Figure 5.19 shows the effect of LUCC on the heat flux from 2010 to 2050, including latent heat flux and sensible heat flux and net radiation.

From Fig. 5.19, we find that the net radiation is the highest heat flux, while the sensible one is the lowest. In Zone IV, the urban expansion area, latent heat flux with urban expansion, and sensible heat flux without urbanization stay at the

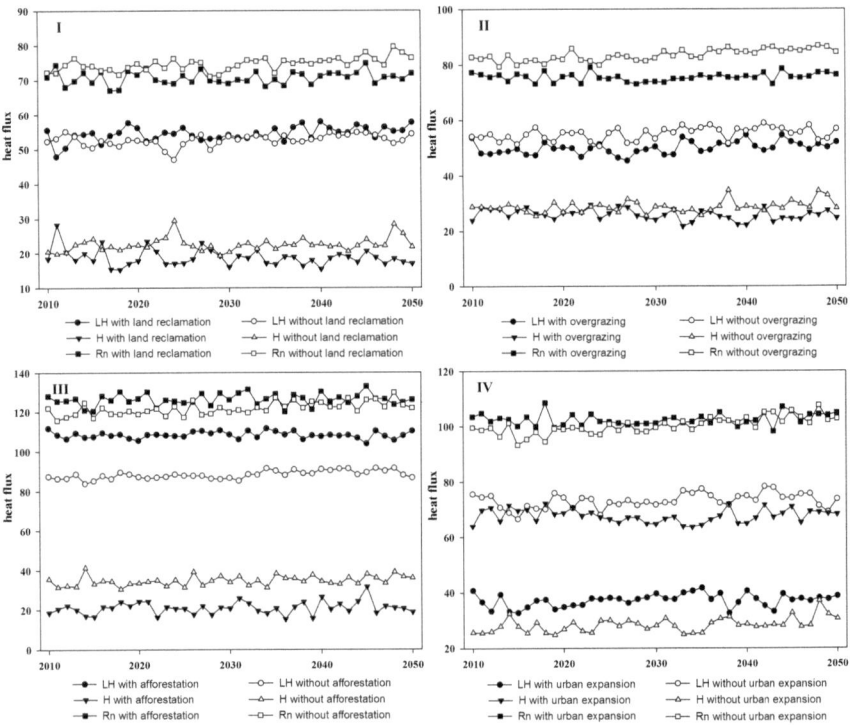

Fig. 5.19 Annual average changes of latent heat flux (LH), sensible heat flux (H), and net radiation (R_n) in the grids with LUCC from 2010 to 2050. *I* reclamation, *II* overgrazing, *III* afforestation, *IV* urban expansion (*unit* W/m²)

bottom. In the former two areas, the differences between the two types of simulation are less than 10 W/m² for latent heat flux, sensible heat flux, and even net radiation. Different from latent heat flux in Zone I, land reclamation will lead to the decrease in the net radiation and sensible heat flux. On the contrary, for the Zone III, latent heat flux has less discrepancy, which is less than 10 W/m², but difference of sensible heat flux between the area with afforestation and area without afforestation is more than 30 W/m², and as for the net radiation, the difference is about 20 W/m². Under the urban expansion, sensible heat flux increases by about 40 W/m², while latent heat flux decreases by about 40 W/m².

Monthly Average Heat Flux

In order to figure out the monthly changes between 2010 and 2050, we count and analyze the three kinds of heat flux in the four case zones on a monthly basis (Table 5.4).

Table 5.4 Monthly average changes of latent heat flux (LH), sensible heat flux (H), and net radiation (R_n) in the grids with LUCC from 2010 to 2050 (*unit* W/m^2)

Month	Zone I			Zone II			Zone III			Zone IV		
	LH	H	R_n	LH	H	R_n	LH	H	R_n	LH	H	R_n
Jan	1.96	−3.28	−3.36	−4.37	−2.91	−7.28	22.12	−14.27	6.70	−34.75	39.72	3.09
Feb	2.29	−3.88	−3.61	−4.84	−2.71	−8.03	20.54	−17.47	2.00	−35.75	38.76	1.13
Mar	1.96	−4.12	−3.80	−4.93	−3.73	−9.09	19.97	−13.36	3.44	−35.74	39.02	0.66
Apr	1.24	−2.93	−4.20	−5.72	−2.99	−8.00	19.21	−15.49	1.73	−36.14	40.91	3.78
May	1.97	−4.12	−4.04	−4.38	−0.32	−4.93	20.99	−11.97	7.38	−33.28	41.05	5.39
Jun	3.17	−3.49	−2.34	−6.11	−1.32	−8.02	19.39	−11.92	5.86	−39.01	39.57	−1.34
Jul	2.09	−2.20	−2.61	−6.35	−2.32	−8.68	19.50	−12.40	5.47	−36.99	38.27	0.08
Aug	0.50	−3.79	−5.78	−4.18	−5.36	−10.34	20.43	−10.25	8.51	−34.65	39.55	3.17
Sep	0.65	−3.85	−4.65	−5.52	−2.91	−7.75	16.60	−11.79	3.35	−39.56	38.90	−2.68
Oct	2.03	−5.36	−4.36	−5.30	−2.15	−7.11	19.67	−12.76	4.20	−36.62	39.65	1.36
Nov	2.17	−4.35	−4.67	−4.86	−3.20	−7.87	20.24	−16.50	3.33	−35.09	38.32	1.04
Dec	2.14	−4.08	−4.67	−5.00	−3.02	−7.27	21.93	−15.20	4.30	−35.57	38.59	2.19
Average	1.85	−3.79	−4.01	−5.13	−2.75	−7.86	20.05	−13.62	4.69	−36.10	39.36	1.49

In the cultivated land reclamation area and overgrazing area, the changes are small with less than 10 W/m^2. Latent heat flux in Zone I will increase by about 1.8 W/m^2 on average, and in June, it will increase by 3.2 W/m^2, which is maximum throughout the whole year. However, sensible heat flux and net radiation will decrease by 3.8 and 4.0 W/m^2 on average, respectively, and the extreme maximum values will appear in August and October with the minimum ones appearing in June and July, respectively. In the Zone II, both types of heat flux will decrease and the smallest change is the change in sensible heat flux. Because of vegetation changes in Zones III and IV, changes in the heat flux present distinctly. Since the forest area increases, latent heat flux and net radiation increase by about 20 and 4.7 W/m^2, respectively, while sensible one will reduce by about 13.6 W/m^2. In contrast, in the Zone IV due to the green reducing, latent heat flux will decrease by 36 W/m^2 with the maximum decrease appearing in September and minimum decrease appearing in May, while sensible one will increase by 39 W/m^2, with minimum and maximum increases of 38 and 41 W/m^2 appearing in July and May, respectively, and the net radiation only changes no more than 1.5 W/m^2 during this period.

Summary

In this study, we investigate the effects of HLCCs on regional climate in typical reclamation, overgrazing, afforestation, and urbanization regions in China with the aid of WRF model. Simulation results show that future HLCC patterns have significant impacts on the surface energy fluxes. From the simulation results, there is a significant variance of albedo under different HLCCs, which suggests that albedo of different underlying surfaces results in spatial variance of the radiation flux. Results show that latent heat flux increases by 2 and 21 W/m^2 in the case of cultivated land reclamation and afforestation, respectively. On the contrary, overgrazing and urban expansion lead to latent heat flux decrease by 5 and 36 W/m^2, correspondingly. Simulation results also indicate that urban expansion has significant effect on sensible heat flux, and it leads sensible heat flux to have an average increase of 40 W/m^2 during 2010–2050. Reclamation, afforestation, and overgrazing have opposite trends, i.e., sensible heat flux decrease. In addition, net radiation would decrease by approximately 4 and 7 W/m^2 in reclamation and overgrazing-dominated region, respectively, and increase by 6 and 3 W/m^2 in afforestation and urbanization regions, separately.

This study also reveals the seasonal pattern of the impacts of HLCCs on surface heat flux. The considerable changes of latent and sensible heat fluxes and net radiation appear in winter and spring, especially in the afforestation and urbanization areas. Furthermore, different HLCCs have various effects on monthly variation of latent and sensible heat fluxes and net radiation. The change tendency of sensible heat flux is consistent with net radiation under various underlying surface types. Urban expansion, afforestation, cultivated land reclamation, and overgrazing have different impacts on latent heat flux, and the effect of urban expansion is by far the most significant of all. The simulated impacts of projected HLCCs on surface energy fluxes in a relatively long run will shed light on screening climate change mitigation strategies through regulating land use activities.

Predicted Impacts of Land Use Change on Surface Temperature in the Typical Areas Around the World

Introduction

LUCC is widely recognized as one of the major contributors to global climate change (Foley et al. 2005). It has long been known that the land use change caused by human activities, e.g., afforestation and deforestation, agriculture practices, and urbanization, can have significant effects on the climate (Foley et al. 2005; Bonan 2008; Kueppers and Snyder 2012). LUCC influences climate by changing the properties of the land surface which is not only the direct heat source of the troposphere, but also one of the main sources of the atmospheric water vapor (Betts et al. 1996). In addition, the land use activities all influence the energy budget and Bowen ratio (the ratio of energy fluxes from one state to another by sensible and latent heating, respectively) of the land surface, the distribution of the precipitation among the soil water, runoff, and evapotranspiration (Phillips et al. 2009; Arora and Montenegro 2011). Thereafter, characteristic changes of the land surface would directly influence the land surface–atmosphere interaction, consequently alter the thermodynamic and dynamic characteristics of the atmosphere, and finally bring out different climate processes and patterns.

Urbanization is an extreme way through which human activities alter underlying surface properties and consequently influence the local climate (Li et al. 2011). The urban heat island effect due to urbanization is a typical example of the influence of LUCC on the regional climate (Shepherd 2005). The urbanization has contributed to 50 % of the increase of land surface temperature in the USA since 1950 (Stone 2009). The widespread impervious surface and the roof and wall of buildings, etc., in the cities all influence the energy flux, circulation of water, and other materials (Stewart and Oke 2012). In addition to the research on the small scale, such as a single city, there is evidence that the impacts of urbanization on climate change are significant at a regional scale. These studies have sufficiently and indisputably demonstrated that urbanization affects climate at regional level.

Forests can affect regional and even multiscale climate by exchanging the planetary energetics, sustaining the hydrologic cycle, and containing and releasing the carbon dioxide and through other physical, chemical, and morphological processes (Bonan 2008), thus playing an indispensable role in balancing the climate and land surface systems. Tropical forest occupies nearly 20 % of total forest land area and accounts for more than 30 % of net primary production (NPP) in the terrestrial ecosystem. It contains about 25 % of the carbon in the terrestrial biosphere and can sequester large amounts of carbon on land surface annually (Bonan 1997, 2008). Though tropical forest could mitigate surface warming through considerable evaporative cooling effect, the temperature would increase if the surface albedo rose to a large extent. Deforestation would offset the water and energy exchange feedback effects by comparing with the reduction of convection and precipitation (Phillips et al. 2009). Therefore, it is no doubt that tropical forest as a significant component would be taken into account for climate change research.

There were 1140 Mha forests converted into cultivated land during 1700–1992 (Klein Goldewijk et al. 2011). The farmland management for adapting to climate change have arisen public concern about the irrigation, no-till agriculture, crop rotation, etc. For example, in turn, irrigation significantly increases the regional moisture to the atmosphere, consequently increasing the regional precipitation (Douglas et al. 2009). On the other hand, irrigation plays a role in reducing the regional temperature and the daily temperature difference (Lobell et al. 2009). Thus, the irrigation area has rapidly increased in the past centuries. According to the history of observation records, the global irrigation area was 8.0 Mha in 1800, increased to 40 Mha in 1900, and then further increased to 2.7 Mha by 2000 (Kueppers et al. 2007). The cultivated land with irrigation is mainly distributed in China, India, Pakistan, Thailand, North America, and the Aral Sea watershed. The cooling effect of irrigation in some regions (e.g., North America, northwest part of India, northeast part of China) is comparable to the warming effect on the magnitude and consequently plays a role in alleviating the climate warming (Kueppers et al. 2007). Besides, the conversion of natural vegetation and cultivated land can have warming or cooling effects on the climate, depending on the type of natural vegetation as well as the established crops (Bounoua et al. 2002). That is to say, different cultivated land management activities can have different climatic effects, and therefore, more researches are needed to identify the climate-related trade-offs associated with future cultivated land management.

Grassland, as an indispensable natural resource, has received more attention than ever before because it plays a crucially important role in sequestrating carbon and preserving biodiversity. As one of the most widespread vegetation types worldwide, grassland covers 15 million km^2 in the tropics (as much as tropical forests) and a further 9 million km^2 in temperate regions (Cleveland et al. 2013), and it accounts for about 25 % of the global land area (Fernandez-Gimenez 2000). On the one hand, as an important land use/land cover, it can provide various ecosystem services such as the provision of the forage, milk, and meat. Besides, grassland also provides some important ecosystem services to regulate the regional climate, e.g., the mitigation of greenhouse gas (GHG) emissions through soil organic carbon (C) and nitrogen (N) sequestration (Barthold et al. 2013). On the other hand, the grassland degradation has declined production of ecosystem on grassland, increasing the frequency of the extreme weathers and natural disasters, such as droughts and fierce freeze-up. As a consequence, they have seriously affected the sustainable development of animal husbandry (Deng et al. 2011).

Since there is great difference in the physical, chemical, and biological characteristics of the land surface in different regions on the Earth, the impacts of the

LUCC on the regional climate systems also vary greatly. For example, the deforestation may lead to the temperature increase in the tropic zone, while it may lead to the regional temperature decrease in the frigid zone (Bonan 2008). Different large-scale LUCCs have effect on the characteristics of regional climate system in different ways, especially the changes of temperature and precipitation (Arora and Montenegro 2011; Degu et al. 2011).

Therefore, in order to completely reflect the relationship between LUCC and the climatic factors, it is necessary to carry out and make comparisons among a series of case studies in different regions and at various spatiotemporal scales and analyze the land surface parameters of the LUCC as well as the climatic response. However, there have been very few relevant researches. Thus, the scientific objective of this study is to estimate and compare the potential impacts of future large-scale LUCC on surface temperature from a global perspective. The WRF model, as the latest generation numerical weather prediction model (Moss et al. 2010), is used as an important tool to quantitatively analyze and compare the possible impacts of the future land use change of different regions on the regional climate. Section "Study Areas" introduces the study areas, especially the characteristics of the dominant land use/land cover and the tendency of LUCC. Section "Data and Methodology" describes the downscaling scheme using WRF-ARW model. The atmospheric forcing dataset and the predicted future land use/land cover in our simulation are also introduced in this section. Section "Results" shows the result of the model performance and outlines the effects of future dominant LUCC on the surface temperature projected by the high-resolution simulation. Finally, Section "Summary" consists of the discussion and conclusion.

Study Areas

In this section, four case studies of potential further land use change as well as their impact on surface temperature are introduced and compared worldwide. In terms of the urbanization impacts in developed megalopolis, the first case study presents some evidences for effects of future urbanization on regional climate based on the simulation of temperature in the Northeast megalopolis of the USA [Fig. 5.20 panel A, panel B (a)]. Further, in order to understand the potential climatological variability caused by future forest vulnerability, we chose Brazilian Amazon region as the study area [Fig. 5.20 panel A, panel B (b)]. The third one chooses India [Fig. 5.20 panel A, panel B (c)] as a typical region of the land reclamation, and the possible impacts of cultivated land reclamation are explored. The possible impacts of the future reclamation of cultivated land are analyzed by forecasting the future cultivated land reclamation and its related changes of temperature. Grassland plays an important role in ecosystem service supply, but studies can be rarely found about the influences of grassland change on climate, especially in Mongolia [Fig. 5.20 panel A, panel B (d)]. Therefore, in the last case study, we simulate the climatological changes caused by future grassland changes in

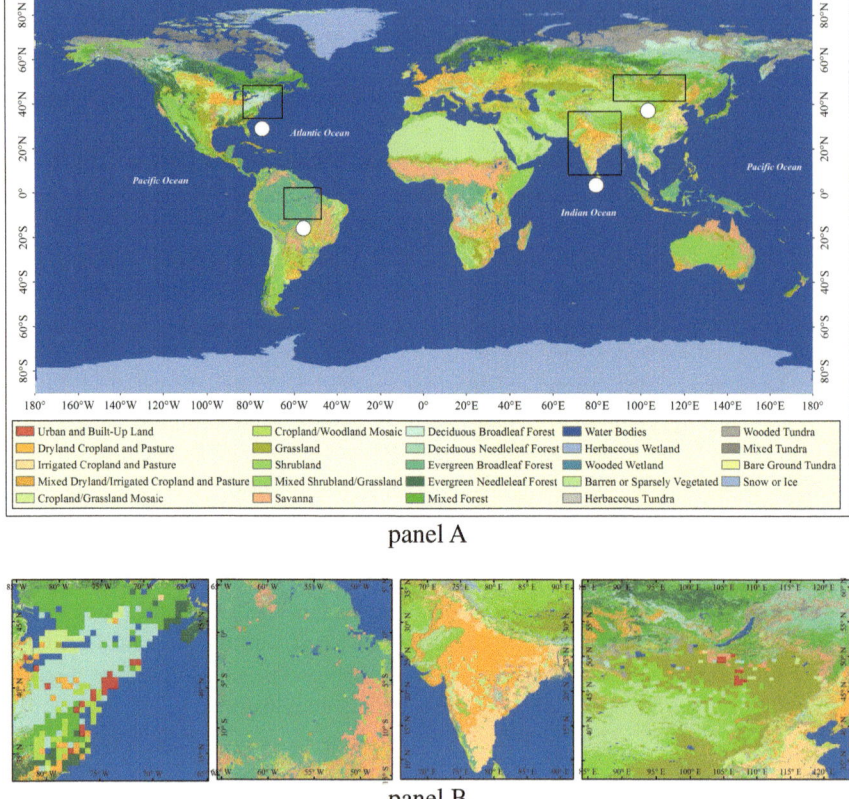

panel A

panel B

Fig. 5.20 Distribution of the study areas (panel **A**) and their land use/land cover (magnified in panel **B**) in 2010

Mongolia along with RCPs of IPCC for the years 2010–2050 (Fujino et al. 2006; Moss et al. 2010).

The Northeast megalopolis is the most populous and largely developed area of the USA (Fig. 5.20 panel A, panel B). A number of cities are distributed in this region, including Baltimore, Boston, Harrisburg, Newark, New York City, Philadelphia, Portland, Providence, Richmond, Springfield, Hartford, and Washington, DC (Powers 2007), and their estimated population could rise up to sixty million by 2025. Thus, this study area is considered as one of the most typical metropolitan area with their economy development and population acceleration to research urbanization effects on climate change.

Amazonia tropical forest (Fig. 5.20 panel A, panel B) constitutes half of the global remaining tropical forests with 6 million km^2, approximately 60 % of which is located in the Brazilian Amazon region (Pawson 2008). The annual average forest clearing rate in Brazil has first increased and then tended to decrease in the past decades. For instance, it increased from approximately 1.3×10^4 km^2/ year in 1990–1994 to over 2.0×10^4 km^2/year in 1995–1996 (Laurance 1998) and then decreased slightly to 1.9×10^4 km^2/year in 1997–2005 (Duchelle et al. 2014). In addition, it has rapidly decreased to 0.7×10^4 km^2/year by 2011 (Davidson et al. 2012). Moreover, these tropical forests would support the massive ecological services to water provisioning and climate regulation; for instance, the 2500 mm annual average precipitation in Amazon reaches could discharge over a trillion m^3 of water into the ocean. In addition, these tropical forests also protect low albedo against radiative forcing effects on keeping lower evapotranspiration rate. Thus, without enough tropical forest, the local residents could not have any settlements with comfortable living conditions in this region (Werth and Avissar 2002), while anthropogenic activities also damage regional ecological system by annual cutting, which have caused the distributions of global climate change (Ramos da Silva et al. 2008). Therefore, land surface changes caused by deforestation would destabilize regional climatic and hydrometeorological variability (Baidya Roy and Avissar 2002) and then induce the climate anomalies such as changing in precipitation and temperature (Schneider et al. 2004).

Cultivation in India, typically, depends upon agricultural irrigation (Fig. 5.20 panel A, panel B). The total area of the cultivated land and pasture reaches around 2.07 million km^2, accounting for 73 % of the study area. There are mainly four kinds of cultivated land in India, which corresponds to the US Geological Survey (USGS) classification, i.e., the dryland cropland and pasture, irrigated cropland and pasture, cropland/grassland mosaic, and cropland/woodland mosaic. The dryland cropland and pasture are mainly allocated in the Gangetic Plain in the north part of India, the Malwa Plateau in the middle part of India, and the northern part of the Deccan Plateau. They occupy nearly 1.18 million km^2, accounting for 56.76 % of the total cultivated land in the study area. The irrigated cropland and pasture is mainly distributed in the northern part of Indus plain, eastern part of Indian Peninsula and the coastal plain in the southern part of India, which accounts for about 32.10 % of the total cultuvated land. The cropland/grassland mosaic and cropland/woodland mosaic only account about 9.82 and 1.32 % of the total cultivated land respectively.

Mongolia is located in the middle of Asia (Fig. 5.20 panel A, panel B) and it is the second largest inland country with a total area of 1.56 million km^2. Most of the country's area is covered by steppes which accounts for more than 70 % of the national area. According to the climate data from Mongolia Meteorological Administration during 1960–2006, the annual average temperature is relatively low in northern part of Mongolia due to its high latitude, which is about -5 °C and even much lower in the winter, while in the southern part, the annual average temperature is stable at 4 °C (Zhang et al. 2013). Besides, the average annual precipitation in Mongolia is 200–300 mm from 1980 to 2006 (Zhang et al. 2007). In the past two decades, the average temperature of Mongolia increased from 1.5 to 2.5 °C by 2–3 times of the world average difference (Yatagai and Yasunari 1994). Because of its large area of grassland cover, the foremost industry in Mongolia is livestock farming, and approximately 30 % of their population is nomadic or semi-nomadic. Thus, grassland resource is crucially important to Mongolia economy. Mongolia grasslands can be mainly divided into five subtypes: forest grassland, typical grassland, mountain grassland, desert grassland, and desert. The high-quality forage grass has been gradually recessing or replacing by inferior plants (bushes, shrubs, etc.) (Barger et al. 2004) that led to many plant species greatly decreasing and then its production gradually being declined as well. For example, the rates of pasture production decline rate in forest grassland, typical grassland, mountain grassland, desert grassland, and desert during 1961–2006 are 40, 52, 39, 33, and 39 %, respectively (Woldemichael et al. 2012).

Data and Methodology

In order to make contribution to analyze and forecast climate change in the future with a global perspective, LUCC in the four study areas during the period from 2010 to 2050 will be firstly identified along with IPCC RCPs. In our study, we simulate the influence of climate change caused by different types of land use change based on the implementation of WRF model with land use/land cover data and atmospheric forcing data.

Downscaling Scheme Using WRF-ARW Model

Land use/land cover data from Landsat TM in 2010 are used as the baseline underlying surface data in this study, while the predicted land use/land cover raster data from 2010 to 2050 are derived from the database of IPCC RCP6.0 (Moss et al. 2010). This database is developed by the Asia-Pacific Integrated Model (AIM) modeling team at the National Institute for Environmental Studies (NIES), Japan. IPCC RCP6.0 is selected as the reasonable land use/land cover patterns in the future because it is a stabilization scenario and the total radiative forcing is stabilized after decades without overshoot by employment of a range of technologies and strategies for reducing greenhouse gas emissions (Fujino et al. 2006). In addition, model output of the latter of IPCC RCP6.0 such as air temperature, specific humidity, sea-level pressure, eastward wind, northward wind, and geopotential height from 2010 to 2050 are used as the atmospheric forcing dataset of WRF-ARW model (Fig. 5.21).

WRF is a next-generation, limited-area, non-hydrostatic, mesoscale modeling system developed by a group of scientists in different institutes (Hernández-Ceballos et al. 2013). It consists of three modes, including preprocessing module of mode (WPS), main module of mode (ARW), and assimilation module of mode

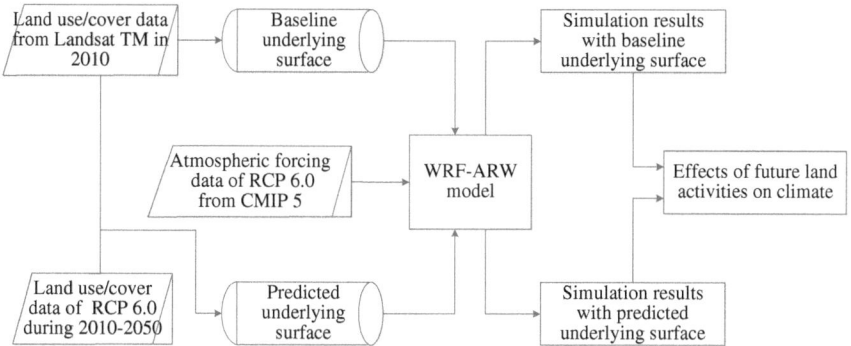

Fig. 5.21 Downscaling scheme using WRF-ARW model for the four case studies

and postprocessing tools of mode data (WRF-VAR). The WRF-ARW (edition 3.3) is used to simulate the impacts of large-scale LUCC on the temperature in this study.

Predicted Changes of Land Use/Land Cover in the Four Typical Areas

In order to effectively investigate and compare the possible future temperature change of the four typical areas, we only analyze the typical land type in the four areas, even though change patterns of all kinds of land use/land cover in 2010–2050 are available in the database. The new urbanization region in the eastern of the USA, the deforestation region in Amazon, the newly increased cultivated land in India, and the grassland region in Mongolia are all extracted and analyzed, separately. All the land use/land cover of these four case study areas is presented in the form of raster data. Thereafter, the change areas (in the form of raster data) of the four types of large-scale LUCC derived from the database of IPCC RCP6.0 are overlaid with the baseline underlying surface data, separately. Consequently, two types of major underlying surface data, baseline underlying surface data directly derived from Landsat TM and predicted underlying surface data by overlaying the typical change area to the baseline underlying surface map, are finally obtained (Fig. 5.22).

In the case study of Northeast megalopolis, both the two underlying surface data are transformed to raster data at the resolution of 30 km × 30 km by resampling. From 2010 to 2050, the land cover conversion is dominated by the conversion from deciduous broadleaf forest to urban and built-up land in the middle part of the study area (Fig. 5.22a). The predicted underlying surface data indicate that the urbanization would continue during 2010–2050.

Figure 5.22b provides the geographic distribution of changed forestry land in Brazilian Amazon. Both the two underlying surface data are resampled at

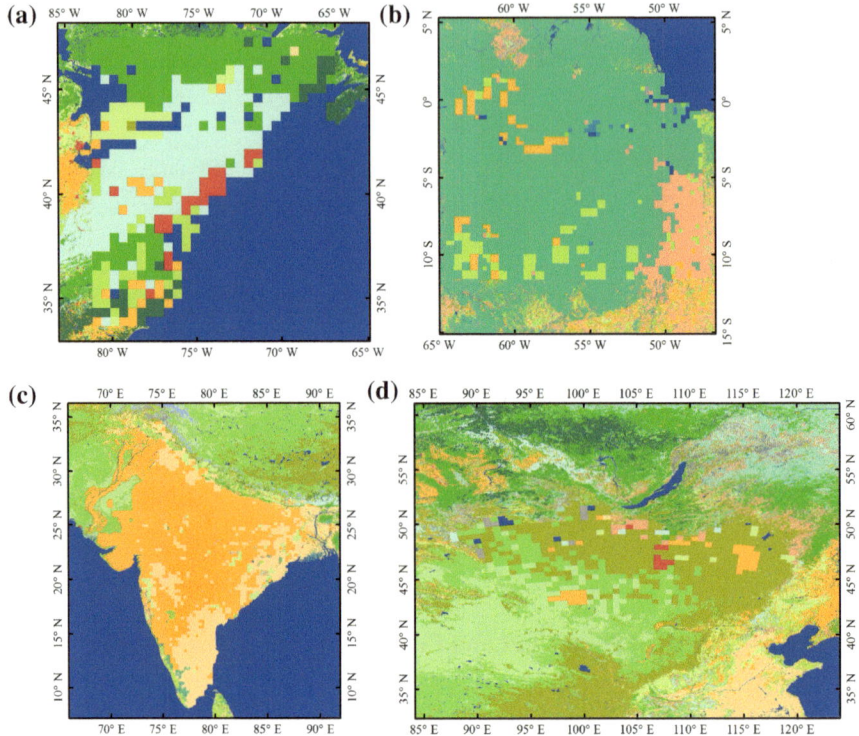

Fig. 5.22 Predicted underlying land use/land cover data in 2050 of the four case studies (**a** Northeast megalopolis of the USA; **b** Amazonia; **c** India; **d** Mongolia)

the resolution of 30 km × 30 km. The degraded and transitioned area is primarily distributed in the periphery of the entire Amazon, as well as along the rivers. Compared to forestry land patterns in panel B (b) of Fig. 5.20, the area in which forest is converted into cropland is principally distributed in the most disturbed and populated area. In addition, the statistical analysis shows that the projected deforested land occupies 15 % of the study area during the period 2010–2050.

In the case study of India, the predicted land use/land cover data in 2050 are obtained by combining the data of the newly increased cultivated land and the former data, and both the two underlying surface data are also resampled at the resolution of 30 km × 30 km (Fig. 5.22c). According to the land conversion data under the AIM performance, the changing area of the cultivated land in the future is around 1.16 million km². It is mainly caused by the increasing demand for the grain due to the population growth, and the area of this kind of conversion will be up to 560 thousand km², accounting for 48 % of the total conversion area. There has been a large area of dryland cropland in India, and more land will be converted into the irrigated land which is of higher productivity with the improvement of the irrigation conditions, development of irrigation techniques, and increase of the grain demand.

Similarly, in the case study of Mongolia, the two underlying land use/land cover data in the downscaling scheme showed in Fig. 5.21 are also transformed to raster data of 50 km × 50 km by resampling. Compared to land use/land cover patterns in panel B (d) of Fig. 5.20, the conversion from grassland to mixed dry-land/irrigated cropland and pasture will dominate in the study area by 2050 (Fig. 5.22d). Besides, it also indicates that there is a strong tendency of conversion from grassland to savanna in this region in future 40 years.

Model Validation

In order to verify the ability of WRF model to simulate the temperature changes in the study areas, we compared the simulation result from WRF model to the observed data of land surface temperature in the case study of India (Fig. 5.23). The simulation results of the monthly average temperature are similar to the observations with a relatively lower error rate which is less than 10 %. Thus, the bias is statistically acceptable because the simulated temperature has roughly a small positive bias due to overestimation in March and September, respectively. Therefore, the result of the model verification indicates that WRF is well behaved to simulate the annual temperature changes and its spatial pattern.

Results

The effects of future large-scale LUCC can be measured by the difference of the simulation results with predicted underlying surface and baseline underlying surface (Fig. 5.21):

$$E_i = R_i - r_i$$

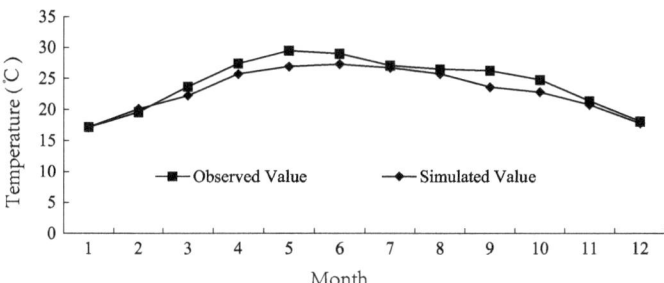

Fig. 5.23 Simulated and observed values of the monthly average near-surface temperature of India in 2010

where i refers to average annual or monthly temperature; E refers to the effects of future predominant land type changes on climate; and R and r are the simulation results of WRF-ARW model with predicted underlying surface and baseline underlying surface, respectively.

On the whole, the surface temperature shows an increasing tendency in these four study areas to some extent (Fig. 5.24). The urban temperature would increase more obviously among the four land types, and we can see that in the USA, where the surface temperature would be significantly increased by severe continuous urbanization over time. Moreover, the spatial pattern of average annual temperature indicates that highly significant spatial heterogeneity of the land surface changes impacts on the temperature changes in Brazilian Amazon and India. In addition, in order to reflect effectively on the tendency of the surface temperature change in the future, in the case study of Northeast megalopolis and Mongolia, we project the result with two periods: 2010–2020 and 2040–2050 for further analysis.

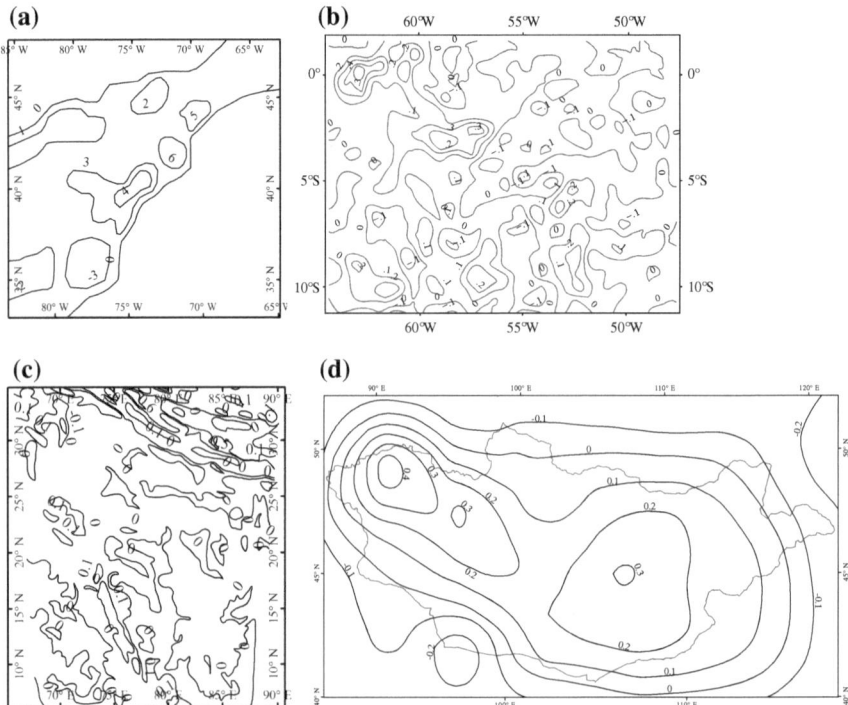

Fig. 5.24 Difference of simulated annual average temperature between 2000 and 2050 (°C) in the four case studies

Simulated Effects of Urbanization in Northeast Megalopolis

According to the simulation results, the urbanization would have effect on average annual temperature in the future (Fig. 5.24a). On the one hand, the largest change in average annual temperature will occur in the expanded urban area during 2010–2050. In some urbanization area, the average annual temperature will even increase by 2–4 °C. On the other hand, the significant average annual temperature decrease will mainly happen in the south of the Northeast megalopolis where the forestry area and its cooling effect should not be ignored. Overall, the results clearly reflect the spatial heterogeneity of the temperature rise caused by urbanization. In addition, the warming effect of urbanization is more obvious in these four case studies. Therefore, to predict the possible influenced area by future urbanization on monthly average temperature, we further analyze by calculating the number of pixels (to reflect the area) in non-urban area with average monthly temperature change exceeding 0.5 °C (Fig. 5.25, panel b).

In terms of the monthly average temperature, the differences of simulated monthly average temperature change indicate that the warming effect of future urbanization will be more significant in the summer than in the winter in new urbanization area (Fig. 5.25a). For example, the urbanization will lead to an increase of monthly average temperature by 3.15 °C in new urbanization area during 2040–2050. There will be also some cooling effect in new urban area in the winter (November, December, and January) during 2010–2020. However, this tendency can be ignored compared to the significant monthly average temperature increase during 2040–2050 and the notable monthly average temperature increase in other months during 2010–2020.

In the non-metropolitan areas, the future urbanization would also drive the monthly average temperature increase because of global climate change, and a tendency of its variety is simulated month to month (Fig. 5.25b). Furthermore, the results show that the influenced area by future urbanization on monthly average temperature would be getting larger in the period from July to December.

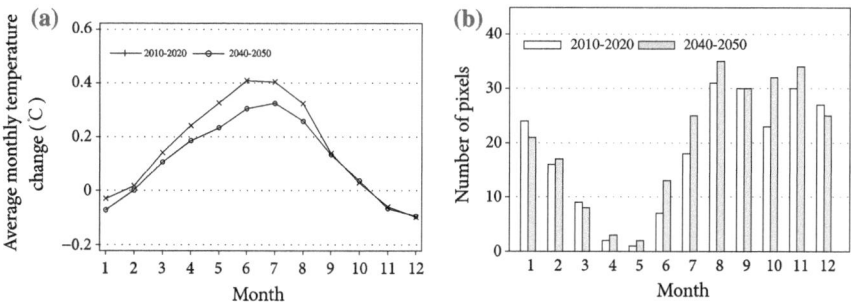

Fig. 5.25 Effects of future urbanization on average monthly temperature in Northeast megalopolis, USA. Panel **a** average monthly temperature changes in urban area and panel **b** number of pixels in non-urban area with average monthly temperature changes exceeding 0.5 °C

Especially in the period of 2040 to 2050, the area that severely influenced by future urbanization on monthly average temperature is more than 6.25×10^4 km^2 from July to December during 2040–2050. Thus, by comparing the number of pixels separately with monthly average temperature change over 0.5 °C during 2010–2020 and 2040–2050, the tendency of urbanization expanding areas could be observed and more notable urbanization effects on monthly average temperature in the future.

Simulated Effects of Deforestation in Brazilian Amazon

The spatial distribution of differences in surface temperature during 2010–2050 is significantly influenced by the forestry land changes (Fig. 5.24b). Obviously, the surface temperature increases over the western region and massive deforestation areas experience a significant increase of temperature. In addition, the spatial pattern of the temperature change during winter corresponds to the land cover to some extent. For instance, the temperature increment was relatively higher in the northwestern and the middle part of the study area, which will mainly experience the land conversion from evergreen broadleaf forest to mixed dryland/irrigated cropland and pasture. Similarly, there is a slight change of the average annual temperature in the south part of the study area, where the land cover is cropland/woodland mosaic.

We also calculate the temperature change variability on different land cover types according to the simulation results (Fig. 5.26). Future temperature in the evergreen broadleaf forest will increase 0.02 °C according to the statistical analysis. As shown in Figs. 5.20 and 5.22b, the land cover change in the study area during 2010–2050 is mainly characterized by the conversion from forestry land to pasture and woodland, and it will lead to a slight change of the surface temperature. This insignificant change of the average annual temperature change can be

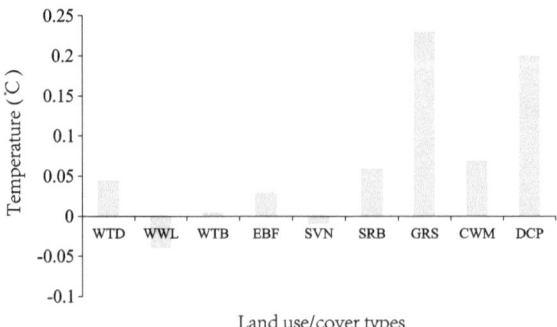

Fig. 5.26 The difference of temperature (°C) of different land covers in Brazilian Amazon in 2050. Note: *DCP* dryland cropland and pasture; *CWM* cropland/woodland mosaic; *GRS* grassland; *SRB* shrubland; *SVN* savanna; *EBF* evergreen broadleaf forest; *WTB* water bodies; *WWL* wooded wetland; *WTD* wooded tundra

explained by the large area of the tropical forests which cover most of the study area. On the contrary, the dryland cropland and pasture as well as the grassland will experience an obvious increment of the average annual temperature change, partly because the relatively small area of the dryland cropland and pasture and the grassland have more uncertainties of the simulation result.

Simulated Effects of Irrigation in India

Simulation result indicates that the future reclamation of cultivated land in India will have some effect on the spatial heterogeneity of the annual average temperature (Fig. 5.24c). Compared to the simulation result with the baseline land surface data, the simulated temperature with the predicted land surface data is higher in the southern and western parts of the study area. Meanwhile, the temperature change is higher in the mountainous area and lower in the plain at the same latitude. Additionally, the simulated result also indicates that the average annual temperature in the inland would show an obvious increase during 2010–2050. However, this warming tendency cannot be found in coastal area.

In order to better understand the relationship between future land conversion and the temperature change, we also calculate the monthly average temperature change in the area where other types of land are converted to irrigated cropland and pasture in summer (Fig. 5.27). Overall, different types of land conversation have different effects on temperature. Specifically, the conversion from evergreen broadleaf forest to irrigated cropland and pasture would lead to an increment of the average monthly temperature by 0.31 °C. Similarly, the conversion from mixed forest to irrigated cropland and pasture will result in an increment of 0.21 °C of the average monthly temperature.

In contrast to these two types of cultivated land reclamation, other kinds of land irrigation have some cooling effect on the surface temperature in summer.

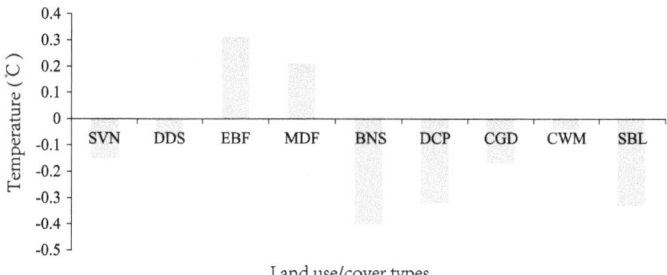

Fig. 5.27 Change of the monthly average temperature due to land conversion from other types to irrigated cropland and pasture in summer in India. Note: *SVN* savanna; *DDS* deciduous broadleaf forest; *EBF* evergreen broadleaf forest; *MDF* mixed forest; *BNS* barren or sparsely vegetated; *DCP* dryland cropland and pasture; *CGD* cropland/grassland mosaic; *CWM* cropland/woodland mosaic; *SBL* shrubland

For example, conversion from barren or sparsely vegetated to irrigated cropland and pasture will lead to a significant decrease of the monthly average temperature by 0.40 °C. Also, the climatic effect of the land conversion from dryland cropland and pasture and shrubland to irrigated cropland and pasture should not be ignored, with a decrement of 0.32 and 0.33 °C, separately. Generally, the different land irrigation activities have different effect on the surface temperature.

Simulated Effects of Grassland Degradation in Mongolia

Overall, the simulation results indicate that grassland degradation in Mongolia also has some effect on the spatial heterogeneity of the annual average temperature (Fig. 5.24d). There would be a slight increase of the surface temperature on the whole region. As for the spatial variation, this increasing tendency of average annual surface temperature would decrease gradually from west to east. In addition, most of the areas (about 70 %) would experience an increase of the annual average temperature, especially the northwestern and central regions of Mongolia, while the increasing tendency can be up to 0.1 °C. Meanwhile, in small part of the southwestern and eastern regions of Mongolia, average annual surface temperature will decrease by about 0.1 °C.

The impact of grassland degradation on surface temperature is more complicated and widespread in the summer than in the winter, in the time period of 2040–2050 than 2010–2020 (Fig. 5.28). Generally, there will be an increasing tendency trend of annual average temperature with an increment of 0.03–0.07 °C during 2010–2020 and 2040–2050. The simulated result indicates that increase tendency of the monthly average temperature from month to month during 2010–2020 and 2040–2050 is basically the same, with obvious seasonal characteristics.

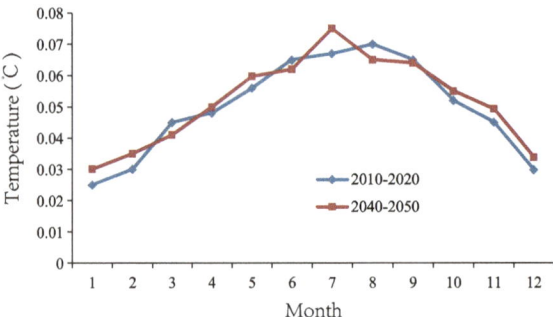

Fig. 5.28 Effects of future grassland degradation on average monthly temperature during two decades in Mongolia

Summary

In this study, the potential impacts of four types of large-scale LUCC (urbanization, deforestation, cultivated land reclamation, and grassland degradation) on surface temperature are analyzed along with IPCC RCPs. For the case study in urbanization in the USA, the land cover change is just characterized by urban expansion, and in the case of grassland degradation in Mongolia, the main land change progress is just the grassland degraded to barren or sparsely vegetated land (one type of land cover in USGS classification which is presented in Fig. 5.20). In contrast to these distinct LUCCs, in the case studies of Brazilian Amazon and India, we observed that there are several land type conversions in the case of irrigation in India and deforestation in Brazilian Amazon. Specifically, the deforestation in Brazilian Amazon includes the land conversion among dryland cropland and pasture, cropland/woodland mosaic, grassland, shrubland, savanna, evergreen broadleaf forest as well as wooded wetland. In addition, the land irrigation in India includes different land conversions from other land types (savanna, deciduous broadleaf forest, evergreen broadleaf forest, barren or sparsely vegetated, shrubland, etc.) to the irrigated cropland and pastureland types.

Urbanization is a predominant LUCC and it is one of the most important human activities that influence the climate system. As more megacities (more than 10 million people) are developing, the climatic effect of this urban expansion pattern should not be ignored. The average annual and monthly temperature change caused by urbanization during 2010–2050 is presented by taking the Northeast megalopolis, USA, as a case study area. The effects of future urbanization on average monthly temperature would vary from month to month and become more and more severe along with urbanization in not only original and new urban area but also non-urban area. The warming effect of future urbanization in original and new urbanization area will be more serious in summer than in winter even if there would be a cooling effect in winter in original urban area. More importantly, the spatial pattern of urban expansion is coordinated with that of the temperature increment; therefore, the results of this study can provide some scientific basis for the urban planning. In addition, more megacities (with more than 10 million people) are developing in developing countries. For example, as one of the biggest developing countries in the world, China has crossed the "urban tipping point," with about 52.57 % of its population now living in cities. The warming effect of urbanization, which is shown in this study, can serve as an alert for the disordered urban expansion.

For climatic changes in the tropical area, a lot of studies have been conducted to explore the influences of vegetation vulnerability on climate variability in Amazon using various methods such as statistical extrapolation and general circulation models (GCMs). However, no debate of reasonable future tropical forest land changes will be complete without taking the Brazilian Amazonia forests as a key component into account in climate researches with high resolution. For the effects of deforestation in the case study of Brazilian Amazon, the land surface changes are quite different from previous studies. Forests are mainly converted into pasture and woodland, distributed along the edge of the study area and river branches. The simulated climatic results caused by these potential future land surface changes show that expanding deforestation would principally trigger the increase of surface temperature in the deforested area. For the deforestation in other parts of the world, such as the boreal forest region, the tendency of the regional annual temperature may be different because of the some other factors such as the different vegetation cover.

As for the cultivated land reclamation, there is still a trend of reclamation of cultivated land in 2010–2050 from the database of IPCC RCP6.0 since the population growth leads the increasing grain demand. The cultivated land reclamation has obvious impacts on the climate in India. Since the different types of reclamation of cultivated land involve different types of LUCC and cropland management modes, there will be some differences in the change of the geophysical parameters such as latent heat flux and sensible heat flux, which will subsequently lead to different effects on the surface temperature in different cultivated reclamation regions. However, with the help of the methodology, the impacts of cropland management practices can be analyzed and emphasized since there will be likely substantial cultivated land change due to the land use management as the society strives to meet growing food demands. For example, China as a large agricultural country plays a very important role in the international grain market and has great demand for grain. As a result, the cultivated land reclamation in northeast China and northern China has lasted for many years.

Among the four case studies, the grassland degradation has some slight effect on the regional temperature in the next 40 years in Mongolia and the effects have spatial and temporal heterogeneity. Annual average temperature will change significantly from April to September, and the maximum change is in July. Additionally, results indicate that the WRF model can well simulate the spatial pattern and change of temperature, although the simulated value is a bit lower than the observed value. The results of this study can also provide an alert for other regions with widespread grasslands, e.g., Inner Mongolia of China, which is located to the south of Mongolia. Compared to urban land and cultivated land, the grassland in the world covers a much larger area and is generally more continuously distributed with a large scale. Therefore, the grassland degradation is analyzed with the resolution of 50 km × 50 km in the case studies of the grassland degradation in Mongolia.

References

Adegoke JO, Pielke RA, Eastman J, Mahmood R, Hubbard KG (2003) Impact of irrigation on midsummer surface fluxes and temperature under dry synoptic conditions: a regional atmospheric model study of the U.S. high plains. Mon Weather Rev 131(3):556–564

Arora VK, Montenegro A (2011) Small temperature benefits provided by realistic afforestation efforts. Nature Geosci 4(8):514–518

Baidya Roy S, Avissar R (2002) Impact of land use/land cover change on regional hydrometeor-ology in Amazonia. J Geophys Res: Atmos 107(D20):LBA 4-1–LBA 4-12

Barger NN, Ojima DS, Belnap J, Shiping W, Yanfen W, Chen Z (2004) Changes in plant func-tional groups, litter quality, and soil carbon and nitrogen mineralization with sheep grazing in an Inner Mongolian grassland. Rangeland Ecol Manage 57(6):613–619

Barthold FK, Wiesmeier M, Breuer L, Frede HG, Wu J, Blank FB (2013) Land use and cli-mate control the spatial distribution of soil types in the grasslands of Inner Mongolia. J Arid Environ 88:194–205

Betts AK, Ball JH (1997) Albedo over the boreal forest. J Geophys Res: Atmos 102(D24): 28901–28909

Betts AK, Ball JH, Beljaars ACM, Miller MJ, Viterbo PA (1996) The land surface-atmosphere interaction: a review based on observational and global modeling perspectives. J Geophys Res: Atmos 101(D3):7209–7225

Biggs TW, Scott CA, Gaur A, Venot J-P, Chase T, Lee E (2008) Impacts of irrigation and anthro-pogenic aerosols on the water balance, heat fluxes, and surface temperature in a river basin. Water Resour Res 44(12):W12415

Bonan G (1997) Effects of land use on the climate of the United States. Clim Change 37(3):449–486

Bonan GB (2008) Forests and climate change: forcings, feedbacks, and the climate benefits of forests. Science 320(5882):1444–1449

Bonan GB, Pollard D, Thompson SL (1992) Effects of boreal forest vegetation on global climate. Nature 359(6397):716–718

Bounoua L, DeFries R, Collatz GJ, Sellers P, Khan H (2002) Effects of land cover conversion on surface climate. Clim Change 52(1–2):29–64

Brovkin V, Ganopolski A, Claussen M, Kubatzki C, Petoukhov V (1999) Modelling climate response to historical land cover change. Glob Ecol Biogeogr 8(6):509–517

Chalita S, Le Treut H (1994) The albedo of temperate and boreal forest and the Northern Hemisphere climate: a sensitivity experiment using the LMD GCM. Clim Dyn 10(4–5):231–240

Chase TN, Pielke RA Sr, Kittel TGF, Nemani RR, Running SW (2000) Simulated impacts of historical land cover changes on global climate in northern winter. Clim Dyn 16(2–3):93–105

Chen F, Dudhia J (2001) Coupling an advanced land surface-hydrology model with the Penn State-NCAR MM5 modeling system. Part I: model implementation and sensitivity. Mon Weather Rev 129(4):569–585

Chen L, Wang J, Fu B, Qiu Y (2001) Land-use change in a small catchment of northern Loess Plateau, China. Agric Ecosyst Environ 86(2):163–172

Cleveland CC, Houlton BZ, Smith WK, Marklein AR, Reed SC, Parton W, Del Grosso SJ, Running SW (2013) Patterns of new versus recycled primary production in the terrestrial bio-sphere. Proc Natl Acad Sci 110(31):12733–12737

Dümenil Gates L, Ließ S (2001) Impacts of deforestation and afforestation in the Mediterranean region as simulated by the MPI atmospheric GCM. Glob Planet Change 30(3–4):309–328

Davidson EA, de Araujo AC, Artaxo P, Balch JK, Brown IF, Bustamante MMC, Coe MT, DeFries RS, Keller M, Longo M, Munger JW, Schroeder W, Soares-Filho BS, Souza CM, Wofsy SC (2012) The Amazon basin in transition. Nature 481(7381):321–328

Debele B, Srinivasan R, Gosain AK (2010) Comparison of process-based and temperature-index snowmelt modeling in SWAT. Water Resour Manage 24(6):1065–1088

Defries RS, Bounoua L, Collatz GJ (2002) Human modification of the landscape and surface climate in the next fifty years. Glob Change Biol 8(5):438–458

Degu AM, Hossain F, Niyogi D, Pielke R, Shepherd JM, Voisin N, Chronis T (2011) The influ-ence of large dams on surrounding climate and precipitation patterns. Geophys Res Lett 38(4):L04405

Deng X-P, Shan L, Zhang H, Turner NC (2006a) Improving agricultural water use efficiency in arid and semiarid areas of China. Agric Water Manage 80(1–3):23–40

Deng X, Huang J, Rozelle S, Uchida E (2006b) Cultivated land conversion and potential agricultural productivity in China. Land Use Policy 23(4):372–384

Deng X, Huang J, Huang Q, Rozelle S, Gibson J (2011) Do roads lead to grassland degradation or restoration? A case study in Inner Mongolia, China. Environ Dev Econ 16(06):751–773

Deng X, Jiang Qo, Zhan J, He S, Lin Y (2010) Simulation on the dynamics of forest area changes in Northeast China. J Geogr Sci 20(4):495–509

Deng X, Zhao C, Yan H (2013) Systematic modeling of impacts of land use and land cover changes on regional climate: a review. Adv Meteorol 2013:11

Douglas EM, Beltrán-Przekurat A, Niyogi D, Pielke Sr RA, Vörösmarty CJ (2009) The impact of agricultural intensification and irrigation on land–atmosphere interactions and Indian monsoon precipitation—a mesoscale modeling perspective. Glob Planet Change 67(1–2):117–128

Duchelle AE, Cromberg M, Gebara MF, Guerra R, Melo T, Larson A, Cronkleton P, Börner J, Sills E, Wunder S, Bauch S, May P, Selaya G, Sunderlin WD (2014) Linking forest tenure reform, environmental compliance, and incentives: lessons from REDD+ initiatives in the Brazilian Amazon. World Dev 55:53–67

Dudhia J (1989) Numerical study of convection observed during the winter monsoon experiment using a mesoscale two-dimensional model. J Atmos Sci 46(20):3077–3107

Feddema J, Oleson K, Bonan G, Mearns L, Washington W, Meehl G, Nychka D (2005) A comparison of a GCM response to historical anthropogenic land cover change and model sensitivity to uncertainty in present-day land cover representations. Clim Dyn 25(6):581–609

Fernandez-Gimenez ME (2000) The role of mongolian nomadic pastoralists' ecological knowledge in rangeland management. Ecol Appl 10(5):1318–1326

Foley JA, DeFries R, Asner GP, Barford C, Bonan G, Carpenter SR, Chapin FS, Coe MT, Daily GC, Gibbs HK (2005) Global consequences of land use. Science 309(5734):570–574

Fu B, Chen L (2000) Agricultural landscape spatial pattern analysis in the semi-arid hill area of the Loess Plateau, China. J Arid Environ 44(3):291–303

Fujino J, Nair R, Kainuma M, Masui T, Matsuoka Y (2006) Multi-gas mitigation analysis on stabilization scenarios using AIM global model. Energy J 343–353

Guan X, Huang J, Guo N, Bi J, Wang G (2009) Variability of soil moisture and its relationship with surface albedo and soil thermal parameters over the Loess Plateau. Adv Atmos Sci 26(4):692–700

Hernández-Ceballos MA, Adame JA, Bolívar JP, De la Morena BA (2013) A mesoscale simulation of coastal circulation in the Guadalquivir valley (southwestern Iberian Peninsula) using the WRF-ARW model. Atmos Res 124:1–20

Hijioka Y, Matsuoka Y, Nishimoto H, Masui T, Kainuma M (2008) Global GHG emission scenarios under GHG concentration stabilization targets. J Glob Environ Eng 13:97–108

Hong S-Y, Dudhia J, Chen S-H (2004) A revised approach to ice microphysical processes for the bulk parameterization of clouds and precipitation. Mon Weather Rev 132(1):103–120

Jin J, Lu S, Li S, Miller NL (2010) Impact of land use change on the local climate over the Tibetan Plateau. Adv Meteorol 2010:6

Klein Goldewijk K, Beusen A, van Drecht G, de Vos M (2011) The HYDE 3.1 spatially explicit database of human-induced global land-use change over the past 12,000 years. Glob Ecol Biogeogr 20(1):73–86

Kueppers L, Snyder M (2012) Influence of irrigated agriculture on diurnal surface energy and water fluxes, surface climate, and atmospheric circulation in California. Clim Dyn 38(5–6):1017–1029

Kueppers LM, Snyder MA, Sloan LC (2007) Irrigation cooling effect: regional climate forcing by land-use change. Geophys Res Lett 34(3):L0370

Kumar A, Dudhia J, Rotunno R, Niyogi D, Mohanty UC (2008) Analysis of the 26 July 2005 heavy rain event over Mumbai, India using the Weather Research and Forecasting (WRF) model. Q J R Meteorol Soc 134(636):1897–1910

Landman W (2010) Climate change 2007: the physical science basis. S Afr Geogr J 92(1):86–87

Laurance WF (1998) A crisis in the making: responses of Amazonian forests to land use and climate change. Trends Ecol Evol 13(10):411–415

Lean J, Warrilow DA (1989) Simulation of the regional climatic impact of Amazon deforestation. Nature 342(6248):411–413

Lee E, Chase TN, Rajagopalan B, Barry RG, Biggs TW, Lawrence PJ (2009) Effects of irrigation and vegetation activity on early Indian summer monsoon variability. Int J Climatol 29(4):573–581

Lee E, Sacks WJ, Chase TN, Foley JA (2011) Simulated impacts of irrigation on the atmospheric circulation over Asia. J Geophys Res: Atmos 116(D8):D08114

Li J, Song C, Cao L, Zhu F, Meng X, Wu J (2011) Impacts of landscape structure on surface urban heat islands: a case study of Shanghai, China. Remote Sens Environ 115(12): 3249–3263

Liu J, Liu M, Zhuang D, Zhang Z, Deng X (2003a) Study on spatial pattern of land-use change in China during 1995–2000. Sci China, Ser D Earth Sci 46(4):373–384

Liu JY, Zhuang DF, Luo D, Xiao X (2003b) Land-cover classification of China: integrated analysis of AVHRR imagery and geophysical data. Int J Remote Sens 24(12):2485–2500

Liu J, Zhang Z, Xu X, Kuang W, Zhou W, Zhang S, Li R, Yan C, Yu D, Wu S, Jiang N (2010) Spatial patterns and driving forces of land use change in China during the early 21st century. J Geog Sci 20(4):483–494

Lobell D, Bala G, Mirin A, Phillips T, Maxwell R, Rotman D (2009) Regional differences in the influence of irrigation on climate. J Clim 22(8):2248–2255

Lobell DB, Bala G, Duffy PB (2006) Biogeophysical impacts of cropland management changes on climate. Geophys Res Lett 33(6):L06708

Mölders N (2008) Suitability of the Weather Research and Forecasting (WRF) model to predict the June 2005 Fire Weather for Interior Alaska. Weather Forecast 23(5):953–973

Mahmood R, Foster SA, Keeling T, Hubbard KG, Carlson C, Leeper R (2006) Impacts of irrigation on 20th century temperature in the northern Great Plains. Glob Planet Change 54(1–2):1–18

Mellor GL, Yamada T (1982) Development of a turbulence closure model for geophysical fluid problems. Rev Geophys 20(4):851–875

Mlawer EJ, Taubman SJ, Brown PD, Iacono MJ, Clough SA (1997) Radiative transfer for inhomogeneous atmospheres: RRTM, a validated correlated-k model for the longwave. J Geophys Res: Atmos 102(D14):16663–16682

Moss RH, Edmonds JA, Hibbard KA, Manning MR, Rose SK, van Vuuren DP, Carter TR, Emori S, Kainuma M, Kram T, Meehl GA, Mitchell JFB, Nakicenovic N, Riahi K, Smith SJ, Stouffer RJ, Thomson AM, Weyant JP, Wilbanks TJ (2010) The next generation of scenarios for climate change research and assessment. Nature 463(7282):747–756

Ozdogan M, Rodell M, Beaudoing HK, Toll DL (2010) Simulating the effects of irrigation over the United States in a land surface model based on satellite-derived agricultural data. J Hydrometeorol 11(1):171–184

Pawson E (2008) Gottmann, J. 1961: Megalopolis. The urbanized northeastern seaboard of the United States. New York: The Twentieth Century Fund. Prog Hum Geogr 32(3):441–444

Phillips OL, Aragão LEOC, Lewis SL, Fisher JB, Lloyd J, López-González G, Malhi Y, Monteagudo A, Peacock J, Quesada CA, van der Heijden G, Almeida S, Amaral I, Arroyo L, Aymard G, Baker TR, Bánki O, Blanc L, Bonal D, Brando P, Chave J, de Oliveira ÁCA, Cardozo ND, Czimczik CI, Feldpausch TR, Freitas MA, Gloor E, Higuchi N, Jiménez E, Lloyd G, Meir P, Mendoza C, Morel A, Neill DA, Nepstad D, Patiño S, Peñuela MC, Prieto A, Ramírez F, Schwarz M, Silva J, Silveira M, Thomas AS, Steege Ht, Stropp J, Vásquez R, Zelazowski P, Dávila EA, Andelman S, Andrade A, Chao K-J, Erwin T, Di Fiore A, Honorio CE, Keeling H, Killeen TJ, Laurance WF, Cruz AP, Pitman NCA, Vargas PN, Ramírez-Angulo H, Rudas A, Salamão R, Silva N, Terborgh J, Torres-Lezama A (2009) Drought sensitivity of the Amazon rainforest. Science 323(5919):1344–1347

Pielke RA Sr, Avissar R, Raupach M, Dolman AJ, Zeng X, Denning AS (1998) Interactions between the atmosphere and terrestrial ecosystems: influence on weather and climate. Glob Change Biol 4(5):461–475

Powers JG (2007) Numerical prediction of an antarctic severe wind event with the weather research and forecasting (WRF) model. Mon Weather Rev 135(9):3134–3157

Puma MJ, Cook BI (2010) Effects of irrigation on global climate during the 20th century. J Geophys Res: Atmos 115(D16):D16120

Ramos da Silva R, Werth D, Avissar R (2008) Regional impacts of future land-cover changes on the Amazon basin wet-season climate. J Clim 21(6):1153–1170

Sacks W, Cook B, Buenning N, Levis S, Helkowski J (2009) Effects of global irrigation on the near-surface climate. Clim Dyn 33(2–3):159–175

Sahu N, Behera S, Yamashiki Y, Takara K, Yamagata T (2012) IOD and ENSO impacts on the extreme stream-flows of Citarum river in Indonesia. Clim Dyn 39(7–8):1673–1680

Schneider N, Eugster W, Schichler B (2004) The impact of historical land-use changes on the near-surface atmospheric conditions on the Swiss Plateau. Earth Interact 8(12):1–27

Seto KC, Shepherd JM (2009) Global urban land-use trends and climate impacts. Curr Opin Environ Sustain 1(1):89–95

Shepherd JM (2005) A review of current investigations of urban-induced rainfall and recommendations for the future. Earth Interact 9(12):1–27

Shiklomanov IA (2000) Appraisal and assessment of world water resources. Water Int 25(1):11–32

Stainforth DA, Aina T, Christensen C, Collins M, Faull N, Frame DJ, Kettleborough JA, Knight S, Martin A, Murphy JM, Piani C, Sexton D, Smith LA, Spicer RA, Thorpe AJ, Allen MR (2005) Uncertainty in predictions of the climate response to rising levels of greenhouse gases. Nature 433(7024):403–406

Stewart ID, Oke TR (2012) Local climate zones for urban temperature studies. Bull Am Meteorol Soc 93(12):1879–1900

Stone B (2009) Land use as climate change mitigation. Environ Sci Technol 43(24):9052–9056

Stubenrauch CJ, Briand V, Rossow WB (2002) The role of clear-sky identification in the study of cloud radiative effects: combined analysis from ISCCP and the scanner of radiation budget. J Appl Meteorol 41(4):396–412

Sud YC, Lau WKM, Walker GK, Kim JH, Liston GE, Sellers PJ (1996) Biogeophysical consequences of a tropical deforestation scenario: a GCM simulation study. J Clim 9(12):3225–3247

Tao Z, Santanello JA, Chin M, Zhou S, Tan Q, Kemp EM, Peters-Lidard CD (2013) Effect of land cover on atmospheric processes and air quality over the continental United States—a NASA Unified WRF (NU-WRF) model study. Atmos Chem Phys 13(13):6207–6226

Werth D, Avissar R (2002) The local and global effects of Amazon deforestation. J Geophys Res: Atmos 107(D20):8087

Wessels KJ, Prince SD, Malherbe J, Small J, Frost PE, VanZyl D (2007) Can human-induced land degradation be distinguished from the effects of rainfall variability? A case study in South Africa. J Arid Environ 68(2):271–297

Woldemichael AT, Hossain F, Pielke R, Beltrán-Przekurat A (2012) Understanding the impact of dam-triggered land use/land cover change on the modification of extreme precipitation. Water Resour Res 48(9):W0954

Wu F, Zhan J, Yan H, Shi C, Huang J (2013) Land cover mapping based on multisource spatial data mining approach for climate simulation: a case study in the farming-pastoral ecotone of North China. Adv Meteorol 2013:12

Xu GC, Kang MY, Jiang Y (2012) Adaptation to the policy-oriented livelihood change in Xilingol Grassland, Northern China. Procedia Environ Sci 13:1668–1683

Yang H, Zhang X, Zehnder AJB (2003) Water scarcity, pricing mechanism and institutional reform in northern China irrigated agriculture. Agric Water Manag 61(2):143–161

Yatagai A, Yasunari T (1994) Trends and decadal-scale fluctuations of surface air temperature and precipitation over China and Mongolia during the recent 40 year period (1951–1990). J Meteor Soc Jpn 72(6):937–957

Yuan Y, Zhao T, Wang W, Chen S, Wu F (2013) Projection of the spatially explicit land use/cover changes in China, 2010–2100. Adv Meteorol 2013:9

Zhang F, Li X, Wang W, Ke X, Shi Q (2013) Impacts of future grassland changes on surface climate in Mongolia. Adv Meteorol 2013:9

Zhang MA, Borjigin E, Zhang H (2007) Mongolian nomadic culture and ecological culture: on the ecological reconstruction in the agro-pastoral mosaic zone in Northern China. Ecol Econ 62(1):19–26

Zhang X, Zhang L, Zhao J, Rustomji P, Hairsine P (2008) Responses of streamflow to changes in climate and land use/cover in the Loess Plateau, China. Water Resour Res 44(7):W00A07

Chapter 6
Land-Use-Oriented Conservation of Ecosystem Services

Jinyan Zhan, Haiming Yan, Chenchen Shi, Yingcheng Liu, Feng Wu and Guofeng Wang

Abstract Ecosystem services suffer from both the human intervention, such as land-use zoning change, and the natural intervention, such as climate change. In this chapter, we first conducted land-use zoning for conserving ecosystem services in the middle reaches of the Heihe River Basin (HRB). Under the background of climate change, regulation services of ecosystem could be strengthened under proper land-use zoning policy to mitigate climate change. We conducted a case study in the middle reaches of the HRB to assess the ecosystem service conservation zoning under the change of land use associated with climate variations. The results show the spatial impact of land-use zoning on ecosystem services in the study area which are significant reference for the spatial optimization of land-use zoning in preserving the key ecosystem services to mitigate climate change. The research contributes to the growing literature in finely characterizing the ecosystem service zones altered by land-use change to alleviate the impact of climate change, as there is no such systematic ecosystem zoning method before. Further, we introduced ecosystem-based adaptation measures for climate change in Qinghai Province. The change of land surface can exert significant influence on the future climate change. We analyzed the effects of herdsmen's adaptation to climate change on the livestock breeding, income, and land surface dynamics with a land surface parameterization scheme. The empirical analysis was first carried out on the impacts of the adaptation measures of herdsmen on their income in the context of the climate change with the positive mathematical programming

J. Zhan (✉) · H. Yan · C. Shi · F. Wu
State Key Laboratory of Water Environment Simulation, School of Environment,
Beijing Normal University, Beijing 100875, China
e-mail: zhanjy@bnu.edu.cn

Y. Liu
College of Resources and Environmental Sciences, China Agricultural University,
Beijing 100083, China

G. Wang
School of Economics and Management, Beijing Forestry University,
Beijing 100083, China

© Springer-Verlag Berlin Heidelberg 2015
J. Zhan (ed.), *Impacts of Land-use Change on Ecosystem Services*,
Springer Geography, DOI 10.1007/978-3-662-48008-3_6

(PMP) model on the basis of the household survey data in the Three-River Source Region, an ecologically fragile area in Qinghai Province, China. Then, the land surface parameterization process is analyzed based on the agent-based model (ABM), which involves the herdsmen's adaptation measures on climate change, and it also provides reference for the land surface change projection. The result shows that the climate change adaptation measures will have a positive effect on increasing the amount of herdsman's livestock and income as well as future land surface dynamics. Some suggestions on the land-use management were finally proposed, which can provide significant reference information for the land-use planning. Finally, we introduced how to build resilience to climate change for conserving ecosystem servers. The ecosystem resilience plays a key role in maintaining a steady flow of ecosystem services and enables quick and flexible responses to climate change, and maintaining or restoring the ecosystem resilience of forests is a necessary societal adaptation to climate change; however, there is a great lack of spatially explicit ecosystem resilience assessments. Drawing on principles of the ecosystem resilience highlighted in the literature, we built on the theory of dissipative structures to develop a conceptual model of the ecosystem resilience of forests. A hierarchical indicator system was designed with the influencing factors of the forest ecosystem resilience, including the stand conditions and the ecological memory, which were further disaggregated into specific indicators. Furthermore, indicator weights were determined with the analytic hierarchy process (AHP) and the coefficient of variation method. Based on the remote sensing data, forest inventory data, so forth, the resilience index of forests was calculated. The result suggests that there is a significant spatial heterogeneity of the ecosystem resilience of forests, indicating it is feasible to generate large-scale ecosystem resilience maps with this assessment model, and the results can provide a scientific basis for the conservation of forests, which is of great significance to the climate change mitigation.

Keywords Land-use zoning · Ecosystem service conservation · Climate change adaptation · Ecosystem resilience · SIZES · Economic analysis · PMP · ABM · AHP

Land-Use Zoning for Conserving Ecosystem Services in the Middle Reaches of the Heihe River Basin

Introduction

Ecosystem services are the natural environment provision and utility the ecosystem provides upon which human beings survival lie (Daily et al. 1997). It is the material basis to maintain the human survival and support earth life system. According to the Millennium Ecosystem Assessment (MA) (MEA 2005), the ecosystem services can be mainly divided into four categories, namely provisioning, supporting, regulating, and cultural services. The degree of human demand influences the sustaining of the ecosystem services (Burkhard et al. 2012). Meanwhile, the services provided by the ecosystem directly affect the human well-being (Jordan et al. 2010; Deng et al. 2013a). Both the natural and human interventions will exert great influence on the ecosystem services, which will hinder the social and economic development and have an impact on the human well-being.

Among those natural interventions of ecosystem services, climate change is likely to affect the abundance, production, distribution, and quality of ecosystems. Therefore, ecosystem services, such as climate stabilization through carbon sequestration, the provision of nonirrigated forage for livestock and wildlife species, the delivery of water which supports fish for commercial and recreational sport fishing, the provision of critical habitat for biodiversity, and many other types of ecosystem services, are likely to be impacted by a changing climate (Shaw et al. 2011). And in return the ecosystem service conservation and restoration under proper land-use zoning policy play a key role in climate mitigation.

As for human interferences that influence the ecosystem services, numerous studies have showed that land-use change associated with climate variations could affect the structure and function of ecosystem and then affect the supply of ecosystem service functions (Watanabe and Ortega 2014). The spatial land-use allocation may affect the provision of a wide range of ecosystem services and is instrumental to their conservation and enhancement (Wang et al. 2012).

Previous studies have shown that, in order to characterize the ecological changes, monitoring landscape pattern changes provides an indirect approach, as such changes would influence a variety of ecological processes and functions (Shrestha et al. 2012). However, ecosystem service zoning, which aims at identifying major regional ecological environment problems and therefore establishes ecosystem service zones, seems to be a more direct way of showing the ecosystem service change associated with land-use change.

Ecological function zoning is one of the hot issues among academia in the field of ecology (Fang et al. 2008). There are various researches of ecosystem service function zoning from different research aspects (Shitikov et al. 2007). However, these researches either focus on only one type of ecosystem or are confined to land-use zoning (Dehring and Lind 2007). The spatial changes of the regional ecosystem service have not been reflected. And case studies of grid scale space dynamic zoning are absent.

A scientifically sound guidance is always needed for planners to reveal the mechanism of the ecological and climatic effect of land-use change and additionally to promote sustainable development. In order to achieve that it is necessary to conduct zoning characterization of the ecosystem service under the comprehensive consideration of land-use/land-cover change and the site condition. While there is lack of a successful model and systematic method to accomplish this goal in the academia, this study addressed the issue by presenting a case study research aimed at empirically exploring how land-use zoning will affect the ecosystem services. Different land-use zoning through the time period from 1988 to 2008 was analyzed and the ecosystem services were identified and zoned by using SIZES. The SIZES can automatically generate the boundaries of the ecosystem service function zone in the case study area after the input of the raster data and therefore visually reveals the temporal and spatial variation characteristics of the regional ecosystem services. Finally, through the spatial and temporal variation of the key ecosystem services generated by SIZES, effects of the land-use zoning on selected ecosystem services are then qualitatively analyzed.

The study area is located in the Heihe River Basin (HRB), China. As one of the largest inland river basins in northeast China, it is characterized by the multiple natural landscapes and is composed of various natural geographic units. As a typical arid area, it comprises almost all the major natural characters. And the living of the residents or farmers there depends largely upon the provision of the ecosystem services. While in the middle reaches, there are 95 % of the cultivated land, 91 % of the population, and over 80 % of the production value of the whole watershed. It is the main agricultural production area of the whole region. However, in recent decades, with the strengthening of human economic activities, a series of ecological environment problems appear, such as grassland degradation, natural forest decrease, species number and glacier area reduction, land desertification and secondary salinization, and water scarcity. These problems are all associated with the land-use/land-cover change and the quality changes of ecological environment. The degraded ecosystem will weaken the ecosystem service function, thus influencing the social-economic development of the whole region. Therefore, we selected the three cities and five counties along the middle reaches as the study area to assess the impact of LUC on ecosystem services. And the method we used will lend some credence to the related research in similar areas.

The following content of this study is divided into four parts, namely methodology, case study, results, and discussion and conclusions. In the methodology part, we will explicitly introduce SIZES. Thereafter, a case study of the middle reaches of the HRB is presented to elaborate the land-use zoning for conserving ecosystem services under the impact of climate change in this ecological fragile region. Major results of this study are stated in the "Results" section. Finally, several discussions are made to conclude this research.

Methodology

We built a system for identifying and zoning ecosystem services (SIZES) to identify the key ecosystem services and their spatial and temporal variation. The system is composed of the following four principal components (PCs): indicator system for ecosystem service evaluation based on analytic hierarchy process (ISESE-AHP), ranking matrix of ecosystem services based on factor analysis (RMES-FA), identification of the core ecosystem services based on principal component evaluation (ICES-PCE), and identification and zoning of ecosystem services based on unsupervised fuzzy clustering (IDES-UFC).

Principle of SIZES

SIZES delimitates the functional zones for regional ecosystem services based on core ecosystem services in the minimum consistent units. The indicator system of ecosystem services in SIZES is developed with the conceptual framework provided by MEA (2003) and the influencing factors at different spatiotemporal scales are accordingly selected. The core ecosystem services, the indicators which comprehensively represent the most important ecosystem services in a functional zone, are identified with factor analysis according to their contribution to the variance of indicators of ecosystem services. The core ecosystem service are interpreted according to the factor loading matrix obtained with the principal component method (PCM) and further calculated on the 1 km × 1 km grid cells. Finally, the functional zones for ecosystem services are delimited with fuzzy clustering analysis and critical areas for ecosystem services are identified. The procedures of SIZES are illustrated more specifically as the following.

1. AHP-Based Indicator System for Ecosystem Services Assessment (ISESA-AHP). An assessment hierarchy is constructed to break down ecosystem services into a series of interrelated factors. Since the regional ecosystem services are influenced by various factors (Shaw et al. 2011), measurable primary indicators for ecosystem services have been selected based on the analytic hierarchy process (AHP) so as to integrate these various influencing factors into one model to determine their interdependencies and the perceived consequences interactively. With a conceptual framework provided by MEA (Gosling 2013), important influencing factors of ecosystem services are elicited according to the driving mechanism of regional ecosystem services. There are three levels in this assessment hierarchy—the overall goal at the top, criteria and subcriteria in the middle, and the measurable primary indicators at the bottom. During the process of selecting primary indicators of regional ecosystem services, three principles are followed: (1) the indicators characterizing ecosystem services are easy to access; (2) the selected indicators should cover all the aspects of regional ecosystem services; and (3) the selected indicators should be coherent with the upper-layer indicators.

2. FA-Based Extraction of the Core Ecosystem Services (ECES-FA). Factor analysis is used to summarize and aggregate the aforementioned primary indicators of ecosystem services into a few PCs that represent certain core ecosystem services. It describes the variance of primary indicators of ecosystem services in terms of potentially fewer comprehensive indicators, that is, PCs. The PCs for a data set are defined as linear combinations of variables that account for the maximum variance within the entire data set and are orthogonal to each other. With the correlation matrix of standardized data of primary indicators of ecosystem services, the eigenvalues, that is the contribution of core ecosystem services to the variance of primary indicators of ecosystem services, are calculated and their corresponding eigenvectors are obtained. The contribution rate of each PC and its cumulative contribution rates are also calculated. It is

assumed that the PCs with higher contribution rate represent better core eco-system services. Only the PCs with the cumulative contribution rate reaching a certain threshold are retained. The threshold is generally set to be above 60 %. Therefore, only the PCs with the cumulative contribution reaching above 60 % are retained and used to represent the core ecosystem services in SIZES.

3. PCM-Based Ranking Matrix of Ecosystem Services (RMES-PCM). The rela-tive importance of primary indicators to the core ecosystem services is ana-lyzed with RMES-PCM, a ranking matrix based on the factor loading matrix obtained with PCM. The factor loading reflects the correlation between one primary indicator and one certain PC. The core ecosystem services are identi-fied and interpreted with the primary indicators with the heavy factor loading; a correlation matrix involving these primary indicators was used as an input for the analysis instead of a covariance matrix, resulting in normalized PCM. Since there may be no representative indicators of the PCs, it is necessary to conduct factor rotation so as to obtain satisfactory PCs. The eigenvalues are used as a criterion for selecting PCs for the factor rotation. Only PCs with eigenvalues above 1 are retained and subjected to varimax rotation for maximizing the cor-relation between primary indicators and PCs (Fezzi et al. 2014). The RMES-PCM is obtained by sorting the primary indicators of ecosystem services according to their factor loadings in descending order. The relative importance of primary indicators of ecosystem services to PCs is ranked to represent cer-tain core ecosystem services.

   With RMES-PCM, the core ecosystem services can be identified and inter-preted according to the most important primary indicators of ecosystem ser-vices. The factor scores of the retained PCs are calculated at each 1 km × 1 km grid cell to spatially represent the core ecosystem services. The factor score can be calculated with many methods, such as the regression estimation method, Bartlett estimation method, and Thomson estimation method. SIZES adopts Bartlett estimation method and spatially represents the core ecosystem services on 1 km × 1 km grid cells in the study area with the factor score.

4. UFC-Based Zonation of Ecosystem Services (ZES-UFC). With the core eco-system services represented at 1 km × 1 km grid cells, the functional zones for ecosystem services are delimitated with unsupervised fuzzy clustering (UFC) algorithm, which is based on the modified fuzzy c-means algorithm and maxi-mum likelihood estimation (MLE). The minimum consistent units (samples) are coherent in the type of core ecosystem services and they are classified into clusters with unsupervised optimal fuzzy clustering algorithm. The core eco-system services of samples within one cluster are more similar to each other than that belonging to different clusters, where the similarity is defined by a distance measure. This partition approach defines the cluster to which each sample belongs, using the elements of the membership matrix. Each sample is attached with different degrees of membership to each cluster. The initial cent-ers of each partition are found with the fuzzy c-means with the Euclidean dis-tance and the fuzzy c-means with the exponential (or MLE) distance are used for precise modeling of the Gaussian mixture in the samples. The number

of clusters is determined by finding the partition with minimum number of Gaussian clusters. Specifically, the number of clusters is determined by beginning with one group at the center of data and adding another group with each iteration process at a location in which membership in the previous groups is small, while examining the partition validity criteria until reaching the maximal number of groups. Finally, the number of clusters providing maximal validity criteria measurements is chosen for optimal partition.

Function Modules Embedded in SIZES

The procedures mentioned above are realized with four function modules in SIZES—module for identifying ecosystem services (MIES), module for ranking ecosystem services (MRES), module for identifying the core ecosystem services (MECES), and module for zoning ecosystem services (MZES).

Module for Identifying Ecosystem Services (MIES)

MIES mainly consists of an indicator system for ecosystem services, representing interaction processes between ecosystem services and influencing factors, and enabling interactive access to SIZES. There are four indicators that fit broadly at the top level in this indicator system, namely supporting, provisioning, regulating, and cultural services. Since the indicators at lower levels may vary among different researches, users of SIZES may recode the program and set these indicators by specifying input parameters and the system will dynamically track the input indicators. Users may also access the online help system, clarifying the underlying assumptions and formal definitions of parameters used in this software. Within MIES of SIZES, each cell (1 km × 1 km) is assigned with multilayer values to represent attributes of each factor (e.g., NPP and land quality). With the 1-km grid pixel data model, MIES can combine, analyze, and interpret the multiple and hierarchical spatial factors efficiently in consideration of maintenance and improvement of ecosystem services.

Module for Extracting the Core Ecosystem Services (MECES)

MECES aims to extract the PCs that represent the core ecosystem services according to established criteria with factor analysis. The core ecosystem services are extracted with the following steps:

1. Assessment matrix for ecosystem services is calculated with

$$X = \left[x_{ij}\right]_{n \times p} = \begin{bmatrix} x_{11} & x_{12} & \cdots & x_{1p} \\ x_{21} & x_{22} & \cdots & x_{2p} \\ \vdots & \vdots & \cdots & \vdots \\ x_{n1} & x_{n2} & \cdots & x_{np} \end{bmatrix} \tag{6.1}$$

where x_{ij} is the value of the jth ($j = 1, 2, L, p$) variable of ecosystem services on the ith ($i = 1, 2, L, p$) grid cell. The vector on the ith ($i = 1, 2, L, n$) grid cell can be presented by. $X_i = (x_{i1}, x_{i2}, L, x_{ip})'$.

2. Sample deviation matrix is constructed with

$$E = \sum_{i=1}^{n} (X_i - \overline{X})(X_i - \overline{X})' \triangleq (e_{kl}) \tag{6.2}$$

where $\overline{X} = \frac{1}{n} \sum_{i=1}^{n} X_i = (\bar{x}_1, \bar{x}_1, \ldots, \bar{x}_1)'$.

3. Sample correlation matrix is calculated as

$$R = \left(\frac{e_{kl}}{e_{kk} e_{ll}} \right), k, l = 1, 2, \ldots, p \tag{6.3}$$

4. The eigenvalues and corresponding eigenvectors are obtained with following steps: The eigenvalue $\lambda_i (i = 1, 2, K, p)$ of the equation $(|\lambda I - R| = 0)$ is obtained with the Jacobi method, and the values are ranked as $\lambda_1 \geq \lambda_2 \geq L \geq \lambda_p \geq 0$. Then $l_i (i = 1, 2, K, p)$, the eigenvector of the corresponding eigenvalue λ_i is calculated. Concurrently, the condition should agree with $l_i = 1$, that is, $\sum_{j=1}^{p} l_{ij}^2 = 1$, where l_{ij} is the jth of vector (l_{ij}). Then, the ith PC is $Z_i = l_i X$.

5. The eigenvalues represent the contributions of these PCs to the variance of primary indicators of ecosystem services. The contribution rate of PCs is calculated with $m_i = \lambda_i / \sum_{j=1}^{p} \lambda_j$. The PCs are sorted according to their contribution rate in descending order and then their cumulative contribution rate is calculated with $\sum_{k=1}^{i} m_i$. When the cumulative contribution rate of the PCs satisfies $\sum_{k=1}^{i} m_i \geq \delta$, where δ is set to be 60 %, the first m PCs is selected to represent the core ecosystem services.

Module for Ranking Ecosystem Services (MRES)

MRES determines the relative importance of primary indicators to the core ecosystem services with RMES-PCM. The factor loading matrix is obtained with the PCM. The factor loading vector is calculated with

$$a_i = \sqrt{\lambda_i} l_i (i = 1, 2 \ldots, p) \tag{6.4}$$

where λ_i is the eigenvalue corresponding to the ith retained PC and l_i is its corresponding orthonormal eigenvector. The factor loading matrix is $A = (a_1, \ldots, a_p)$. The varimax rotation is conducted for maximizing the correlation between primary indicators and PCs for making the representative variables of PCs more prominent. The RMES-PCM is then obtained by sorting the primary indicators of ecosystem services in the factor loading matrix according to their factor loadings in descending order. With RMES-PCM, the core ecosystem services can be identified and interpreted according to the primary indicators of ecosystem services with the factor loading. Finally, SIZES calculates the factor scores of core ecosystem services with Bartlett method and spatially represents the core ecosystem services on 1 km × 1 km grid cells with their factor scores.

Module for Zoning Ecosystem Services (MZES)

MZES utilizes UFC analysis to delimitate and generate functional zones for ecosystem services. The steps of delimitating functional zones for ecosystem services with UFC analysis are as follows:

1. Calculate the average value of all data records:

$$P = \frac{1}{m} \sum_{i=1}^{m} X_i \tag{6.5}$$

   and set the number of clusters (c) to be one, i.e., $c = 1$.
2. Let $U_{ki-\text{old}} = 0$, ($U_{ki-\text{old}} = \text{previous} U_{ki}$);
   The basic fuzzy c-means algorithm (FCM):
3. Calculate the Euclidean distance (for all i, k):

$$d^2(P_k, X_i) = [(P_k - X_i)^T \cdot (P_k - X_i)] \tag{6.6}$$

4. Calculate the degree of membership (U_{ki}) of all data records in all clusters (for all i, k):

$$U_{ki} = \frac{1}{d^2(P_k, X_i)} \bigg/ \sum_{j=1}^{c} \frac{1}{d^2(P_j, X_i)} \tag{6.7}$$

5. Compute the new set of cluster centers, $P_{1 \to k}$ (for all k):

$$P_k = \sum_{i=1}^{m} U_{ki}^2 X_i \bigg/ \sum_{i=1}^{m} U_{ki}^2 \tag{6.8}$$

6. If $\max_{k,i} \left[|U_{ki} - (U_{ki-\text{old}})| \right] > \varepsilon$, then $U_{ki-\text{old}} = U_{ki}$, go to step 3, or else a loop is started and continues step 6.
   The modified fuzzy c-means algorithm based on MLE (FMLE):
7. Compute the MLE distance:
   The sum of memberships within the kth cluster (the a priori probability of the kth cluster):

$$A_k = \frac{1}{m} \sum_{i=1}^{m} U_{ki} \tag{6.9}$$

   The fuzzy covariance matrix of the kth cluster:

$$F_k = \frac{\sum_{i=1}^{m} U_{ki}(P_k - X_i)(P_k - X_i)^T}{\sum_{i=1}^{m} U_{ki}} \tag{6.10}$$

The fuzzy hypervolume of the kth cluster:

$$H_k = [\det(F_k)]^{1/2} \tag{6.11}$$

Then the MLE distance (for all i, k) is:

$$d^2(P_k, X_i) = \frac{H_k}{A_k} \exp[(P_k - X_i)^T \cdot F_k^{-1} \cdot (P_k - X_i)/2] \tag{6.12}$$

8. Repeat steps 4 and 5 using the MLE distance.

9. If $\max_{k,i} \left[|U_{ki} - (U_{ki-\text{old}})| \right] > \varepsilon$, $U_{ki-\text{old}} = U_{ki}$ go to step 7; if not, continue.

10. Compute the partition and average density validity criteria for choosing the optional number of clusters. For instance, the partition density criterion definition is:

$$\text{PD}(c) = \frac{\sum_{j=1}^{c} A_j}{\sum_{j=1}^{c} H_j} \tag{6.13}$$

11. Create an imaginary center that has the same distance (very far) from all data records. For instance, by choosing the initial distance to the new cluster as:

$$d^2(P_{k+1}, X_i) = 10 \cdot \text{variance } (X) \ (i = 1, \dots, m) \tag{6.14}$$

12. Let $c = c + 1$. If $c \leq c_{\max}$ ($c_{\max}$ is the maximum feasible number of clusters in the data), $U_{ki-\text{old}} = 0$ and go to step 4; or else, the UFC algorithm is completed.

The matrices U and P created in step 9 are the required outputs of the UFC algorithm.

Case Study in the Heihe River Basin

Study Area

The HRB, with an altitude between 2000 and 5500 m, is one of the largest inland river basins in Western China, which stretches over 821 km and covers an area of 142.9 thousand km^2. It is characterized by the multiple natural landscapes of ice, river, lake, oasis, and desert and is composed of the natural geographic units of high mountain snow-ice zone, plain oasis zone, and Gobi desert zone. As a typical arid area, it comprises almost all the major natural characters and therefore possesses various ecosystem services. And the living of the residents or farmers depends largely upon the provision of the ecosystem services. In the middle reaches, the runoff has been fully used, while 95 % of the cultivated land, 91 % of the population, and over 80 % of the production value of the whole watershed gathered here. It is the main agricultural production area of the whole region. Both human and natural interventions will exert great influence on the ecosystem services, which will hinder the social and economic development and thus impact the human well-being.

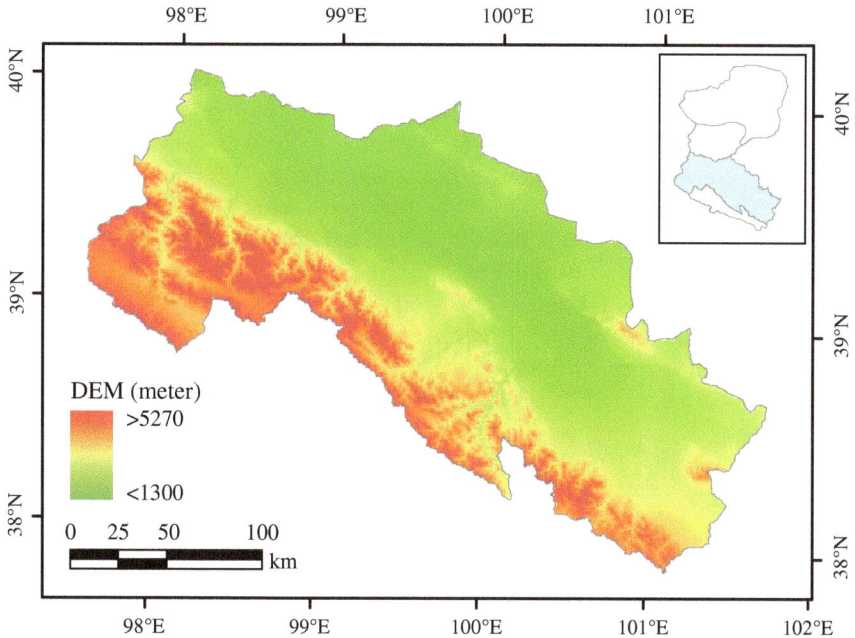

Fig. 6.1 Location and geographic boundaries of the middle reaches of the HRB

In this situation, climate change plays a key role in altering the provision service of the middle reaches and the region depends largely on the regulating services of ecosystem to mitigate this change. Therefore, investigating how the land-use zoning could conserve the ecosystem service and thus counter the impact of climate change is of great significance in this area. So, we selected the three cities and five counties along the middle reaches as the study area (Fig. 6.1).

Identifying the general characteristics of the regional ecological environment is of great significance to the rational regional ecosystem service function identification. In the middle reaches, the key ecological environment problems are land desertification and salinization and water pollution. Besides, due to the lack of monitoring and control measures, the condition is getting worse. The reduction of grassland area caused by the massive land reclamation as well as overgrazing leads to the grassland degradation. The unreasonable industrial structure not only limited the regional economy development, but also worsened the supply and demand contradictions of the water resources, leading to further problems such as land desertification and salinization.

Data Selection

The selection of the ecosystem services is based on the characteristics of the study area mentioned above and is under the guidance of MA, which has been systematically elaborated in Sect. Principle of SIZES, "(1) AHP-based indicator system for ecosystem services evaluation" part. In this study, we divided the ecosystem services in the middle reaches of HRB into three major categories, namely provisioning services, regulating services, and supporting services. And in each category, there are subcategories and further items, each being relevant to the land-use change process that characterizes this study; thus, an index system of the key ecosystem services of the HRB was built (Tables 6.1 and 6.2). Given the fact that the cultural service function is difficult to quantify, this research did not take the cultural function into consideration.

In this study, the information about land-use change was derived from the remote sensing images of the HRB in the years of 1988, 1955, 2000, and 2008 through the human–computer interactive interpretation (Fig. 6.2). And the information can be used as the basic data for analyzing the different land-use zoning.

Data Processing

Our study uses a land-cover classification data set developed by the Chinese Academy of Sciences (CAS) (Liu et al. 2003) as the background data for spatially identifying the ecosystem services. The data set is derived from Landsat TM/ETM

Table 6.1 Key ecosystem service indicators in the HRB

Categories	Sub-categories	Indicators and description
Provisioning services	Food	Cultivated land percentage %
	Freshwater	Water area percentage %
	Fuelwood	Forestry area percentage %
Regulating services	Climate regulating	Air temperature: mean annual temperature (0.1 °C)
		Temperature departure
		Precipitation departure
		Cumulated temperature (>10 °C): annually cumulated temperature of daily mean air temperature above 10 °C (0.1 °C)
		Sunshine hours: sunshine hours rectifying the spatial variability of solar radiation
		Sunshine hours departure
	Species diversity	Landscape diversity index(Shannon diversity index)
Supporting services	Soil formation	Loam proportion %
		Slope: terrain slope derived from DEM (0.01 degrees)
	Nutrient cycling	Soil phosphorus content %
	Primary production	Net primary production g cm^{-2} year^{-1}
	Grassland area	Grassland percentage %

Table 6.2 Descriptive statistics of key ecosystem service indicators in the Heihe River Basin

Variable	Unit	Mean	Std. Dev.	Min	Max
Cultivated land	%	5.52	1873.59	0	100
Water area	%	4.49	1616.67	0	100
Grassland	%	16.79	3091.04	0	100
Forestry area	%	1.12	660.01	0	100
Air temperature	°C	0.51	2.54	0	0.96
Temperature departure	°C	0	2.52	−5.59	4.27
Precipitation departure	mm	0	75.04	−137.93	213.55
Cumulated temperature	°C	2479.10	733.29	461.29	3942.56
Sunshine hours	h	8.33	0.35	7.31	9.30
Sunshine hours departure	h	−2.85e−15	0.31	−1.00	0.76
Landscape diversity index	–	−37.81	116.46	−460.52	460.52
Loam proportion	%	22.50	6.37	4	34
Slope	0.01°	4.72	5.53	0	60.57
Soil phosphorus content	%	0.07	0.01	0.02	0.08
Net primary production	g cm-2 year-1	97.54	112.16	0	897

Note *Total observation: 161584

images of bands 3, 4, and 5 with a spatial resolution of 30 m × 30 m. And the classification contains 6 types of land uses/land covers, including cultivated land, forestry, grassland, water area, built-up area, and unused land (Liu et al. 2010).

Raster Data Processing

The identification of the key ecosystem service function is based on the raster data model. The indicators of the ecosystem service function are characterized by multi-source and multi-scale. Therefore, the 1-km area percentage data (Deng et al. 2010a, 2011a) (Fig. 6.3) is an expressive way of data integration and fusion in monitoring, forecasting, and analyzing the regional ecosystem service changes. The procedure to generate the 1-km area percentage data set includes four steps. The first step is to generate a vector map of land-cover and land-use change during the study period at a scale of 1:100,000 based on the remote sensing Landsat TM/ETM data. The second step is to generate a 1-km FISHNET vector map georeferenced to a boundary map of the study area at a scale of 1:100,000. Each cell of the generated 1-km FISHNET vector map owns a unique ID. The third step is to overlay the vector map of land-cover and land-use change with the 1-km FISHNET vector map. This is done by aggregating converted areas in each 1-km grid identified by cell IDs of the 1-km FISHNET vector map in the TABLE module of ArcGIS software. Finally, the area percentage vector data are transformed into grid raster data to identify the conversion direction and intensity. The design of working flow insists on no loss of area information. Without special notification, the statistical area of cultivated land according to the GRID data is survey area by satellite remote sensing data, which can be called "gross area."

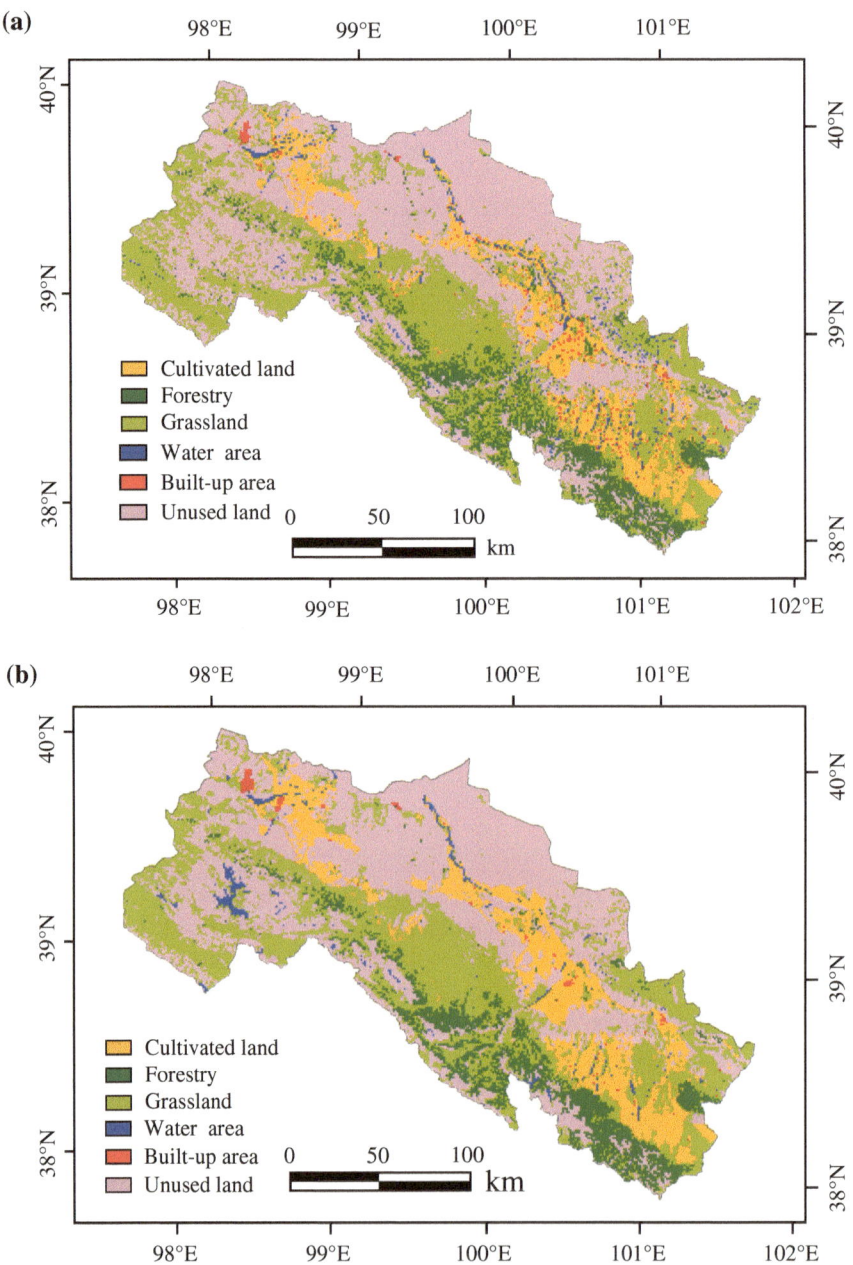

Fig. 6.2 **a** Land-use patterns of middle reaches of the HRB in 1988 and **b** land-use patterns of middle reaches of the HRB in 2008

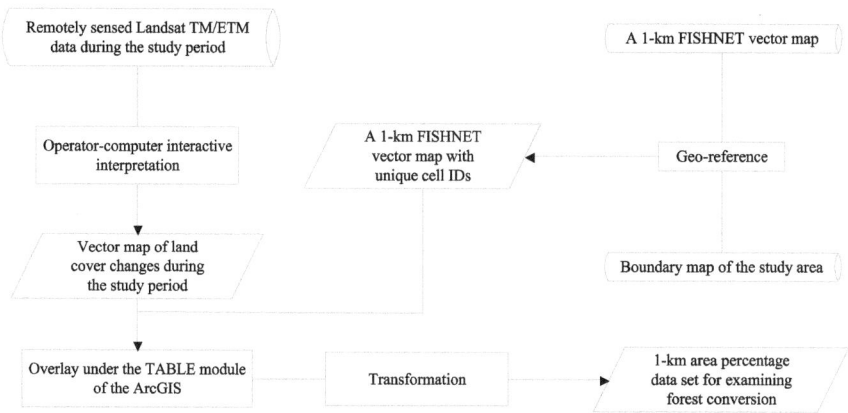

Fig. 6.3 Flowchart of generating 1-km area percentage data set of land-use change

Indicator Data Processing

Among the indicators that are needed in the identification of the key ecosystem services (Fig. 6.4), the raster data of the net primary production are provided by the CAS. The data sources and processing methods of other indicators are listed below. (1) Food, water, and fuelwood supply: in order to precisely characterize the supply of grain, water, and fuelwood in each cell, in this research, we use the percentage of the cultivated land, water area, and forestry to indicate these provisions. The 1-km area percentage data processing method is shown above in Fig. 6.3. (2) Climate variables: All the climatic variables are generated from the site-based observations from the China Meteorological Administration. The spline interpolation algorithm is employed to make the surface data of climatic variables acquired at observation stations (Price et al. 2000). The values for the climatic variables during simulation period are estimated using the space-time stochastic model (Hutchinson 1995). (3) Landscape diversity index: We selected the Shannon diversity index to measure the landscape diversity index (Zhou et al. 2014).

The Shannon diversity index (H) is expressed as (Jost 2006):

$$H = \sum_i p_i \ln(p_i) \qquad (6.15)$$

where p_i is the share of land-use type i in total area of the landscape.

(4) Soil data: Soil data were from multiple sources. Firstly, the specific soil survey points were calibrated on local administrative division base map, based on carefully comparing the information on National Secondary Soil Survey data and 1:250,000 topographic maps. Secondly, an Albert projection value for specific soil survey point was assigned onto the map and generated a soil survey base map. After that, soil attribute data were linked onto soil survey base map by using the Kring interpolation algorithm and generated spatial soil data encompassing soil type and physicochemical characters. Finally, these soil data were

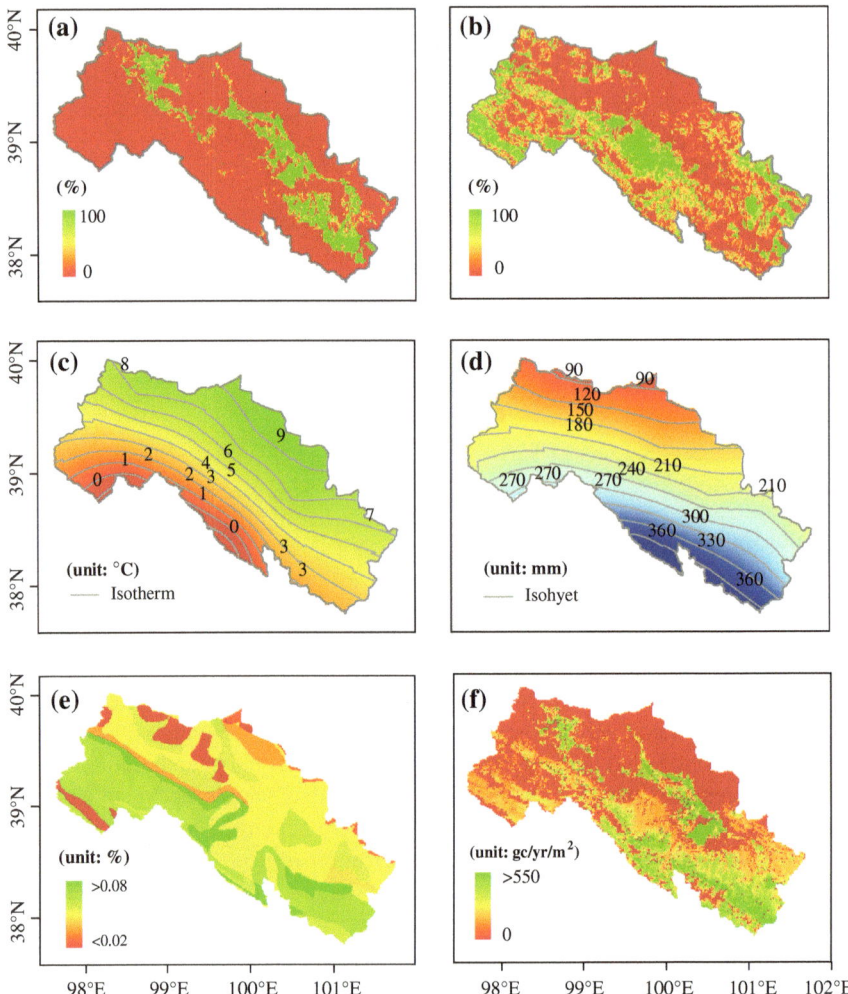

Fig. 6.4 Spatial dispersion and characterization of the ecosystem service indicators, **a** cultivated land percentage, **b** grassland area percentage, **c** air temperature, **d** precipitation, **e** soil phosphorus content, and **f** net primary production

saved as ArcGIS Grid format, with soil type, soil depth, soil organic content, and soil texture information. (5) Elevation data: The elevation data were derived from China 1:250,000 digital elevation models, with a spatial resolution of 50 m. Average slope information was extracted on each 1 km × 1 km plot by GIS spatial analyzing function; then it was used as basic information for land suitability evaluation.

Results

Land-Use Change

Land-use changes, which are mainly caused by human activities, are the major driving force of the ecosystem service changes. The land-use changes in the study area from 1988 to 2008 (Table 6.3) were characterized by a general increase of cultivated land, with a decrease in forestry, grassland, and water area. Built-up area recorded the most significant expansion, being the result of rapid urbanization. Among all six types of land use, the unused land accounts for the largest proportion in all of the four study time periods, which is composed mainly of alp and desert and concentrated mainly in the north part of the region, where the major ecosystem service there is the climate regulating services.

As for the transformation of land-use types (Table 6.4) characterizing the ecosystem structure changes, we can see that the increased cultivated land mainly comes from grassland and used land. During this process, though the ecosystem service of the cultivated land is strengthened, the grassland's ecosystem services, which are also crucial in providing provision and supporting services, are impaired. Therefore, in land-use zoning, it is important to consider not only the amount of the land-use conversion but also the direction of the conversion among various land-use types.

Table 6.3 Land-use change during 1988–2008 (*Unit* km^2)

Land-use type	1988	1995	2000	2008	Change rate during 1988–2008 (%)
Cultivated land	4650	4290	4814	5194	11.70
Forestry	3509	3575	3538	3505	−0.11
Grassland	13382	13642	13186	13269	−0.84
Water area	739	1075	750	701	−5.14
Built-up area	390	365	419	466	19.49
Unused land	17718	17446	17686	17239	−2.70

Note *Change rate during (R) 1988–2008 is calculated based on following formula: $R = (B − A)/A \times 100\%$; A represents the area of each type of land in 1988, and B represents the area of each type of land in 2008

Table 6.4 Land-use transformation matrix during 1988–2008 (*Unit* km^2)

	1	2	3	4	5	6
1		1.84	42.61	2.16	23.41	4.06
2	24.99		14.63	0.06	0.34	2.95
3	345.92	6.46		3.71	7.15	10.28
4	45.23	0	5.44		0.98	5.35
5						
6	201.16	2.77	41.41	12.99	31.83	

Note *1 cultivated land, 2 forestry, 3 grassland, 4 water area, 5 built-up area, and 6 unused land

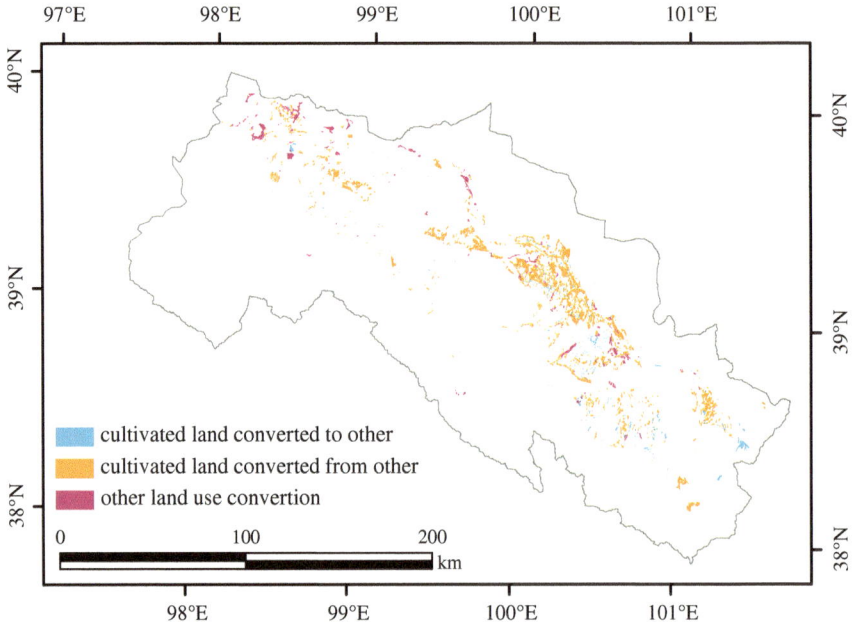

Fig. 6.5 Spatial land-use change during 1988–2008

The spatial land-use change (Fig. 6.5) mainly occurred in the mid-north, mid-central, and mid-south area, which could explain the boundary shifting of the ecosystem service zone in these areas. Through studying the historical land-use zoning policy and its impact on ecosystem services, we aim at optimizing the spatial planning of land use to conserve ecosystem services under the impact of climate change.

Ecosystem Service Zoning Change

In identifying and zoning of ecosystem services, firstly, we applied the AHP to build an indicator system (Table 6.1) with the conceptual framework provided by MEA for ecosystem service assessment with 15 indicators that proved to be the major driving forces of ecosystem services in the middle reaches of the HRB. The data of these indices were then spatially represented with the 1-km grid pixel data model. Then using a correlation matrix, factors with eigenvalues >1 were retained, and factors were subjected to varimax rotation to maximize the correlation between factors and measured attributes. The magnitude of the eigenvalues was used as a criterion for interpreting the relationship between ecosystem services and factors. Afterward, to summarize and aggregate the selected ecosystem service attributes, a principal component evaluation was performed on the aforementioned variables. We assumed that PCs receiving a high contribution best represented the core ecosystem services. Therefore, we retained only the PCs with a cumulative contribution greater than 60 %. An UFC algorithm was then performed using values for the retained PCs to develop the functional zones for ecosystem services, with subsequent valuing of each 1-km grid cell after the iterations reaching their threshold. Finally, we grouped data in clusters, where within each cluster the data exhibit similarity. Besides, in this study, the center of gravity of each service zone in each time is generated in the ArcGIS to analyze the shifts of center of gravity (Fig. 6.6).

In 1988 and 1995, there are five major clusters, namely climate regulating function, food provision function, nutrient cycling supporting function, soil supporting function, and supporting function. The name of each cluster is determined according to the above-mentioned principal component analysis within each cluster. Similarly, four clusters have been identified in 2000 and 2008, which are climate regulating function, food provision function, soil supporting function, and supporting function.

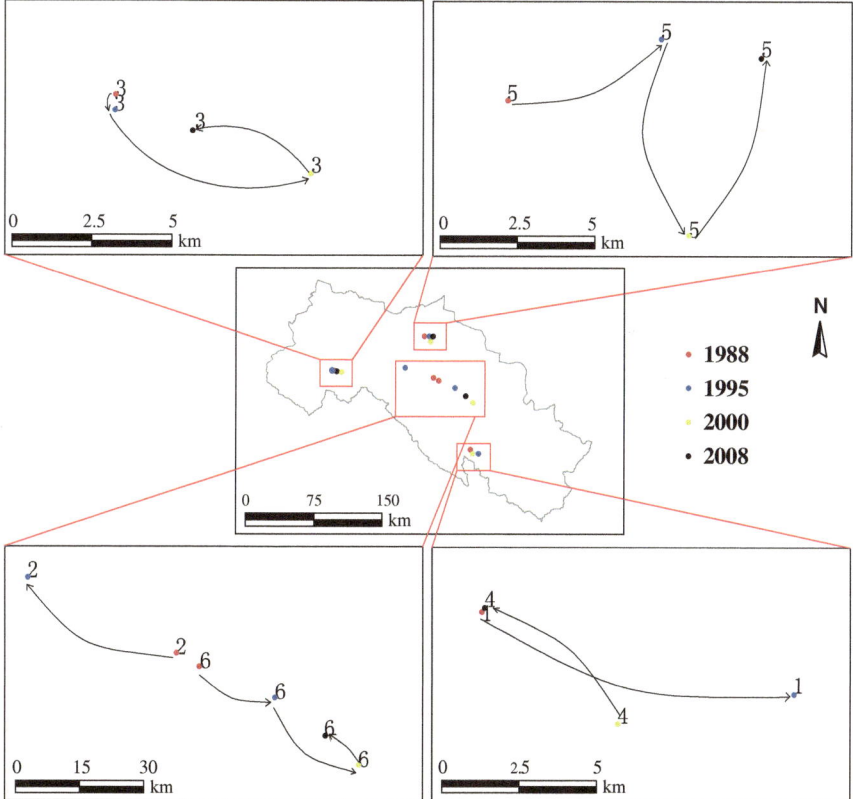

Fig. 6.6 Spatial shifts of the center of gravity of ecosystem service zones. Note: *1* supporting function; *2* nutrient cycling supporting function; *3* soil supporting function; *4* fuelwood provision function; *5* climate regulating function; *6* food provision function

Each functional zone involves one kind of ecosystem services, which is displayed with different zoning boundaries. The functional zones for ecosystem service obtained with SIZES comprehensively reflect the spatial clusters of ecosystem services in the middle reaches of the HRB. There were some uncertainties, but the result obtained with the SIZES still reflected the conditions of local ecosystem services in the HRB; the result indicated the boundaries of functional zones for ecosystem services were conspicuously different from the administrative boundaries (Fig. 6.7). The functional zones generally expand across several counties; there is generally one major kind of ecosystem services as well as several kinds of secondary ecosystem services in each county.

The ecosystem service zoning results (Fig. 6.7) showed that, during 1988–2008, the middle reaches maintained the climate regulating function, which is the major function of the area mainly distributed in the north region. It plays a key role in mitigating climate change. The boundaries of the climate regulation service zone changed slightly and the center of gravity (Fig. 6.6) changed the most among all the other services in this region, which is a result of the changing in land-use structure and climate and its own fragile ecological condition in the northern part.

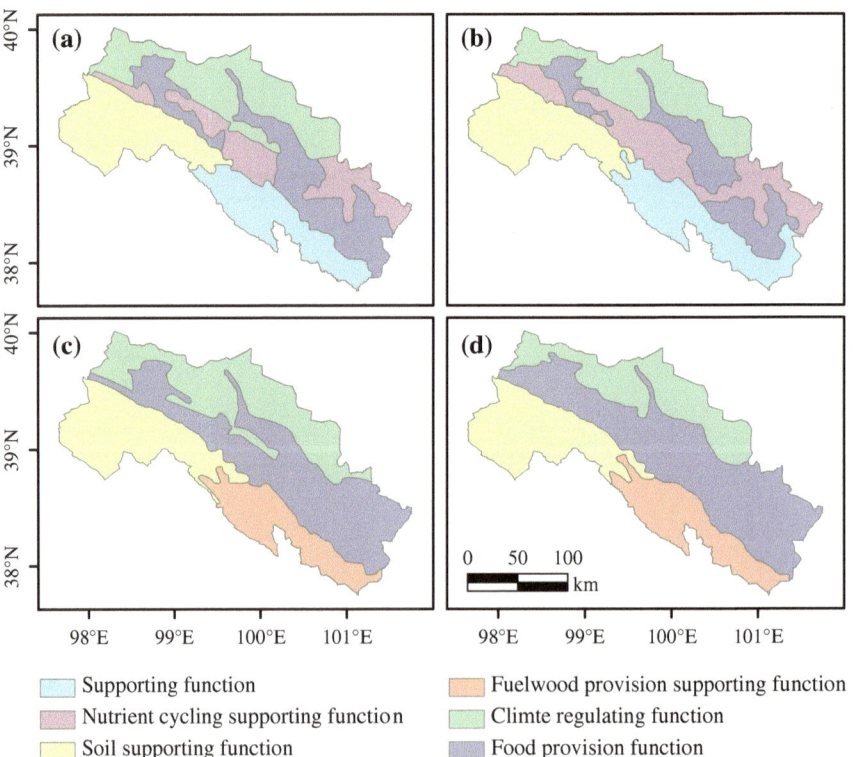

Fig. 6.7 Ecosystem service zoning of middle reaches of the HRB in the years of **a** 1988, **b** 1995, **c** 2000, and **d** 2008

The food provision function zone is mainly concentrated in the cultivated land area. From 1988 to 1995, the food provision function is decreasing in response to the cultivated land degradation during the same time period. And by 2008, this provision increased with the expansion of the cultivated land area. The center of gravity also shifted between each two of the study time period, though being the slightest shift among others. In 2000, a large area of fuelwood supply function zone occurred in the southern part, which is considered as the result of forestry land increase.

The supporting function showed an overall declining trend characterized by boundary shrinking, which is the result of the net primary production loss and the decreased soil function due to the forestry and grassland reduction. In the southern part, the shift from supporting function to fuelwood provision function after 2000 is the result of the forestry increase in this region; fuelwood provision took place as the major ecosystem service.

Summary

This study used a model to identify and zone the ecosystem services to explore the land-use zoning on ecosystem services in climate mitigation. The research contributes to the growing literature in finely characterizing the ecosystem service zones via land-use zoning to mitigate the impact of climate change, as there is no such systematic ecosystem zoning method before.

The application of SIZES proves capable of automatically identifying the boundary of ecosystem service function zone. The method of ecosystem service spatial clustering at watershed scale can be applied in other regions. Though in our study in the middle reaches of HRB we focused on the climate regulation ecosystem services, our method can be further applied in other regions to explore their specific typical ecosystem services. And the results will provide scientific basis for the formulation of regional ecosystem service development planning and ecosystem adaptive management decision. However, the spatial mapping capability of the software needs to be further strengthened.

As one of the largest inland river basins, the provision of the ecosystem services in the HRB is of great significance to the production and life of the local residents. And the regulating services are of crucial importance for this ecological fragile region to counter the impact of climate change. In this research, the impact of land-use change on ecosystem services was explored in the middle reaches of the HRB. Through the identification of the ecosystem service, we recognize that, in the middle reaches of the HRB, it is important to protect the provision and regulating services, especially given the fragile ecology in this region. We identify that it is a direct and effective way of land-use zoning to protect the key ecosystem service and raise the provision and regulating function.

Identifying the major regional ecological environment problems and thereafter establishing the prioritized ecosystem conservation zone are of practical significance. In the middle reaches of the HRB, the major provision function of ecosystem is food provision function, which is mainly concentrated in the cultivated land area and is strengthened with the expansion of cultivated land during the studied time period. In 2000, a large area of fuelwood provision function zone occurred in the southern part, which is considered as the result of forestry land increase. The supporting services mainly occurred in the grassland and water area and show a general trend of degradation, as a result of the net primary production loss and the decreased soil function due to the forestry and grassland reduction. The major regulation service in this study area is climate regulation services, mainly concentrated in the northern desert, Gobi, and grassland area. The conservation of major ecosystem services should be prioritized. And land-use zoning serves as an effective way to protect the key ecosystem services.

Ecosystem-Based Adaptation to Climate Change in Qinghai Province

Introduction

Climate change poses great threats to the human society and natural environment. In order to adapt to climate change, social agents would take some adaptation measures that have some synergistic effects on the dynamics of land surface. However, land surface dynamics would also react on the regional climate condition. Anthropogenic adaptation measures on climate change and variability may directly or indirectly influence the dynamic land surface change (Smithers and Smit 1997), which will have feedback effects on climatic conditions (Smithers and Smit 1997; Feddema et al. 2005). One of the effects of climate change on land surface dynamics reflects on the land surface dynamics response to climate change in the Alpine region. Some human decisions can somehow mitigate the climate change by influencing CO_2 concentration according to previous studies that were conducted with the empirical or numerical methods (Feddema et al. 2005; Wickham et al. 2013), while some adaptation measures will indirectly change the climatic conditions through the intermediate impacts mainly derived from the change of land surface properties (Bounoua et al. 2002; Feddema et al. 2005). The Intergovernmental Panel on Climate Change (IPCC) Third Assessment Report has updated a framework of interrelationship between human adaptation design and implementation and climate change (Smit and Pilifosova 2003). However, previous studies have rarely analyzed the potential impacts of human's adaptation measures on land surface change which will result in destabilizing the climatic conditions in return (Smithers and Smit 1997).

There are still various difficulties in parameterizing the effects of human adaptation measures on climate change for climate models' performance through the land surface dynamics. First, the current researches on the adaptation measures are mostly focused on the cost-benefit analysis, while there is less concern about effects of land surface change on climate which take adaptation measures into consideration (de Bruin et al. 2009). Second, various views and preferences should be taken into consideration when implementing the systematic simulation for land surface during the implementation process of adaptation, which involves the policies, economy, social agents, and many stakeholders (Smit et al. 2000). Third, the research object of adaptation measures is the ecosystem that is open, complex, and full of uncertainties, the knowledge about it is still limited, and therefore there are still some uncertainties in the implementation of adaptation measures (Adger et al. 2009). To sum up, there are still a lot of difficulties in the land surface parameterization scheme in the climate models; for example, the parameters are complex; data are difficult to capture and quantify, and so forth, while those are the indispensable contents for the study of climate models.

The case study in typical ecofragile regions is helpful to comprehensively understand the importance of adaptation measures in land surface dynamics. In this study, the Three-River Source Region in Qinghai Province, which has a fragile ecoenvironment and relatively simple industry structure, was selected as the study area. The future land surface was projected on the basis of the herdsmen's adaptation measures to climate change. An agent-based model (ABM) was developed, in which the adaptation measures would be considered in land surface projection for further climate models' performance in the ecologically fragile areas.

Study Area

Basic Information

The Three-River Source Region, which is the headstream of Yangtze River, Yellow River, and Lancang River, is between 31 39′–36 12′N and 89 45′–102 23′E and in the southern part of Qinghai Province, China. It plays an important role as the ecological barrier for the ecological environment security and sustainable development of regions in the middle and lower reaches of Yangtze River and Yellow River in China as well as Southeast Asian countries. In the recent 30 years, the mean annual temperature is 3.96 °C, with an interannual increment of 0.04 °C, while the average annual precipitation is 500 mm, with a small interannual growth rate of 0.17 mm. The annual average precipitation gradually decreases from east to west on a whole. In addition, the frequency and intensity of climatic disasters also tended to increase in the past decades. The frequency of droughts showed a decreasing trend after 1961, but the duration time of droughts has extended, which exert great adverse impacts on husbandry production. The temperature has increased significantly since 1998, and the winter snowfall and frequency of snow disaster have also increased. Meanwhile, there is an obvious periodical change in the storm; that is, the storm frequency in Qinghai Province has increased notably since the 1990s, especially in the eastern husbandry area. The grasslands occupy the largest area in the study area, reaching 2.665×10^5 km^2 and accounting for about 71.50 % of the total land area of the study area. While the unused lands account for 16.50 %, the forestlands and waters account for 9.20 and 2.10 %, respectively. In the past 20 years, the cultivated lands and unused lands expanded by 46.2 and 0.64 %, respectively, while the forestlands and grasslands have decreased by 0.30 and 0.28 %, respectively. In 2009, the population was 7.5×10^4, with the urbanization rate reaching 15.73 %. Among the 16 counties in the study area, more than half of them are key counties for poverty alleviation and development work, where the poor accounted for 63 % of the total population. The animal husbandry is the main economic base for survival and development of pastoral minorities, the output value of which accounts for 57.28 % of the total agricultural output.

Herdsmen's Adaption to Climate Change

The adaptation of the farming and animal husbandry to climate change has been regarded as one of the key research spots, and adaptation measures are of great importance for mitigating the adverse effects of climate change (Deng et al. 2013b). The Three-River Source Region is an important base for the development of animal husbandry in China, and it is very sensitive to climate change. The local social agents have various measures to adapt to climate change, which were preliminarily classified mainly from the perspective of technology, engineering, management, and policy on the basis of achievements of previous researches on the adaptation measures on climate change (Smit and Skinner 2002). The herdsmen's adaptation measures will have different effects, the spatiotemporal heterogeneity of which is mainly reflected by the resources endowment, population, climatic conditions, different social agents, cost-benefit, and other hidden effects that are not easy to be captured (Smit et al. 2000; Lobell et al. 2008).

Herdsmen's decision on the adaptation measures can greatly affect the land surface dynamics (Adger et al. 2003). There may be twofold impacts of these adaptation measures on the land-use dynamics. Herdsmen's choice of the adaptation measures on climate change can affect their income, and the change of income will in turn affect the herdsmen's choices to adapt to the different cost, then it will further affect the decision-making behavior of the social agents. It has been discussed how economic benefits affect decision-making behavior of agents in the previous works, with economic benefits represented with the cost of adaptation measures and its potential benefits in the ABM model (Izquierdo et al. 2004; Polhill et al. 2010). We captured the economic benefits of climate change adaptation measures with the positive mathematical programming (PMP) model. In addition, this study also involved some adaptation measures that may have impact on land surface dynamics directly, for example, the effective artificial maintenance on grassland, which can prevent the degradation from grassland to desert or even wilderness.

The adaptation measures, which are the most typical and most concerned by the herdsmen, were selected from each category based on the household survey (Table 6.5). It shows that the major adaptation measures in the Three-River Source

Table 6.5 Descriptions of various adaptions to climate change

Categories	Adaption	Descriptions	Effects
Engineering measure	Artificial maintenance of grassland	Combining captive breeding with stocking	The captive breeding is conductive to the protection of natural vegetation, while recuing physical consumption of livestock, contribute to the growth of livestock, and directly increase the benefit. Meanwhile, a large-scale captive breeding has the potential to change the original pattern of land use (Dong et al. 2002)
Technical measure	Pest control	Combing biological control and chemical control	It is helpful to control the desertification and degradation of grassland, protect grassland resources, maintain the ecological balance of grassland, and increase the income of herdsman (Gillies and Pierce 1999)
Management measure	The early warning mechanism of snow disasters	Early warning of monitoring of snow disaster and layout of snow equipment Management and emergency plan of snow disaster	Early warning work of disaster can help farmers effectively and make them take measures in time to reduce the loss of disasters. But the setting of monitoring facilities and monitoring point will increase the area of built-up land (Einstein and Sousa 2007)
Policy measure	Fiscal subsidies	Subsidy policy of livestock breeding, subsidy of good seeds of artificial planting forage seed, policy of production subsides, subsidy policy of animal husbandry machinery purchase, subsidy policy of grazing prohibition for nurture	Policy measures can avoid herdsman's blind pursuit of profit maximization which may destroy the ecological balance. On the premise of guaranteeing the herdsman's income, the government subsidy can help the implication of policy and guide farmer and herdsman to maintain the reasonable pattern of land use (Wang et al. 2013)

Region mainly include the pest control, artificial maintenance of grassland, early warning mechanism of snow disasters, and fiscal subsidies, which represent the impacts of adaptation measures from the perspective of technology, engineering, extreme weather prevention and management, and fiscal subsidies, respectively. During the survey interview, the herdsmen were asked about their preferences for adaptation measures on climate change, choice probabilities, and thinking ways, based on which the behavior rules were defined (Kelly and Adger 2000) and lay the foundation for the land surface parameterization scheme in the climate simulation. Additionally, the ABM model is used to simulate the effects of interaction between different social agents under particular conditions.

Data Collection and Processing

The household survey data used in this study were obtained from the questionnaire survey, which was implemented during August and October 2012 in three counties of the Three-River Source Region. First, three counties were selected from Tibetan Autonomous Prefecture of Golog, Tibetan Autonomous Prefecture of Huangnan, and Tibetan Autonomous Prefecture of Hainan. The selection of the sample counties was mainly based on three factors, that is, per capita grassland area, population density, and per capital GDP. Second, nine towns were selected in the three-sample counties with stratified random sampling. Then, two towns with the most and least grassland area were selected from the 9 towns, and thereafter two villages were selected in each sample town according to the employment situation in sectors except the animal husbandry. Finally, 10 herdsmen families were selected as the sample families from the herdsmen families with the large, medium, and small grassland area in each sample village. The questionnaire survey was carried out on 200 herdsman families, and the response rate of the questionnaires was 100 %, with 193 valid questionnaires and a ratio of valid questionnaire of 96.50 %. The contents of the questionnaire are as follows: population, land area (mainly including the grassland area), livestock production, sale of animal products, investment into the prevention and control of the damage from rats, subsidies granted for policy considerations, employment situation of sectors except animal husbandry, income from animal husbandry, and so forth.

In addition, the supplementary data consist of the basic geographic information data which are supported by Data Center for Resources and Environmental Sciences, Chinese Academy of Sciences (RESDC); the data mainly include land-use data, DEM data, and topographic location data. The socioeconomic data are obtained from the Qinghai Statistical Yearbook, China Agriculture Yearbook, and China Animal Industry Yearbook. The natural environmental data and climate data are collected from the meteorological stations.

Land Surface Parameterization Scheme

Estimation Measures of the Impacts of Herdsmen's Adaption

It is necessary to quantitatively analyze the economic benefits of the herdsmen's decision-making behavior about adaptation measures in order to better provide reference for the land surface parameterization scheme in the ABM model. There are many methods for quantitative assessment of adaptation measures, and this study used the PMP model proposed by Howitt (1995). This model is somehow a popular approach to analyze the policy-oriented environmental and economic problems, using even scarcely available information. The method has great flexibility and conforms to the basic hypothesis of economics of diminishing marginal returns, and these advantages support wide application of PMP model in policy effect analysis (Heckelei et al. 2012).

There are mainly three steps to analyze the cost and benefit of the adaptation measures for climate change (Izquierdo et al. 2004). First, the constraint conditions were set in the linear programming model according to the observed values of the baseline period, and the dual prices (shadow prices) of the constraint conditions were further calculated with the linear programming model. Second, the slope of the average cost function in the objective function of the PMP model was calculated according to the dual price of the constraint conditions. Third, the corresponding parameters in the model were adjusted according to the need of the assessment of specific adaptation measures, and the parameters were obtained and then put in the objective function to obtain the optimal solution of the nonlinear programming for the investment into adaptation measures.

Finally, in order to analyze the impacts of the adaptation measures on the income of herdsmen, the PMP model was established on the basis of the household survey data of the herdsmen families as follows. The objective function is as follows:

$$\text{TMG} = \max \sum (p_i y_i + \text{sub} - (\alpha_i + \gamma_i/2x_i))x_i \tag{6.16}$$

Constraint conditions are as follows:

$$\sum x_i \times l_i \leq L \tag{6.17}$$

$$\sum x_i \leq \text{Area} \tag{6.18}$$

$$\sum x_i \times q_i \leq I \tag{6.19}$$

$$
\begin{aligned}
y_i = {} & \beta_0 + \beta_1 \text{dtpro} + \beta_2 \text{Gld} + \beta_3 \text{POP} + \beta_4 \text{NPP} + \beta_5 \text{dtwater} + \beta_6 \text{dtroad} \\
& + \beta_7 \text{rain\_cv} + \beta_8 \text{temp\_cv} + \beta_9 t_0 + \beta_{10} t_{10} + \beta_{11} \text{conrodent} \\
& + \beta_{12} \text{aprotect} + \beta_{13} \sup \text{int} + \mu
\end{aligned} \tag{6.20}
$$

where TMG is the herdsmen family income from livestock breeding; y_i is the number of livestock per unit area of grassland; x_i is the area of grasslands for the livestock; p_i is the selling price of the livestock; α_i is the cost per unit area of grassland; γ_i is the shadow price; l_i is the labor input into each livestock; q_i is the capital income per unit area of grassland; dtpro is the distance to capital of province; Gld is the area of grassland; POP is population; NPP is the net primary production; dtwater is the distance to water area; dtroad is the distance to road; rain_cv is the coefficient of variance of rain; temp_cv is the coefficient of variance of temperature; t_0 is the cumulative temperature above 0 °C; t_{10} is the cumulative temperature above 10 °C; conrodent is the rodent control; aprotect is the manual maintenance of grasslands; supint is the grazing intensity supervision; μ is the residual error.

In the PMP model mentioned previously, the objective function is defined as the maximization of the income of the herdsmen's family from livestock breeding for the following reasons. First, these adaptation measures implemented in the ecologically fragile areas aim to reduce the livestock loss of the herdsmen, and they consequently have positive significant impacts on the income of the herdsmen's family from livestock breeding. Second, there is a very low proportion of the income from crop production and other agricultural industries except animal husbandry in the total income of the herdsmen's family since the animal husbandry is the dominant industry in all the three counties where we do household survey in the Three-River Source Region. Third, this study is mainly focused on the impacts of these adaptation measures on the income of the herdsmen's family.

The income per unit area of grassland equals to the product of capital income per unit area of grassland q_i and the area of grasslands for grazing the livestock x_i; (6.17) expresses the constraint conditions of the labor; (6.18) expresses the constraint conditions of the grassland area; that is, the sum of the area of grasslands for grazing the livestock cannot exceed the total land area of the herdsmen's family; (6.19) is the constraint conditions of the capital; and (6.20) provides the number of livestock per unit area of grassland under specific condition. Considering that in the system are mainly the herdsmen who graze the cattle and sheep and will not invest the income from other sources into animal husbandry, so the upper limit of total amount of investment into the livestock production should be set as the total income from livestock.

Parameterize the Adaption in Land Surface Projection

According to different agents, the parameters of main decision-making behavior in the ABM model are different to simplify and simulate the adaptive behavior choice and the influence of each agent on climate change. Different agent has different preference for various adaptation measures, resulting in the different land surface dynamics effect. Therefore, this study classified and extracted the agents in the study area and then determined their preference for adaptation measures according to the investigation and parameterized them.

Firstly, classify the counties and cities in Three-River Source Region into three categories according to the analysis of their social and economic development, natural environment characteristics, and resource endowment characteristics; then, the herdsmen's agents that are extracted from each category mainly include herdsmen's agent, half-herdsmen's agent, and restrictive herdsmen's agent. Secondly, design and determine the rules of agents' behavior through on-the-spot investigation about herdsmen's willingness to take climate change adaptation measures and other social-economic and ecological environment consciousness, combined with the cost-benefit analysis of the adaptive measurements.

In addition, through the survey, it is found that in the Three-River Source Region, the output of grasslands was relatively higher compared with that of other land-use types. The grassland is also the main ecosystem in the study area, and its ecosystem service functions, such as livestock pasturing, water conservation, water purification, and climate regulation, are indispensable to local individuals and communities. Therefore, the relative social agents are more willing to preserve and restore the grassland.

Land Surface Dynamics Response to Adaptation for Climate Change

Economic Analysis of Adaption Measures

This study analyzed the economic benefits of adaptation measures for climate change with PMP model and explored the impacts of different adaptation measures on the livestock production and herdsmen's income based on the different context of climate change. The results show that the livestock production and the economic benefits have a spatial heterogeneity and scale effects (Fig. 6.8).

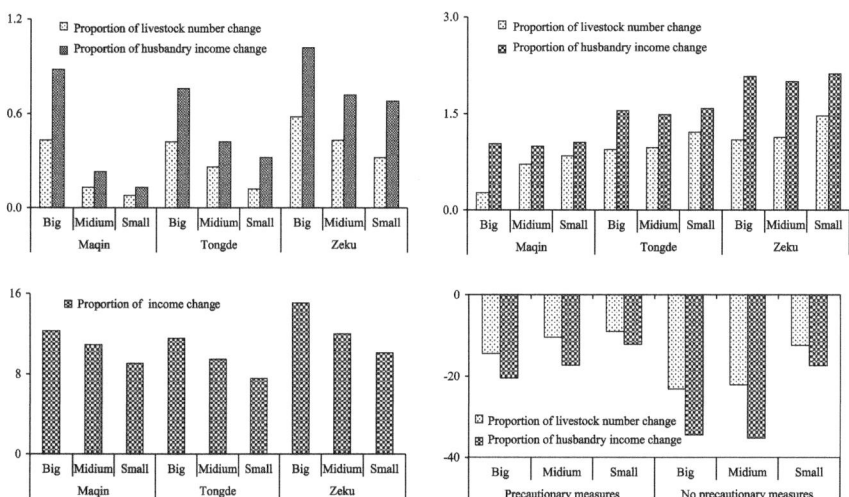

Fig. 6.8 Mathematical analysis of herdsmen's adaption to climate change including the technical measures, engineering measures, extreme weather adaption, and government subsidies, from three-level pasture scales, such as large (0.67–4.33 km^2), medium (0.13–0.67 km^2), and small (0.01–0.13 km^2)

Livestock number and animal industry income in the same county are correlated positively, which means that animal industry income would increase with the increase of livestock number. Under the conditon with high temperature change, the adaptive technical measures will make the livestock number and the income of animal husbandary positively related. With the decrease of farm size, the income increment of livestock number and animal industry income would be declined gradually. Thus, we may deduce that the effect of the measure and the size of farm have a good positive correlation, and adaptive measure could increase herdsmen's income, but a more obvious impact on economic benefits of large-scale pastures. The implementation of non-engineering adaptation measures plays 0.8–1 % of the effect on the animal husbandry income of the herdsman after estimating based on survey data. It is also found that the impact of non-engineering adaptation measures on income of animal husbandry is different due to the pasture size. Under the high effect of precipitation change condition, increment of livestock number would increase gradually with the decrease of grassland scale in the same county. However, the variation of animal industry income would decrease firstly and then increase with the decline of grassland scale, and the income increment of small-scale grassland is the biggest. Hence, engineering measure may develop the livestock production of the herdsman and the income of animal industry. The utility of engineering measure is related positively to grassland scale, but utility effect of water resource in small-scale grassland would be bigger, thus engineering measure has a bigger impact on livestock number and animal industry income in small-scale grassland. Abnormal climate may lead to the large-scale reduction on livestock number and animal industry income; however, herdsman's income may reduce one-third of loss under the condition of early warning mechanism. At the same time, the risk management is divided into 3 levels, and we evaluated whether they develop the management of early warning at the government level to adapt to climate change using PMP model. Simulation analysis shows that the establishment of early warning mechanism can avoid about 12 % of economic loss in the large-scale grassland. The loss of small-scale grassland would be the minimum compared with the other two kinds of grassland when analyzed from the scale of grassland, which means that small-scale grassland has a good flexibility, and it can adjust the scale of production timely to minimize the losses. Government subsidy has a strong positive relationship with different scale of grassland and income of animal industry in the same county.

In conclusion, the four types of adaptation measures, which start from the character of climate impact factor in different typical ecologically fragile areas and can analyze filed sample data as well as getting the results, all can develop the livestock number and income of industry in different degrees from empirical analysis of PMP model. Therefore, positive adaptive measure has a positive effect on the herdsman to adapt to climate change.

Land Surface Dynamics

According to the simulation results based on ABM model, the land surface in the Three-River Source Region shows some temporal and spatial characteristics

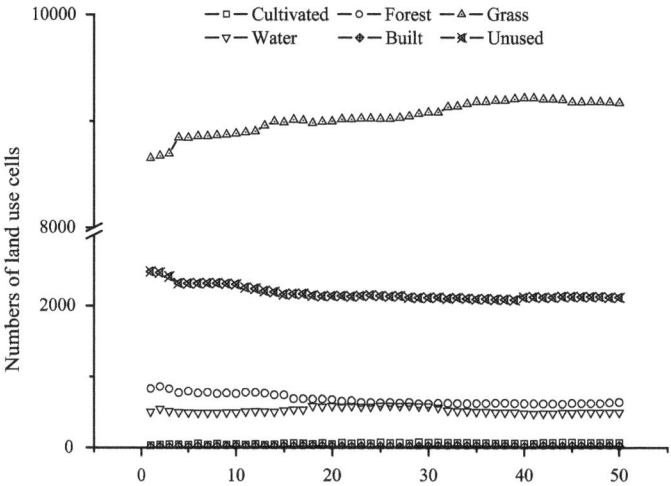

Fig. 6.9 The underlying land surface structure in the next 50 years. The grid cell size is 2 km × 2 km

during the next 50 years. Generally, the grassland area is increasing to a certain extent with an increment of 6.04 %, and the unused land gets moderate development and utilization, so the unused lands area experiences a reduction, with a rate of 14.64 %. Besides, the forestland area has some decrease and other lands use type almost maintain the same (Fig. 6.9).

In the next 50 years, the area of changed land accounted for 6.65 % of the total land area, which is dominated by the conversion from unused lands and forestlands to grasslands, accounting for 2.99 and 1.81 % of the total land area, respectively. Specifically, as to the transformed lands, 74.92 % of forestlands and 80.6 % of unused lands are converted to grasslands; in contrast, the increasing of grassland area mainly comes from the forestland and unused land, with 34.39 and 56.67 %, respectively. The dramatic changes in grassland area are mainly because farms mainly pursue profit maximization and focus on considering the economic benefits of land use on the decision-making process. Grassland area would increase gradually and other land types would show a decreasing trend according to the simulation of the next 50 years. It means that the utilization way of land by farmers and herdsmen becomes simplex gradually, and dominant land-use type is more and more obvious. Since single type of land will inevitably lead to the instability of ecological system, farmers and herdsmen should change the traditional way for adapting the nature, utilize adaptation measures rationally, not just pursue the revenue maximization of livestock yield and income of animal husbandry, but fully consider the rationality of land-use pattern.

The distribution of land surface change simulated by ABM is shown in Fig. 6.10. The land-use change mainly occurred in the central and east part of the study area; the east part was mainly dominated by the conversions from forestlands to grasslands and from grasslands to unused lands, while the central part was

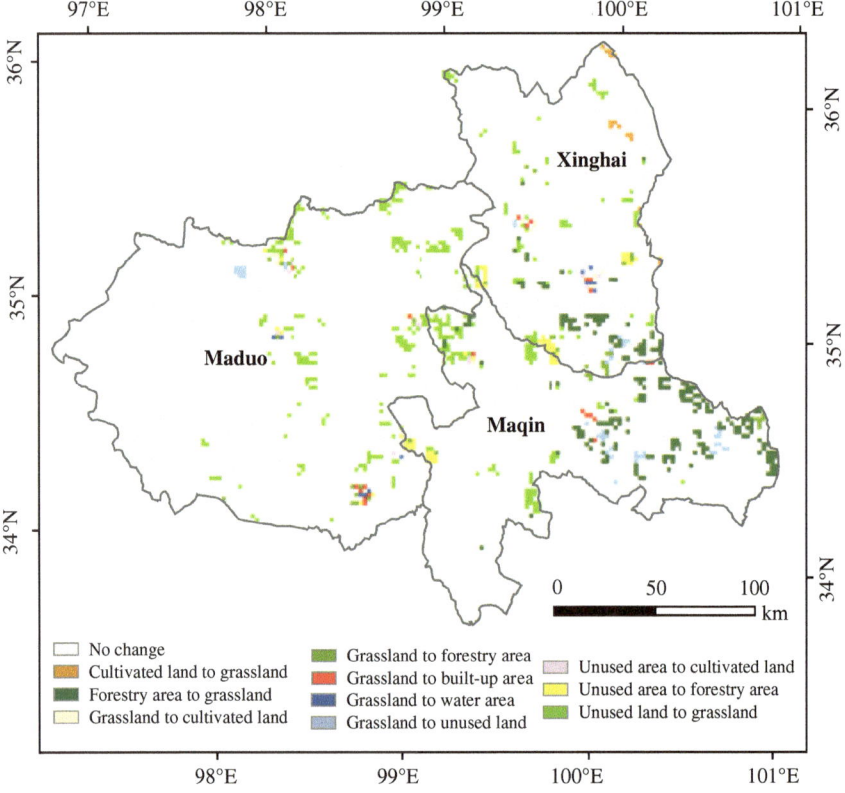

Fig. 6.10 The spatial pattern of land surface change in the next 50 years. Only large proportion of land surface change is showed in this picture

manifested by the development and utilization of unused land, thus showing the conversion from unused lands to grasslands. That is because of the restively low altitude and flat terrain in the eastern area which are relatively easy for reclaiming of unused land, and there is restively high precipitation which is advantageous to the pasture production and livestock farming. These favorable factors make the herdsman develop animal husbandry by reclaiming forestland and unused land to increase the income of animal husbandry.

Summary

This study analyzed the economic benefits of the adaptation measures on climate change with the PMP model based on the survey data in the Three-River Source Region, and then projected the land-use change with the ABM model. This method may provide some reference for the research on the climate simulation with different underlying land surface. The simulation result shows that the four kinds of adaptation measures will have different impacts on the development of the animal husbandry and economic benefits. Within a certain range, the larger the scale of grassland is, the more the income is. Therefore, the scale of pastures should be maintained at a certain appropriate range, and the government should continue to promote the return of the farmland to grassland and ecological migration projects. Besides, it is urgent to carry out more research on monitoring of ecological environment and ecosystem restoration, promote the comprehensive management of the ecological environment, and control the destruction of the ecological environment caused by herdsmen. According to the analysis with the ABM model on the basis of the herdsmen's adaptation measures to climate change, we could know that the adaptation measures will have synthetic impacts on the evolution of the underlying surface and promote the development and utilization of unused land and conversion to the forestlands and grasslands. The grassland increased by approximately 6.04 %, while the unused land decreased by 14.64 %. According to the simulation result of the land use in Three-River Source Region during the next 50 years, we can find that all the land-use types except grassland show a decreasing trend under the condition of adaptation measures. It means that the land utilization way of farmers and herdsmen gradually becomes simplex, and the predominance of grassland is more and more obvious. This suggests that herdsmen will ignore the rationality of land use when they excessively pursue the maximization of benefit.

The herdsmen are the main agent of economic activities and the basic unit of decision-making on the adaptation to climate change in the grazing districts, the decision-making behaviors of which have profound impacts on the land surface dynamics. The animal husbandry is the main economic source in the study area, and the adaptation measures on climate change will affect the economic income of herdsmen, and further influence the land surface dynamics since the herdsmen's income will have effects on their decision-making on the adaptation measures and consequently lead to different land-use change. This study analyzed the impacts of adaptation measures on climate change, but the land-use classification is relatively rough, and the resolution is low, and there are many other factors influencing the land surface dynamics; therefore, it is necessary to implement some further researches in the future.

Although there are some uncertainties in the result, it can still provide value reference information for the land-use management. Based on the result mentioned previously, some measures and suggestions of land use were put forward from all aspects. It is necessary to implement a number of measures in the overgrazing zones, for example, the reasonable land-use planning, intensive development of land, control of excessive reclamation of grassland, policy of returning grazing land to grassland, and grazing prohibition. Meanwhile, the government should provide food and economic compensation to the farmers and herdsmen so as to avoid the conversion from grassland to desert and other unfavorable land-use types due to excessive degradation. Meanwhile, it is urgent to strengthen the prevention and control of pests, reduce the loss of grassland, and promote the utilization of existing grassland so as to further control the reclamation of grassland. Moreover, it is also urgent to establish the early warning mechanism of natural disasters in view of the increased frequency and intensity of natural disasters and their direct impacts on the herdsmen's economic benefit. The forest accounts for a relatively small proportion of the total area of the Three-River Source Region; however, it provides abundant ecosystem services; therefore, it is necessary to make effort to manage and protect the forest in order to guarantee the sustainable development.

Building Resilience to Climate Change for Conserving Ecosystem Servers

Introduction

There is ample evidence of the ecological impacts of climate change, which may pose considerable challenges to the terrestrial ecosystems and change the provided ecosystem services in the future (Prato 2012), and the forest ecosystem resilience plays an important role in maintaining the desirable ecosystem states that allow these ecosystem services to be delivered, and enables quick and flexible responses to climate change (Thompson et al. 2009). For example, forests are the major reservoirs of terrestrial biodiversity and contain about 50 % of the global terrestrial biomass carbon stocks, emissions from deforestation and degradation remain a significant source of annual greenhouse gas emissions into the atmosphere, and therefore the conservation, appropriate management, and restoration of forests will make a significant contribution to climate change mitigation (IPCC 2007). Resilience is the capacity of an ecosystem to withstand external pressures and return to its predisturbance state over time; the loss of ecosystem resilience indicates that ecosystems are prone to the shifts to undesirable states in which the ecosystem services needed by humans can no longer be delivered, and maintaining or restoring the forest ecosystem resilience is often cited as a necessary societal adaptation to climate change (Brand 2009). However, forest ecosystem resilience has continually declined at the regional scale and even global scale due to climate change and human disturbance (Folke et al. 2002). The quantitative assessment of ecosystem resilience can provide a scientific basis for the forest resource management and conservation and therefore is of great significance to the maintenance of critical ecosystem services (Strickland-Munro et al. 2010).

The concept of resilience has been widely used, and there have been some ecological theories that attempt to explain the mechanism of resilience through a variety of models (Gallopín 2006), for example, "species richness-diversity" (MacArthur 1955), "functional redundancy" (Bellwood et al. 2003), "keystone species hypothesis" (Walker 1992), and "resilience-productivity hypothesis" (Moore et al. 1993). However, these theories are generally based on the concept of species populations as the basic functional unit and therefore fail to capture the importance of the interactions among individual organisms in the ecosystem. Among the current theories, the theory of dissipative structures seems particularly suitable for investigating the dynamics of structural change and resilience of ecosystems. It shows that the open and self-organizing systems maintain their structural order by keeping their internal state far from thermodynamic equilibrium through active exchanges with their environment. Those dissipative structures are in principle stable as long as the exchanges with the environment are maintained and the continuous perturbations are absorbed within the framework of the given dynamic regime (Nichols and Prigogine 1989). The theory of dissipative structures provides a scientific theoretical framework for explaining the mechanism of the ecosystem resilience; however, there have been very few researches on the quantitative measurement of ecosystem resilience on the basis of this theory; more in-depth research should be carried out on how to more scientifically and accurately assess the ecosystem resilience with reasonable indicators of the ecosystem resilience.

Resilience can be measured in terms of change in a system level property and function following perturbation, and the perturbation can be simulated. In previous research, the ecosystem resilience was generally measured by the rate of return of the ecosystem state after disturbance or the maximum disturbance that the ecosystem can absorb before shifting to another state (Carpenter et al. 2001a). Currently, researchers generally select one key indicator associated with the ecosystem, for example the key species and vigor of the ecosystem, and they then simulate the time for the key indicator to return from the stressed state to the normal state (Tr) and the maximum stress that the ecosystem can withstand (MS) with models such as the CENTRURY model and the GAP model (Costanza 2012). Ecosystem resilience can be represented by the values of MS, 1/Tr, or MS/Tr (Bennett et al. 2005). This approach assumes that the dynamics of the ecosystem can be understood by analyzing a few key variables, which is termed the "rule of hand" (Brand 2009). However, the concept of "rule of hand" is limited and relatively unrepresentative because it is impossible to represent the complete recovery of ecosystem function by the recovery of only a few key variables (Schmitz 2000). In addition, although it is in principle possible to measure ecosystem resilience by fitting a dynamic model to time series, this approach imposes extraordinary data requirements. It is usually difficult to obtain the data that can meet such requirements in practice (Carpenter et al. 2001a).

It is more plausible to measure resilience in terms of the factors influencing resilience. The literature reported a number of factors that influence the ecosystem resilience, for example, the diversity within functional groups and variability of habitats (Franklin and MacMahon 2000). However, these factors have not previously been considered comprehensively by researchers. The operational indicators of resilience have received little attention in the literature, and there is no consensus-based view of how to measure resilience or even of the exact nature of resilience (Gibbs 2009). Rosset and Oertli assessed the resilience of species to warming with five ecological and biogeographical metrics and explained their theoretical basis (Rosset and Oertli 2011); however, this approach may be relatively biased since these researchers simply used equal weights for each metric.

This study aims to develop a conceptual framework for the spatially explicit assessment of the forest ecosystem resilience based on the theory of dissipative structures, and the rest of this study is organized as follows. The second part presents a brief overview of the study area and explains how the indicator system was constructed, how the indicator weights were determined, and how the resilience index was calculated. Besides, this part also shows the data used in this study and how they were processed. The third part presents the results and discusses the underlying reasons for the spatial heterogeneity of the forest ecosystem resilience, and the final part concludes.

Data and Methodology

Study Area

Yongxin County is representative of the subtropical mixed conifer and broad-leaved forest area in the Poyang Lake watershed, with a forest area of 143,980 ha and a forest coverage rate of 65.6 %. It is located between 26°47′–27°14′N and 113°50′–114°29′E, in the upper and middle reaches of the Heshui River which is the largest secondary tributary of Poyang Lake (Fig. 6.11). The northern and southern parts of Yongxin County primarily include mountains and hills, where there are a lot of forests, whereas the central part of the county includes hills and plains, most of which are covered by cultivated land. It is the subtropical monsoon climate in this region, with an annual average temperature of 18.2 °C and an annual average precipitation of 1530.7 mm, and the zonal vegetation is the evergreen broadleaved forest, but the existing forests primarily consist of Pinus massoniana and *Cunninghamia lanceolata* (*C. lanceolata*), and the current state of the forests is the result of both the long-term human disturbance and restoration and the natural recovery under the influence of the regional natural background (Deng et al. 2011b).

Data and Processing

The data used primarily include the forest inventory data, remotely sensed data, and statistical data. The forest inventory data in 2009 were obtained from the Yongxin Forestry Bureau, including the forest resource distribution map, data for 707 sample plots, and data for 41603 sample trees. We obtained the forest distribution map by integrating the forest form map with the Landsat Thematic (TM) image covering the study area. Besides, we interpolated the sample plot data and sample tree data into 30 m × 30 m grid data with the Kriging method to obtain the

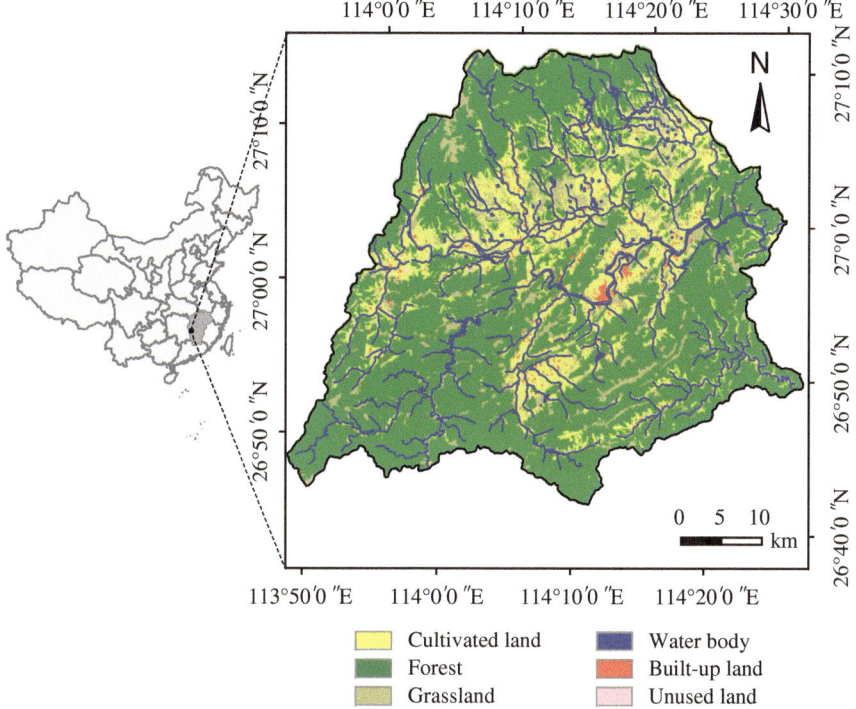

Fig. 6.11 Location of Yongxin County and the distribution map of forests

data of specific indicators of the internal memory, such as the average stand age, average DBH growth rate, and mature dominant tree density (Deng et al. 2010b). The external memory was then calculated on the basis of the internal memory. Besides, the factor influencing the availability of the external memory was indicated with the proximity index, which was calculated on the basis of the forest distribution map. In addition, the data of indicators of the stand conditions were prepared with the remote sensing data, observation data, and statistical data. For example, the climatic data were from the observation stations in Jiangxi Province and Hunan Province maintained, and the original data were interpolated into 30 m × 30 m grid data using the gradient plus inverse distance squares method and then extracted the part of Yongxin County. The soil data were extracted from the forest inventory data of the second nationwide general soil survey and were interpolated into 30 m × 30 m grid data with the Kriging method (Wu et al. 2013). The terrain data were obtained with the 30 m × 30 m digital elevation model (DEM) data. The distance to the nearest river was obtained from the 1:250,000 topographic maps of Jiangxi Province. The population data were obtained from China Population Statistics Yearbook 2010 and were spatially disaggregated into 30 m × 30 m grid data with the spatial disaggregating method (Ye et al. 2010).

Model Development

The forest ecosystem resilience is quantitatively measured with its influencing factors on the basis of the theory of dissipative structures in this study. First, some definitions related to the ecosystem resilience were clarified, which lay the foundation for the ecosystem resilience assessment. Then based on the theory of dissipative structures, a hierarchical indicator system was constructed according to the factors influencing the forest ecosystem resilience, most of which have some impacts on the energy and material flows between the forest ecosystem and the environment. Thereafter, the indicator weights were determined with the combined weighting method, including the AHP and coefficient of variance (CV) method (Wolfslehner et al. 2005). Finally, the resilience index of forests was calculated as the weighted sum of these indicators at the grid scale.

Definition of Ecosystem Resilience, State, and Scale

It is necessary to make some definitions related to the ecosystem resilience so as to make the results comparable. First and foremost, it is necessary to clarify the definition of ecosystem resilience since the literature offers various definitions of ecosystem resilience and includes a controversy about the existence of different ecosystem states (Nichols and Prigogine 1989). The current viewpoints can be summarized into ecological resilience and engineering resilience, and forests are engineering resilient in the sense that they may in time return to their predisturbance state and maintain approximately the original species composition (Thompson et al. 2009). Therefore, the definition of the engineering resilience has been adopted in this study, referring to the capacity of an ecosystem to absorb disturbance and return to its predisturbance state following a perturbation (Folke 2006).

Secondly, it is crucial to specify the ecosystem state of interest since the forest ecosystems have multiple states under which different ecosystem services may be delivered (Carpenter et al. 2001a). This study has only focused on the resilience under the current state, which is defined with the dominant tree species (Thompson et al. 2009), and it is assumed that there will be no state transformation during the study period. In addition, it is necessary to specify the scale in the resilience assessment since the different scales may lead to different assessment results (Jentsch et al. 2002), and when viewed over an appropriate time span, a resilient forest ecosystem is able to maintain its "identity" in terms of taxonomic composition, structure, ecological functions, and process rates (Thompson et al. 2009). As for the spatial scale, this study proposes to measure the absolute and relative conditions (e.g., space, environmental characteristics, and resource availability) at the patch scale and to analyze questions of resilience at the multipatch scale (Jentsch et al. 2002), and the 30-m resolution patch has been used to analyze the resilience in this study. As for the timescale, the forest ecosystem resilience is analyzed at the annual scale, and it is assumed that no ecosystem state transformations would occur during the study period since the ecosystem generally fluctuates near equilibrium and remains essentially stable during a given period.

Indicator System

A hierarchical model was developed with factors influencing the forest ecosystem resilience according to the theory of dissipative structures. The theory of dissipative structures shows that the forest ecosystem as a system of dissipative structures will soon collapse on condition that there is no input of energy and material (Scarborough and Burnside 2010), and the ecosystem resilience depends on both the favorable stand conditions and the biological and ecological resources in the ecosystem, all of which influence the input of energy and material into the forest ecosystem. Therefore, the resilience index that evaluates the forest ecosystem resilience was disaggregated into the stand conditions and the ecological memory, which were finally disaggregated into simple indices that are measurable and can be observed in the field.

The indicators of the stand conditions include the terrain, soil, climate, water conditions, and human disturbance, which interact with each other and jointly affect the resilience of the forest ecosystem (Zirlewagen et al. 2007). The terrain has obvious impacts on the factors required for plant growth, for example, the water and soil nutrients. For example, there is richer soil fertility and seed bank in the lower slope position than the middle and upper slope positions, and the seed germination rate at the lower slope position is higher (Martin and Ogden 2002), and the terrain indicators in this study include the aspect, slope, and slope position. Besides, the resilience at the local level depends on the ability of the landscape to maintain infiltration, water storage capacity, and nutrient cycles, all of which are threatened by soil loss and structural change (Carpenter et al. 2001b). In this study, we selected the soil depth, humus horizon depth, and loam quantity for use as soil structure indices and used the humus quantity, soil nitrogen (N) quantity, soil phosphorus (P) quantity, and soil potassium (K) quantity as indices of soil fertility. In addition, the climate primarily influences photosynthesis, respiration, and other ecosystem processes through medium-term and long-term temperature, radiation and wetness and consequently exerts great impacts on the plant growth (IPCC 2007), and water conditions also have significant influence on plant growth through influencing the availability of water (Shi et al. 2012). The climate indices used in this study included the annual accumulated temperature above 10 °C, the annual precipitation, and the annual hours of sunshine, and the distance to the nearest river was used as the indicator of the water conditions. What is more, there is still controversy over the identification of humans as a component of natural ecosystems, but human beings have altered the resilience of ecosystems (Dublin et al. 1990); therefore, the population density that is closely related with human activities has been used as the indicator of human disturbance in this study.

The ecological memory include the internal memory within the foci forest patch and external memory in the neighbor patches of the foci forest patch (Nyström and Folke 2001; Bengtsson et al. 2003), which were further represented with more specific indicators. The internal memory includes the species that survive within the disturbed area (i.e., the biological legacies) and the remaining dead organic structures that serve as foci for regeneration and allow species to colonize (i.e., the structural legacies) (Nyström and Folke 2001). The biological legacies include the seed, vegetation materials, and animal communities, while the structural legacies provide critical protective cover, habitat, and food and nutrient sources for a variety of organisms and influence geomorphic processes such as erosion and the deposition of sediments (Nyström and Folke 2001). First, the seed bank is the material basis for the natural regeneration of forests (Bengtsson et al. 2003). Since it is difficult to measure the soil seed bank in a large area, we used the factors influencing the soil seed bank as the specific indicators, including the density of the mature dominant trees and the stand canopy (Khattak et al. 2008). Besides, the ecosystem resilience resides in both the diversity of the drivers and the number of passengers who are potential drivers, which are of different significance to the ecosystem (Berkes and Jolly 2002), and the selected indictors of the diversity of the drivers include the species number of the dominant trees and the subdominant trees and the grass and shrub canopy. In addition, ecophysiological characteristics of the vegetation also contribute to ecosystem resilience (Thompson et al. 2009), the indicators of which include the average stand age and the average growth rate of the diameter at breast height (DBH) in this study. What is more, structural legacies provide critical protective cover that allows species to colonize (Franklin and MacMahon 2000), and the litter depth was used as the indicator of structural legacies in this study. More importantly, the recovery of an ecosystem from disturbance requires an area that is sufficiently large and abundant internal memory to guarantee a rapid reorganization, and a larger and less fragmented forest ecosystem is more resilient (Thompson et al. 2009). In this study, the patch area and perimeter-to-area ratio were used to represent the patch size and patch shape, respectively.

Ecosystem reorganization requires both the internal memory within the disturbed patch and the external memory within neighboring patches, which provides seed flows among forest patches and influence the species composition and facilitate the resilience of disturbed patches (Wolfslehner et al. 2005). Since it is difficult to measure long-distance seed dispersal, we measured the external memory by the density of mature dominant trees within neighboring patches of corresponding patch types whose edges are within a specified distance of the focal patch (1000 m in this study). In addition, many plant species are dispersal limited and are influenced by various factors such as the distance to the seed sources and the availability of dispersal agents (Ehrlén and Eriksson 2000), and therefore, we have used a proximity index that combines the area of the neighboring patch and the distance to the focal patch as the indicator of the factors that influence the availability of external memory (Table 6.6).

Calculation of the Resilience Index

There are many methods to synthesize the basic indicators into one index, such as the fuzzy comprehensive evaluation method, Delphi method, and comprehensive index method (Borja et al. 2008). This study used the classic comprehensive index method to calculate the resilience index of the forest ecosystem. First, the assessment indicators of the forest ecosystem resilience were assigned with different

Table 6.6 The assessment system and weights of specific indicators

Medium-level indicators	Bottom indicators	W_{AHP}	W_{CV}	Weights
Stand conditions				
Terrain	Slope position	0.0123	0.0386	0.0254
	Slope	0.0088	0.0209	0.0148
	Aspect	0.0257	0.0417	0.0337
Climate				
Temperature	Annual average temperature	0.0138	0.0503	0.0320
	Cumulative temperature above 10 °C	0.0292	0.0543	0.0417
Precipitation	Annual precipitation	0.0356	0.0289	0.0322
Solar radiation	Annual sunshine hour	0.0102	0.0467	0.0284
Water condition	Distance to the nearest river	0.0194	0.0075	0.0134
Soil				
Soil structure	Soil depth	0.0063	0.0177	0.0120
	Humus depth	0.024	0.0168	0.0204
	Loam quantity	0.0035	0.00001	0.0017
Soil fertility	Soil organic quantity	0.0045	0.0073	0.00591
	Soil nitrogen quantity	0.0065	0.0016	0.00406
	Soil phosphorus quantity	0.0036	0.00001	0.0018
	Soil kalium quantity	0.0044	0.0016	0.00301
Human disturbance	Population density	0.0237	0.0273	0.0255
Ecological Memory				
Internal memory				
Biological legacies	Species number of dominant and subdominant trees	0.0459	0.0824	0.0641
	Grass-shrub canopy	0.0145	0.0308	0.0226
	Average stand age	0.0293	0.0106	0.0199
	Average DBH growth rate	0.0206	0.0342	0.0274
	Mature dominant tree density	0.0568	0.0487	0.0527
	Vegetation canopy	0.0171	0.0257	0.0214
Structural legacies	Litter depth	0.0828	0.0384	0.0606
Patch area and shape	Patch area	0.2487	0.1001	0.1744
	Perimeter area ratio	0.0749	0.0058	0.0403
External memory and factors influencing its availability	Density of mature dominant tree within neighbor patches of corresponding types	0.0552	0.0665	0.0608
	Proximity index	0.1227	0.1949	0.1588

W_{AHP} and W_{CV} refer to the weights determined with the analytic hierarchy process (AHP) and coefficient of variation method (CV), respectively. The weights was calculated as the average of W_{AHP} and W_{CV}

weights; then, the weighted sum of the assessment indices was calculated with a spatial overlay of these indices. The resilience of the forest ecosystem is calculated with the following formula:

$$\text{resilience}_j = \sum_{i=1}^{n} w_i x_i \ (i = 1, 2, 3 \ldots n) \tag{6.21}$$

where resilience_j is the resilience index of the jth assessment unit; w_i is the weight of the ith assessment index; x_i is the value of the ith assessment index, which is normalized with the extreme value method; and n is the number of the assessment indices.

Besides, there are many methods to determine the index weights, which can be generally classified into the subjective weighing method and objective weighing method, both of which have some disadvantages (Rooney and Bayley 2010). The combinatorial weighing method was used to determine the indicator weights to reduce the possible errors in this study. The indicator weights were first determined with AHP and the coefficient of variation method and then combined as follows (see (6.21)) in order to take full advantage of expert knowledge and give full consideration to specific conditions of the study area:

$$w_i = \frac{w_j^{(1)} \times w_j^{(2)}}{\sum_{j=1}^{m} \left[w_j^{(1)} \times w_j^{(2)} \right]} \tag{6.21}$$

where $w_j^{(1)}$ is the weight vector obtained with the coefficient of variation method; $w_j^{(2)}$ is the weight vector obtained with AHP; and j is the number of assessment indices. The final index weights are listed in Table 6.6.

Results and Discussion

Results

The indicator system was finally established and indicator weights were determined (Table 6.6), with which the forest ecosystem resilience in the study area was calculated and was further divided into five levels (Fig. 6.12), with the thresholds of the resilience index determined with the natural breaks method (Table 6.7).

The forests with the highest resilience were mainly located in the middle of the southwestern mountain area. The result indicated that the forest ecosystem resilience in Yongxin County ranged from 0.1803 to 0.6919, with an average value of 0.3821. Only 40.4 % of the forests were above the average resilience level, and the forest ecosystem resilience in the study area was not very high on the whole. Besides, the forest ecosystem resilience is generally above the average level occurred in the southwestern mountain area, the southeastern mountain area, and the northern mountain area, while the resilience was generally below the

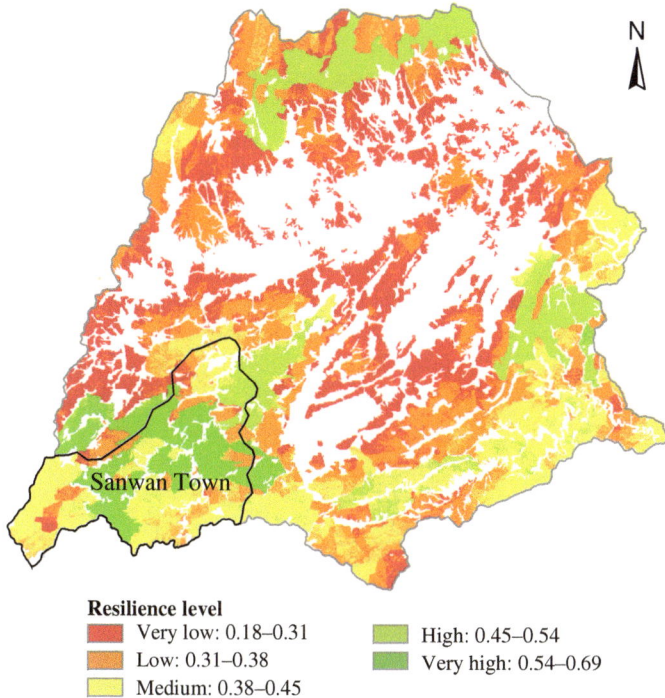

Fig. 6.12 Spatial pattern of the forest ecosystem resilience level in Yongxin County and the location of Sanwan Town where Sanwan National Forest Park is located

Table 6.7 Threshold of forest resilience level and total area at each resilience level

Resilience level	Threshold of resilience levels	Total area (ha)	Area percent (%)
Very low	0.18–0.31	35835.93	25.60
Low	0.31–0.38	44688.15	31.90
Medium	0.38–0.45	30632.22	21.90
High	0.45–0.54	18448.29	13.20
Very high	0.54–0.69	10386.63	7.40

average level in the middle plain and hill area. The map of the forest ecosystem resilience level clearly showed the spatial heterogeneity of the pattern of resilience (Fig. 6.12). The forests reaching a high or very high resilience level generally occurred around the center of the southwestern mountain area and the eastern part of the southeastern mountain area, and they extended from east to west in the northern mountain area. The forests with a medium level of resilience generally occurred in the area of the southern boundary, whereas the forests with a low or very low level of resilience were generally located in the central plain and hill area. The forests in the southwestern mountain area generally reached or exceeded a medium level of resilience, with a large area reaching the very high resilience level. The forests only reached a medium or low level of resilience in the southeastern mountain area except in the eastern part and the area near the administrative boundary, where the forests reached a high level of resilience. In the northern mountain area, only the forests extending from east to west, whose type species was *C. lanceolata*, reached a medium or high level of resilience. The remainder of the forests in the northern mountain area only reached a low or very low level of resilience. In contrast, the forests in the central area were almost entirely at a low or very low level of resilience.

The total forest area at each resilience level was furthermore summarized (Table 6.7). In total, 57.5 % of the forests were at a low or very low level of resilience; 21.9 % reached a medium level of resilience. Only 20.6 % reached or exceeded a high level of resilience, indicating the resilience of the forest was not good. Besides, only five forest types reached the very high resilience level, among which the *C. lanceolata* forest accounted for the largest area proportion of forests with very high resilience, with the total area of 10381.2 ha reaching the very high resilience level. In addition, many forest types reached the high resilience level, while only the *C. lanceolata*, other sclerophyllous forest, and the aceae has relatively large area with high resilience, reaching 16,457.6, 1,951.02, and 12.42 ha, respectively. In summary, most forest types only reached a medium level of resilience, only the forest types with a large total area, such as *C. lanceolata*, tended to have large areas exceeding the medium resilience level.

The key factors influencing the resilience of the forest ecosystem varied among different areas, these spatial patterns of which may provide significant information for formulating appropriate forest resource management measures. For example, in the southwestern mountain area, the stand condition was not favorable due primarily to the terrain, which makes this area susceptible to soil erosion. However, the rich ecological memory found in this area substantially offsets this disadvantage. The patch area, with a weight of 0.1744, is one of the most important indicators at the bottom level, and the patch area of most forest types was generally very large and the fragmentation was not serious in this area. By comparison, the stand condition is very favorable in the southeastern mountain area, with higher temperature and more precipitation that had favorable effects on the forests. However, the internal memory in this area only reached the medium level; only very few parts were rich in external memory. In the northern mountain area, especially in its western portion, the stand condition is not very good, and the ecological memory was only rich in a *C. lanceolata* forest extending from east to west due primarily to the abundant internal memory. However, the ecological memory of other forest types was very poor in this area, primarily due to the serious fragmentation. In the central plain and hill area, where there are primarily plain and low hills, the stand condition is very favorable for humans as well as forests. But, this area has a long history of human disturbance, and the forests were seriously fragmented, leading to the very poor ecological memory.

The current spatial heterogeneity of the forest ecosystem resilience is the result of the cumulative effects of long-term human intervention and the influence of natural conditions (Zhan et al. 2010). The human disturbance plays a subordinate role in influencing the forest ecosystem resilience on the whole; however, it still may be the dominant influencing factor at the local scale. For example, there is still very serious human disturbance in the plain area, and the human activities still play a dominant role in influencing the forest ecosystem resilience, especially in the plain area in the middle part of the study area, where there is a high population density and a lot of forests have been reclaimed for cropland. In fact, the accumulative effects of historical human activities make great contribution to the current spatial pattern of the species composition of the forests. For example, the zonal vegetation is the evergreen broad-leaf forest, which has been seriously damaged during the historical period, and the gradual recovery of the local forests is due primarily to the reconstruction and conservation since the 1980s. As a result, *C. lanceolata*, which has been widely used in the afforestation projects, has accounted for the largest proportion of the forests at present. Overall, the result objectively reflects the spatial pattern of the forest ecosystem resilience in the study area, indicating that it is a practical approach to spatially measure the ecosystem resilience with its influencing factors.

Discussion

This study indicates that the ecosystem resilience can be quantitatively measured with its influencing factors; however, more efforts should be made to further explore how to more scientifically and accurately measure the ecosystem resilience since there are various challenges in both the spatially explicit assessment of the ecosystem resilience and its application. First and foremost, more efforts should be made to explore the theory and assessment of the ecosystem resilience. Although previous researches have provided some methods to measure the ecosystem resilience, there remains an urgent need for an operational tool for assessing and mapping the ecosystem resilience, and it is necessary to make more efforts to select and integrate the resilience indicators according to the research object and data availability based on firm theoretical foundations. Besides, methods for spatially assessing many resilience factors have not yet been developed, and the comparative study is crucial since the ecosystem resilience cannot always be directly observed (Rowlands et al. 2012). In addition, since there may be multiple indicators to measure the ecosystem resilience, it is of great importance to explore how to integrate these indicators. The resilience was represented with weighted sum of indices of its influencing factors in this study; the result may be not very accurate since there may be some nonlinear relationship between the ecosystem resilience and its influencing factors, but it still provides a useful method to the spatially explicitly assess the ecosystem resilience.

There is also a great challenge to operationalize the resilience concept for ecosystem management in a dynamic world; one of the major challenges for progressing resilience-based management lies in successful application. In fact, the general resilience principles have been influencing the way of the ecosystem management and conservation and have been consciously or unconsciously applied in the eco-system management. For example, the biosphere reserves have been generally demarcated into core area(s), buffer area(s), and transition area(s) according to guidelines of the United Nations Educational, Scientific and Cultural Organization (Batisse 1985). The demarcation of the biosphere reserves into three areas can promote the ecosystem resilience, although the ecosystem managers may have been not aware of that. For example, in fact there is a forest park in Sanwan Town where the forest ecosystem resilience is the highest, that is, Sanwan National Forest Park, which has a forest coverage rate of 90.5 % (Fig. 6.12). This forest park was initially established in order to promote the development of the local tourism and pursue more economic benefit rather than improve the ecosystem resilience; however, it has unexpectedly increased the resilience of local forests. Therefore, establishing forest parks may be an effective way to conserve ecosystems and maintain the desirable ecosystem state.

Summary

Drawing on principles of the ecosystem resilience highlighted in the literature, this study selected the indicators that capture the influencing factors of the ecosystem resilience and quantitatively assessed the forest ecosystem resilience in Yongxin County. The result indicates that it is feasible to generate large-scale ecosystem resilience maps with this assessment model and spatially explicitly identify the areas essential to the ecosystem conservation. Besides, the result shows that there is significant spatial heterogeneity of the forest ecosystem resilience in the study area, which can provide a scientific basis for the local forest resource management and conservation to maintain critical ecosystem services and adapt to climate change. But, it is still necessary to make further improvement in the future research since there are still some controversies on the selection and integration of the specific indicators of ecosystem resilience, and it is possible to more accurately measure the ecosystem resilience if high-resolution data in a large area are available. Although there are still some uncertainties, the results still can provide a scientific basis for the conservation of the forests, which is of great significance to the climate change mitigation.

References

Adger WN, Dessai S, Goulden M, Hulme M, Lorenzoni I, Nelson DR, Naess LO, Wolf J, Wreford A (2009) Are there social limits to adaptation to climate change? Clim Change 93(3–4):335–354
Adger WN, Huq S, Brown K, Conway D, Hulme M (2003) Adaptation to climate change in the developing world. Prog Dev Stud 3(3):179–195
Batisse M (1985) Action plan for biosphere reserves. Environ Conserv 12(01):17–27

Bellwood DR, Hoey AS, Choat JH (2003) Limited functional redundancy in high diversity systems: resilience and ecosystem function on coral reefs. Ecol Lett 6(4):281–285

Bengtsson J, Angelstam P, Elmqvist T, Emanuelsson U, Folke C, Ihse M, Moberg F, Nyström M (2003) Reserves, resilience and dynamic landscapes. AMBIO J Hum Environ 32(6):389–396

Bennett E, Cumming G, Peterson G (2005) A systems model approach to determining resilience surrogates for case studies. Ecosystems 8(8):945–957

Berkes F, Jolly D (2002) Adapting to climate change: social-ecological resilience in a Canadian western Arctic community. Conserv Ecol 5(2):18

Borja A, Bricker SB, Dauer DM, Demetriades NT, Ferreira JG, Forbes AT, Hutchings P, Jia X, Kenchington R, Marques JC (2008) Overview of integrative tools and methods in assessing ecological integrity in estuarine and coastal systems worldwide. Mar Pollut Bull 56(9):1519–1537

Bounoua L, DeFries R, Collatz GJ, Sellers P, Khan H (2002) Effects of land cover conversion on surface climate. Clim Change 52(1–2):29–64

Brand F (2009) Critical natural capital revisited: ecological resilience and sustainable development. Ecol Econ 68(3):605–612

Burkhard B, Kroll F, Nedkov S, Muller F (2012) Mapping ecosystem service supply, demand and budgets. Ecol Ind 21:17–29

Carpenter S, Press M, Huntly N, Levin S (2001) Alternate states of ecosystems: evidence and some implications. In: Ecology: achievement and challenge: the 41st symposium of the British ecological society sponsored by the ecological society of America held at Orlando, Florida, USA, 10–13 April 2000, Blackwell Science

Carpenter S, Walker B, Anderies JM, Abel N (2001b) From metaphor to measurement: resilience of what to what? Ecosystems 4(8):765–781

Costanza R (2012) Ecosystem health and ecological engineering. Ecol Eng 45:24–29

Daily GC, Alexander S, Ehrlich PR, Goulder L, Lubchenco J, Matson PA, Mooney HA, Postel S, Schneider SH, Tilman D (1997). Ecosystem services: benefits supplied to human societies by natural ecosystems. Ecological Society of America, Washington (DC)

de Bruin K, Dellink R, Ruijs A, Bolwidt L, van Buuren A, Graveland J, De Groot R, Kuikman P, Reinhard S, Roetter R (2009) Adapting to climate change in The Netherlands: an inventory of climate adaptation options and ranking of alternatives. Clim Change 95(1–2):23–45

Dehring CA, Lind MS (2007) Residential land-use controls and land values: zoning and covenant interactions. Land Econ 83(4):445–457

Deng X, Huang J, Huang Q, Rozelle S, Gibson J (2011a) Do roads lead to grassland degradation or restoration? A case study in Inner Mongolia, China. Environ Dev Econ 16(06):751–773

Deng X, Huang J, Uchida E, Rozelle S, Gibson J (2011b) Pressure cookers or pressure valves: do roads lead to deforestation in China? J Environ Econ Manage 61(1):79–94

Deng X, Jiang QO, Su H, Wu F (2010a) Trace forest conversions in Northeast China with a 1-km area percentage data model. J Appl Remote Sens 4(1):041893–041893–041813

Deng X, Li Z, Huang J, Shi Q, Li Y (2013a) A revisit to the impacts of land use changes on the human wellbeing via altering the ecosystem provisioning services. Adv Meteorol 2013:8

Deng X, Zhao C, Yan H (2013) Systematic modeling of impacts of land use and land cover changes on regional climate: a review. Adv Meteorology 2013

Deng X, Yin F, Uchida E, Rozelle S (2010b) A complementary measurement of changes in China's forestry area using remote sensing data. IJLS 3:1–12

Dong S, Zhou C, Wang H (2002) Ecological crisis and countermeasures of the Three Rivers Headstream Regions. J Nat Res 17(6):713–720

Dublin HT, Sinclair AR, McGlade J (1990) Elephants and fire as causes of multiple stable states in the Serengeti-Mara woodlands. J Anim Ecol 1147–1164

Ehrlén J, Eriksson O (2000) Dispersal limitation and patch occupancy in forest herbs. Ecology 81(6):1667–1674

Einstein H, Sousa R (2007) Warning systems for natural threats. Georisk 1(1):3–20

Fang Q, Zhang L, Hong H, Zhang L, Bristow F (2008) Ecological function zoning for environmental planning at different levels. Environ Dev Sustain 10(1):41–49

Feddema JJ, Oleson KW, Bonan GB, Mearns LO, Buja LE, Meehl GA, Washington WM (2005) The importance of land-cover change in simulating future climates. Science 310(5754):1674–1678

Fezzi C, Bateman I, Askew T, Munday P, Pascual U, Sen A, Harwood A (2014) Valuing provisioning ecosystem services in agriculture: the impact of climate change on food production in the United Kingdom. Environ Res Econ 57(2):197–214

Folke C (2006) Resilience: the emergence of a perspective for social–ecological systems analyses. Glob Environ Change 16(3):253–267

Folke C, Carpenter S, Elmqvist T, Gunderson L, Holling CS, Walker B (2002) Resilience and sustainable development: building adaptive capacity in a world of transformations. AMBIO: J Hum Environ 31(5):437–440

Franklin JF, MacMahon JA (2000) Messages from a mountain. Science 288(5469):1183–1184

Gallopín GC (2006) Linkages between vulnerability, resilience, and adaptive capacity. Glob Environ Change 16(3):293–303

Gibbs MT (2009) Resilience: what is it and what does it mean for marine policymakers? Marine Policy 33(2):322–331

Gillies C, Pierce R (1999) Secondary poisoning of mammalian predators during possum and rodent control operations at Trounson Kauri Park, Northland, New Zealand. New Zealand J Ecology 23(2):183–192

Gosling SN (2013) The likelihood and potential impact of future change in the large-scale climate-earth system on ecosystem services. Environ Sci Policy 27, Supplement 1(0):S15–S31

Heckelei T, Britz W, Zhang Y (2012) Positive mathematical programming approaches–recent developments in literature and applied modelling. Bio-based Appl Econ 1(1):109–124

Howitt RE (1995) Positive mathematical programming. Am J Agric Econ 77(2):329–342

Hutchinson MF (1995) Stochastic space-time weather models from ground-based data. Agric For Meteorol 73(3–4):237–264

IPCC (2007) Climate change 2007: the physical science basis. Agenda 6(07):338

Izquierdo LR, Gotts NM, Polhill JG (2004) Case-based reasoning, social dilemmas, and a new equilibrium concept. J Artif Soc Soc Simul 7(3)

Jentsch A, Beierkuhnlein C, White PS (2002) Scale, the dynamic stability of forest ecosystems, and the persistence of biodiversity. Silva Fennica 36(1):393–400

Jordan SJ, Hayes SE, Yoskowitz D, Smith LM, Summers JK, Russell M, Benson WH (2010) Accounting for natural resources and environmental sustainability: linking ecosystem services to human well-being. Environ Sci Technol 44(5):1530–1536

Jost L (2006) Entropy and diversity. Oikos 113(2):363–375

Kelly PM, Adger WN (2000) Theory and practice in assessing vulnerability to climate change and facilitating adaptation. Clim Change 47(4):325–352

Khattak AB, Zeb A, Bibi N (2008) Impact of germination time and type of illumination on carotenoidcontent, protein solubility and in vitro protein digestibility of chickpea (*Cicer arietinum* L.) sprouts. Food Chem 109(4):797–801

Liu J, Liu M, Zhuang D, Zhang Z, Deng X (2003) Study on spatial pattern of land-use change in China during 1995–2000. Sci China Ser D Earth Sci 46(4):373–384

Liu J, Zhang Z, Xu X, Kuang W, Zhou W, Zhang S, Li R, Yan C, Yu D, Wu S, Jiang N (2010) Spatial patterns and driving forces of land use change in China during the early 21st century. J Geog Sci 20(4):483–494

Lobell DB, Burke MB, Tebaldi C, Mastrandrea MD, Falcon WP, Naylor RL (2008) Prioritizing climate change adaptation needs for food security in 2030. Science 319(5863):607–610

MacArthur R (1955) Fluctuations of animal populations and a measure of community stability. Ecology 36(3):533–536

Martin TJ, Ogden J (2002) The seed ecology of Ascarina lucida: a rare New Zealand tree adapted to disturbance. NZ J Bot 40(3):397–404

MEA (2005) Ecosystems and human well-being. Island Press, Washington DC

Moore JC, De Ruiter PC, Hunt HW (1993) Influence of productivity on the stability of real and model ecosystems. Sc New York Washington 261:906

Nichols G, Prigogine I (1989) Exploring complexity: an introduction. WH Freeman [PJ]

Nyström M, Folke C (2001) Spatial resilience of coral reefs. Ecosystems 4(5):406–417

Polhill JG, Sutherland L-A, Gotts NM (2010) Using qualitative evidence to enhance an agent-based modelling system for studying land use change. J Artif Soc Soc Simul 13(2):10

Prato T (2012) Increasing resilience of natural protected areas to future climate change: a fuzzy adaptive management approach. Ecol Model 242:46–53

Price DT, McKenney DW, Nalder IA, Hutchinson MF, Kesteven JL (2000) A comparison of two statistical methods for spatial interpolation of Canadian monthly mean climate data. Agric For Meteorol 101(2–3):81–94

Rooney RC, Bayley SE (2010) Quantifying a stress gradient: an objective approach to variable selection, standardization and weighting in ecosystem assessment. Ecol Ind 10(6):1174–1183

Rosset V, Oertli B (2011) Freshwater biodiversity under climate warming pressure: identifying the winners and losers in temperate standing waterbodies. Biol Conserv 144(9):2311–2319

Rowlands G, Purkis S, Riegl B, Metsamaa L, Bruckner A, Renaud P (2012) Satellite imaging coral reef resilience at regional scale. A case-study from Saudi Arabia. Mar Pollut Bull 64(6):1222–1237

Scarborough VL, Burnside WR (2010) Complexity and sustainability: perspectives from the ancient Maya and the modern Balinese. Am Antiq 75(2):327–363

Schmitz OJ (2000) Combining field experiments and individual-based modeling to identify the dynamically relevant organizational scale in a field system. Oikos 89(3):471–484

Shaw MR, Pendleton L, Cameron DR, Morris B, Bachelet D, Klausmeyer K, MacKenzie J, Conklin D, Bratman G, Lenihan J, Haunreiter E, Daly C, Roehrdanz P (2011) The impact of climate change on California's ecosystem services. Clim Change 109(1):465–484

Shi Q, Zhan J, Wu F, Deng X, Xu L (2012) Simulation on water flow and water quality in Wuliangsuhai Lake using a 2-D hydrodynamic model. J Food Agric Environ 10(2):973–975

Shitikov VK, Vykhristyuk LA, Pautova VN, Zinchenko TD (2007) Comprehensive ecological zoning of the Kuibyshev Reservoir. Water Resour 34(4):450–458

Shrestha MK, York AM, Boone CG, Zhang S (2012) Land fragmentation due to rapid urbanization in the phoenix metropolitan area: analyzing the spatiotemporal patterns and drivers. Appl Geogr 32(2):522–531

Smit B, Burton I, Klein RJ, Wandel J (2000) An anatomy of adaptation to climate change and variability. Clim Change 45(1):223–251

Smit B, Pilifosova O (2003) Adaptation to climate change in the context of sustainable development and equity. Sustain Dev 8(9):9

Smit B, Skinner MW (2002) Adaptation options in agriculture to climate change: a typology. Mitig Adapt Strat Glob Change 7(1):85–114

Smithers J, Smit B (1997) Human adaptation to climatic variability and change. Glob Environ Change 7(2):129–146

Strickland-Munro JK, Allison HE, Moore SA (2010) Using resilience concepts to investigate the impacts of protected area tourism on communities. Ann Tourism Res 37(2):499–519

Thompson I, Mackey B, McNulty S, Mosseler A (2009) Forest resilience, biodiversity, and climate change. In: A synthesis of the biodiversity/resilience/stability relationship in forest ecosystems. Secretariat of the convention on biological diversity, Montreal. Technical Series

Walker BH (1992) Biodiversity and ecological redundancy. Conserv Biol 6(1):18–23

Wang M, Zhang X, Yan X (2013) Modeling the climatic effects of urbanization in the Beijing–Tianjin–Hebei metropolitan area. Theoret Appl Climatol 113(3–4):377–385

Wang S-H, Huang S-L, Budd WW (2012) Integrated ecosystem model for simulating land use allocation. Ecol Model 227:46–55

Watanabe MDB, Ortega E (2014) Dynamic emergy accounting of water and carbon ecosystem services: a model to simulate the impacts of land-use change. Ecol Model 271:113–131

Wickham JD, Wade TG, Riitters KH (2013) Empirical analysis of the influence of forest extent on annual and seasonal surface temperatures for the continental United States. Glob Ecol Biogeogr 22(5):620–629

Wolfslehner B, Vacik H, Lexer MJ (2005) Application of the analytic network process in multi-criteria analysis of sustainable forest management. For Ecol Manage 207(1):157–170

Wu F, Zhan J, Yan H, Shi C, Huang J (2013) Land cover mapping based on multisource spatial data mining approach for climate simulation: a case study in the farming-pastoral ecotone of North China. Adv Meteorol 2013

Ye J, Yang X, Jiang D (2010) The grid scale effect analysis on town leveled population statistical data spatialization. J Geo-Inf Sci 12(1):40–47

Zhan J, Shi N, He S, Lin Y (2010) Factors and mechanism driving the land-use conversion in Jiangxi Province. J Geog Sci 20(4):525–539

Zhou K, Huang J, Deng X, van der Werf W, Zhang W, Lu Y, Wu K, Wu F (2014) Effects of land use and insecticides on natural enemies of aphids in cotton: first evidence from smallholder agriculture in the North China Plain. Agric Ecosyst Environ 183:176–184

Zirlewagen D, Raben G, Weise M (2007) Zoning of forest health conditions based on a set of soil, topographic and vegetation parameters. For Ecol Manage 248(1):43–55